Climate in Asia and the Pacific

ADVANCES IN GLOBAL CHANGE RESEARCH

VOLUME 56

For further volumes:
http://www.springer.com/series/5588

Michael J. Manton • Linda Anne Stevenson
Editors

Climate in Asia and the Pacific

Security, Society and Sustainability

Editors
Michael J. Manton
School of Mathematical Sciences
Monash University
Melbourne, VIC, Australia

Linda Anne Stevenson
Asia-Pacific Network for Global Change Research
Kobe, Japan

ISSN 1574-0919
ISBN 978-94-007-7337-0 ISBN 978-94-007-7338-7 (eBook)
DOI 10.1007/978-94-007-7338-7
Springer Dordrecht Heidelberg New York London

Library of Congress Control Number: 2013945585

Printed on acid-free paper

Springer is part of Springer Science+Business Media (www.springer.com)

Foreword

The 1992 Rio Declaration, emerging from the UN Conference on Environment and Development, stated that: "Human beings are at the centre of concerns for sustainable development. They are entitled to a healthy and productive life in harmony with nature." Twenty years later, Heads of State and Government renewed their commitment to sustainable development at the Rio+20 UN Conference on Sustainable Development, with an outcome document titled *The Future We Want.* It noted that "to ensure the promotion of an economically, socially, and environmentally sustainable future for our planet, and for present and future generations" will require concrete and urgent action.

Advancing sustainable human development becomes very difficult as the world moves towards the edges of its planetary boundaries: the need for urgent action is real. This can be most clearly demonstrated with respect to climate change, where experts warn that the world must stay under a 2 °C increase in temperature threshold above pre-industrial levels – beyond which it is believed there would be catastrophic and irreversible change to our climate.

The impacts of global warming are already being felt through the depletion of natural resources; more frequent natural disasters, from flooding to heat waves and droughts; and changes in ecosystem dynamics.

The Executive Secretary of the UN Framework Convention on Climate Change, Christina Figueres, has said that "climate change has become the amplifier and multiplier of every crisis we are facing – be it human heath, population growth, the strain on water, food and other resources, or energy insecurity."

What is unfair is that the world's poorest countries, including Small Island Developing States, which have contributed little to greenhouse gas emissions, are among the most vulnerable to the consequences of climate change. Within countries, the poor rely disproportionately on natural resources for their livelihoods, and in urban settings often live in hazard-prone slums or remote locations. Their economic and overall well-being stands to be more directly impacted by the changing climate.

The costs of inaction on climate change are increasingly clear as extreme weather inflicts loss of life and livelihoods and the destruction of property and infrastructure around the world. The Asia-Pacific has suffered disproportionately in recent years, with 45 % of the world's natural disasters in the last three decades occurring in the region, leading to significant losses in human life and GDP.

Climate in Asia and the Pacific: Security, Society, and Sustainability, reviews the current understanding of trends around climate change in the Asia-Pacific, the impact on natural and human systems across the region, and strategies to mitigate and adapt to these impacts. It looks at the relationship between climate change and urbanization – highlighting the vulnerabilities of mega-cities and of unsustainable urbanisation practices. It explores the impact on human security, with a particular focus on food and water security; disaster risk; broader societal concerns – from human health to the needs of vulnerable remote communities; and looks at sustainable energy options for the region and future directions for climate research.

As the Asia-Pacific has not only many of the world's most climate-exposed territories, but also hosts more than half of the world's population, including nearly 900 million of those who are extremely poor, this report is relevant for all those interested in how our changing climate impacts on development.

UNDP's 2012 Asia-Pacific Human Development Report, *One Planet to Share: Sustaining Human Progress in a Changing Climate*, argued that while growth in Asia is important for the world economy and has contributed to poverty reduction in the region, the challenge now is to reduce the emissions intensity of that growth while simultaneously improving the lives of people – including through access to clean energy, and meeting unfinished development agendas.

This publication can provide valuable input into the discussion on how the region can follow a sustainable development path, which fulfils the urgent human needs of today while preserving a habitable planet for future generations. The diverse expertise of the various contributors and the broad scope of issues addressed are valuable for academic and practitioner communities alike.

The goal is clear: to reduce poverty while staying within the boundaries set by nature. To do that we need knowledge and evidence to support better policies, and we need political leadership willing to act.

United Nations Development Programme Helen Clark
New York, USA

Foreword

Actionable Climate Information for Regional and Global Development

It is only in the last 10,000 years that we have moved to our modern society with its critical dependence on mechanized agriculture and exploitation of natural resources at an ever increasing rate using modern technology to improve the quality of humans' life on Earth. Every day climate and weather variability and changes shape the global commerce and development, including the natural environment and its biodiversity on which society depends for water, food and other ecosystem services for our comfort and well being.

Earth's climate and weather not only influence food and water supplies but they also have major impacts on human health, tourism, energy and transport, thus the global economy and society at large. While our increasing mastery of technology and exploitation of energy reserves has given us some ability to adapt to climate variations, the burgeoning global population, increasing urbanisation and the increasing demand on Earth's natural resources means that we are also becoming increasingly vulnerable to changes in climate/weather, particularly through extreme events such as floods, droughts, high heat waves, and other climate phenomena. These impacts are felt by all nations around the world, but the ability to respond to them and the resiliency to recover from their adverse impacts is very limited in most developing regions and nations. There is a growing demand by public and private sectors for timely access to reliable science-based information about climate variability and change, and their potential impacts on people, and natural and managed ecosystems.

The national and international global environmental research and development programmes established during the past several decades (e.g. Earth System Science Partnership, ESSP) have made great strides towards understanding the functioning of Earth's climate system, its natural variability, and human induced changes. Sustained observations of the atmosphere, oceans, terrestrial ecosystems and the polar regions together with development of computer-based models have played a major role in these efforts. Indeed, the revolutionary progress in computation and telecommunication technologies during the recent decades have been instrumental in representing realistically the natural processes in the Earth system models, and to increase the resolution of smallest grid cells represented in these models while we have continued to improve the complexity of Earth's climate system for longer periods (multiple decades to centuries) into the future. Another major success during this period has been the establishment of coordinated international mechanisms for synthesis and translation of best available scientific knowledge about the state of Earth's climate system in a form that is useful to policy decision makers through the Intergovernmental Panel on Climate Change (IPCC) and other similar bodies for atmospheric ozone, biodiversity, water resources, energy, etc. In short, we have managed to make progress on advancing the science of climate change together with an effective approach to using the resulting knowledge for environmental policy decisions and a wide range of other applications around the world.

The need for climate information is now growing rapidly beyond the environmental policy domains by all sectors of the world economy that must consider both the risks and opportunities associated with climate change and variability on seasonal, decadal and longer time scales for day-to-day management activities (e.g. transport and tourism) and long-range planning (e.g. investments in infrastructure) for national, regional and global development. This implies a greater need for more sophisticated models that mimic realistically the behaviour of the entire Earth system at greater time and space resolutions, hence a demand for coordination of research activities across multiple scientific disciplines, more powerful computers, and greater capacity for translation and communication of the resulting information and knowledge to a wide range of users. The tasks of coordination, integration, synthesis of scientific information and effective communication of the results to managers and decision makers have to be carried out by entities at the national, regional and global level. Development and dissemination of "actionable" science-based climate information requires a symbiotic relationship between producers and users of this information. Such a partnership will ensure integration of users' need into the research agenda together with timely and effective access to the research results for decision makers. The international initiatives such as the Global Framework for Climate Services (GFCS) sponsored by the United Nations system and their partners and the Future Earth: Research for Global Development coordinated by the International Council for Science (ICSU) and its Alliance partners are intended to promote a more effective dialogue between providers and users of science-based climate/environment information. The regional organizations such as the Asia-Pacific Network for Global Change Research (APN), development banks and non-governmental organizations are expected to play a major role in implementation of these global initiatives, especially in the development of scientific and technical capabilities and networks that are essential for their success. The papers presented in this monograph describe excellent examples, case studies and projects on how to forge such alliances between providers and users of science-based information for development purposes. Lessons learned from such efforts that are captured in these papers will be invaluable for successful implementation of the GFCS and Future Earth in the ensuing decades.

World Climate Research Program, Ghassem R. Asrar
Geneva, Switzerland

Foreword

Climate in Asia and the Pacific: Security, Society and Sustainability provides a comprehensive description and discussion of the complex and interactive phenomena of climate related global change in the Asia Pacific region. Its chapters address the current state of knowledge from the fields of climate science, environmental science, sociology, technology development, public health, and security policy with an insightful look at issues of governance that are central to managing the impacts of climate change on human and national security. It draws upon these multiple fields of knowledge to consider strategies for mitigating and adapting to those impacts.

This phenomenon of climate related global change is unprecedented in human history in its complexity and in its potential threat to sustainable and secure living on planet Earth. Global warming and its impacts are interactive with other twenty-first century trends including population growth; urbanization; economic development and resource demands; and the generation of waste products and their release into the air, water, seas, and landscapes. Another layer of complexity is added by the political context, where developed and developing nations perceive different and sometimes conflicting interests in a dynamic regional security environment. The present volume does an excellent job of unfolding many of the layers of that complexity.

Climate change is an issue of the global commons because the green house gases, which are the anthropogenic contribution to global warming spread across the atmosphere and the oceans, whatever their point of origin. It is a regional issue because its environmental impacts on water and food supplies, on coastal infrastructure, or on biodiversity are not contained by national boundaries. It is a national issue because mitigation and adaptation must be managed by the executive agencies of sovereign nations, and it is a local issue because its impacts are felt locally.

Because our institutions for governance are the product of our past experience, they are often better reflections of past needs for managing human affairs than of emerging needs. Complex problems are by their nature unpredictable, subject to unexpected consequences and possible tipping points. As Stewart Brand is reported to have said, "Dealing with climate change…involves a level of global cooperation that has never happened, and the mechanisms for that are not in sight" (Achenbach 2012).

Climate change is a slow motion crisis. First recognized by the climate science and meteorological communities about 50 years ago, awareness of the emerging problem slowly spread among environmental research and policy communities, until the 4th Assessment Report of the Intergovernmental Panel on Climate Change (IPCC) brought a shared Nobel Peace Prize for "*for their efforts to build up and disseminate greater knowledge about man-made climate change, and to lay the foundations for the measures that are needed to counteract such change*"

Those measures are the domain of governance, and the goal of good governance must be a secure and sustainable society for peoples and nations across the Asia Pacific region and around the globe. Climate change is a challenge – perhaps an existential challenge – to the people of our region. But it also represents an opportunity to work together across scientific and traditional knowledge communities, across government agencies, across public and private sectors, and through regional organizations for security and for economic development. It represents an opportunity to share knowledge and to collaborate across all of these groups and perspectives to manage the complex problems of climate related global change, to ensure a secure and sustainable environment for our children and posterity. The authors of this book have provided us with a good platform to work from.

Asia Pacific Center for Security Studies J. Scott Hauger
Honolulu, USA

References

Achenbach, J. (2012 January 2). Spaceship earth: A new view of environmentalism. Washington Post.http://articles.washingtonpost.com/2012-01-02/national/35439231_1_planet-climate-change-civilization-and-nature/2.

The Nobel Peace Prize. (2007). Nobelprize.org. 22 Jan 2013 http://www.nobelprize.org/nobel_prizes/peace/laureates/2007/.

Preface

Following a mandate from its governing body, the Intergovernmental Meeting (IGM), the Asia-Pacific Network for Global Change Research (APN) produced a synthesis report of all of the activities it had conducted under one of its four broad themes of global environmental change - climate change and climate variability. The Synthesis Report – *Climate in Asia and the Pacific: A Synthesis of APN Activities*, summarised more than 55 regional research and capacity building projects that the APN had conducted under this theme since 1998.

Positive feedback following wide distribution of the synthesis report prompted the need, and decision, to complement the report with a book explaining the current status of climate change and climate variability in the Asia-Pacific region; future directions in the area and overarching issues. It was agreed among the authors that the foci of the book be security (food, water and energy); society (urban and remote communities; human health and governance) and sustainability (low carbon development and ecosystem services).

The first chapter of the book addresses a number of key questions that relate to our current understanding of the interactions between climate, natural ecosystems and human communities across Asia and the Pacific. The analysis presented in subsequent chapters addresses these questions and provides recommendations for a number of future directions in research needed to better understand and manage the risks associated with climate change and variability in the region. The final chapter summarises the findings presented in the book and provides an overall picture of future needs for climate research in Asia and the Pacific. Finally, we suggest a number of overarching issues that should be taken into account in future considerations of climate interactions across the region.

Immediately following the publication of the Synthesis Report, an authors' workshop for the present book convened in October 2011 kick-starting a gathering and 16-month coordination of the work of 31 authors from broad backgrounds in global environmental change.

We are immensely grateful to the contributing authors, who are not only leaders in their field but most are contributing authors to the Fifth Assessment Report of the

Intergovernmental Panel on Climate Change (IPCC AR5). We also acknowledge and appreciate comments by Francisco Werner in the preparation of the manuscript. Finally, there are four people that deserve special acknowledgement and who worked particularly hard in the background to realise the timely publication of the book. I would like to thank Ratisya Radzi, my right arm through the entire process, as well as fellow Editor, Michael J. Manton; Lead Author, Michael James Salinger; and APN Communications & Development Officer, Xiaojun Deng. All of you went above and beyond the call of duty in your work towards this book and the fruits of your labour, I am sure, will not go unnoticed.

Asia-Pacific Network for
Global Change Research
Kobe, Japan

Linda Anne Stevenson

Contents

1 **Introduction** ... 1
Michael J. Manton

2 **Climate in Asia and The Pacific: Climate Variability and Change** ... 17
Michael James Salinger, Madan Lall Shrestha, Ailikun, Wenjie Dong, John L. McGregor, and Shuyu Wang

3 **Climate and Urbanization** ... 59
Peter Marcotullio, Richard Cooper, and Louis Lebel

4 **Climate and Security in Asia and the Pacific (Food, Water and Energy)** ... 129
Lance Heath, Michael James Salinger, Tony Falkland, James Hansen, Kejun Jiang, Yasuko Kameyama, Michio Kishi, Louis Lebel, Holger Meinke, Katherine Morton, Elena Nikitina, P.R. Shukla, and Ian White

5 **Climate and Society** ... 199
Kanayathu Koshy, Linda Anne Stevenson, Jariya Boonjawat, John R. Campbell, Kristie L. Ebi, Hina Lotia, and Ruben Zondervan

6 **Climate and Sustainability** ... 253
Rodel Lasco, Yasuko Kameyama, Kejun Jiang, Linda Peñalba, Juan Pulhin, P.R. Shukla, and Suneetha M. Subramanian

7 **Future Directions for Climate Research in Asia and the Pacific** ... 289
Michael J. Manton and Linda Anne Stevenson

Index ... 309

Contributors

Ailikun Institute of Atmospheric Physics, Monsoon Asia Integrated Regional Study (MAIRS) IPO, Chinese Academy of Sciences, Beijing, China

Jariya Boonjawat Southeast Asia START Regional Centre (SEA START RC), Chulalongkorn University, Bangkok, Thailand

John R. Campbell Te Whare Wānanga o Waikato, The University of Waikato, Hamilton, New Zealand

Richard Cooper Southeast Asia START Regional Centre (SEA START RC), IW LEARN, Chulalongkorn University, Bangkok, Thailand

Wenjie Dong State Key Laboratory of Earth Surface Processes and Resource Ecology, College of Global Change and Earth System Science, Beijing Normal University, Beijing, China

Kristie L. Ebi ClimAdapt, LLC, Los Altos, USA

Tony Falkland Island Hydrology Services, Hughes ACT, Australia

James Hansen The International Research Institute for Climate and Society (IRI), Columbia University Lamont Campus, Palisades, NY, USA

Lance Heath Climate Change Institute (CCI), The Australian National University, Canberra, ACT, Australia

Kejun Jiang Energy Research Institute, National Development and Reform Commission, Beijing, China

Yasuko Kameyama Centre for Global Environmental Research, National Institute for Environmental Studies, Tsukuba-City, Ibaraki, Japan

Michio Kishi Graduate School of Fisheries Sciences, School of Fisheries Sciences, Hokkaido University, Hakodate, Hokkaido, Japan

Kanayathu Koshy Centre for Global Sustainability Studies (CGSS), Universiti Sains Malaysia, Penang, Malaysia

Rodel Lasco World Agroforestry Centre (ICRAF), Khush Hall, IRRI, Laguna, Philippines

Louis Lebel Unit for Social and Environmental Research (USER), Faculty of Social Sciences, Chiang Mai University, Chiang Mai, Thailand

Hina Lotia Programme Development Department, Leadership for Environment and Development (LEAD), Islamabad, Pakistan

Michael J. Manton School of Mathematical Sciences, Monash University, Clayton, VIC, Australia

Peter Marcotullio Department of Geography, Hunter College, New York, NY, USA

John L. McGregor CSIRO Marine and Atmospheric Research, Aspendale, VIC, Australia

Holger Meinke Tasmanian Institute of Agriculture, University of Tasmania, TAS, Hobart, Australia

Katherine Morton International Relations, Research School of Pacific Asian Studies, The Australian National University, Canberra, ACT, Australia

Elena Nikitina EcoPolicy Research and Consulting (EcoPolicy), Moscow, Russia

Linda Peñalba Institute of Governance and Rural Development, College of Public Affairs,University of the Philippines Los Baños, Laguna, Philippines

Juan Pulhin Department of Social Forestry and Forest Governance, College of Forestry and Natural Resource, University of the Philippines Los Baños, Laguna, Philippines

Michael James Salinger University of Auckland, Auckland, New Zealand

Madan Lall Shrestha Nepal Academy of Science and Technology, Khumaltar, Lalitpur, Nepal

P.R. Shukla Public Systems Group, Indian Institute of Management, Vastrapur, Ahmedabad, Gujarat, India

Linda Anne Stevenson Asia-Pacific Network for Global Change Research, Kobe, Japan

Suneetha M. Subramanian United Nations University Institute of Advanced Studies (UNU-IAS), 6F International Organizations Center, Yokohama, Japan

Shuyu Wang Institute of Atmospheric Physics, Chinese Academy of Sciences, Beijing, China

Ian White The Fenner School of Environment Society, The Australian National University, Canberra, ACT, Australia

Ruben Zondervan Earth System Governance Project, Lund University, Lund, Sweden

Chapter 1
Introduction

Michael J. Manton

Abstract Variations in climate in the Asia-Pacific region play a major role in the development of natural ecosystems and of human societies. Furthermore, human activities place additional stresses on natural and societal systems and climate change is now considered a significant factor in these increases. The book documents the climate of the region and interactions of the climate with both the environment and societies in the region. The book emphasizes the impacts of climate change as well as strategies to mitigate and adapt to those impacts. A number of aspects of climate in the region that capture interactions between climate and natural and human systems are considered and include climate variability and change, climate and urbanization, climate and security, climate and society, and climate and sustainability.

The book draws on published results in the scientific literature and the analysis presented highlights key climate-related issues for Asia and the Pacific. Subsequent chapters of the book include important issues such as: *Climate variability and change* – large-scale climate systems, trends in mean climate, trends in extreme climate events across Asia and the Pacific, challenges and opportunities for modeling the climate, current projections for future climate under climate change; *Society and urbanization* – trends in urbanization, interactions between urban areas and climate, climate hazards and vulnerabilities for urban areas, climate change mitigation and adaptation strategies for urban areas; *Food, water and energy security* – meeting future needs for rice and wheat across Asia, food from fisheries, water security, and balancing energy demands with reduced GHG emissions; *Governance and sustainability* – institutional arrangements to address the impacts of climate change, prospects for remote communities under climate change, effects of climate change on human health, low carbon development pathways, and ecosystem services to enhance the adaptive capacity of communities.

Keywords Asia-Pacific • Climate change and variability • Climate research • Sustainability

---

M.J. Manton (✉)
School of Mathematical Sciences, Monash University, Clayton, VIC 38000, Australia
e-mail: michael.manton@monash.edu

M.J. Manton and L.A. Stevenson (eds.), *Climate in Asia and the Pacific: Security, Society and Sustainability*, Advances in Global Change Research 56,
DOI 10.1007/978-94-007-7338-7_1, © Springer Science+Business Media Dordrecht 2014

1.1 Climate in Asia and the Pacific

Asia and the Pacific is the major region of the world for rapid economic and social development in the twenty-first century. For this development to be sustainable, the growth needs to account for the effects of climate variability and change, as well as various socio-economic factors. This book aims to review the current understanding of the climate of Asia and the Pacific, its impact on natural and human systems across the region, and strategies adopted to mitigate and adapt to these impacts. From this analysis, we are able to identify significant research and development issues that need to be considered in the future. This book complements the report 'Climate in Asia and the Pacific: A Synthesis of APN Activities' (Manton et al. 2011) prepared by the Asia-Pacific Network for Global Change Research (APN), which draws together the climate-related work supported by the APN over the last 15 years.

The region of interest is focused on monsoon Asia and the western Pacific Ocean (Fig. 1.1). It extends from Pakistan in the west to Hawaii and French Polynesia in the east. The latitudinal extent is broadly defined by the influence of the Asian-Australian monsoon, but some issues extend into the more temperate areas of northern Asia. The topographical features of the region vary from the mountains of the Himalayas to the small islands of the Pacific. The variation in climate regime across the region is correspondingly vast, with tropical climates in the Pacific and deserts in continental Asia. The local climate has a significant impact on human and natural systems of each region. For example, both natural ecosystems and human communities have adapted to the seasonal cycle of the monsoon across much of Asia. Similarly the sea-surface temperature (SST) patterns of the Pacific, associated with the El Niño – Southern Oscillation (ENSO), lead to interannual climate variations that affect the lives of many communities, especially in Pacific islands. Other SST patterns in the Indian Ocean, associated with the Indian Ocean Dipole (IOD), impact the climates of Asia.

Fig. 1.1 World map highlighting member countries of the APN (Source: APN)

In addition to the large geographical variations across Asia and the Pacific, there are major differences in the cultures and socio-economic features of the communities of the region. The communities of monsoon Asia include remote mountain groups and mega-cities. All these communities have vulnerabilities to climate variability and change. While many communities have adapted over centuries to the natural variations of climate, the phenomenon of global climate change often brings new hazards about which we have limited knowledge. The time scale for mitigating and adapting to these hazards is short, and so new policy frameworks are being developed to understand and manage the associated risks.

1.2 Scope of Analysis

In this book we consider five broad aspects of climate in Asia and the Pacific that capture the interactions between climate and both natural and human systems. These are:

- Climate variability and change
- Climate and urbanization
- Climate and security
- Climate and society
- Climate and sustainability.

To focus the analysis, which draws on published results in the scientific literature, we first highlight some of the key climate-related issues for Asia and the Pacific. More detailed discussions of each of these issues are presented in the subsequent chapters of the book.

1.2.1 Large-Scale Climate Systems of Asia and the Pacific

The Asian-Australian monsoon influences the lives of about 60 % of the world's population through its major seasonal variations. The movement of the monsoon is essentially a result of the seasonal migration of the sun and the temperature contrast between the oceans and continental land masses. These interactions lead not only to large spatial variations in the monsoon across Asia but also to substantial temporal variability on scales from sub-seasonal to decadal. Indeed, the natural variability of the monsoon makes it difficult to detect significant trends in its characteristics.

The onset of the Indian monsoon is characterized by the transition of the zone of high precipitation (inter-tropical convergence zone – ICTZ) from the equator to about 15 °N at the end of May. This transition is followed by a more gradual progression north, with subsequent return southward in September. The initial onset of the East Asian monsoon is characterized by the establishment of the Meiyu-Baiu-Changma front in May, with further development in June (Goswami et al. 2006).

Since the early studies of (Walker 1924), it has been recognized that the behavior of the monsoon is affected by the east–west circulation across the Pacific of the Walker cell, which is in turn associated with the ENSO phenomenon. This relationship leads to inter-annual and longer-term variability in the monsoon.

The climate of the Pacific is dominated by the easterly trade winds that are driven by the meridional Hadley cell. The upward arm of the Hadley cell is delineated by the rain clouds of the ICTZ. These clouds form as the warm moist air in the Hadley cell is lifted before flowing in the upper troposphere to higher latitudes and subsequently sinking in the dry belts of the sub-tropics. In the South Pacific, the rising arm of the Hadley cell is also apparent from the clouds of the South Pacific Convergence Zone (SPCZ), which is associated with the interaction between the trade winds and mid-latitude disturbances in the prevailing westerly winds.

The seasonal variations in the ICTZ and SPCZ generate the annual cycle of the climate of the Pacific. Inter-annual variability in the climate is greatly affected by the interaction of the Walker cell with the upper ocean that produces the ENSO phenomenon. In the Indian Ocean, some distinct patterns in the sea-surface temperature are known as the Indian Ocean Dipole (IOD), which is correlated with inter-annual variations in climate in parts of Asia and Australia. Correlations are also found between inter-decadal variations in climate of the region and a sea-surface temperature pattern of the Pacific Ocean known as the Inter-decadal Pacific Oscillation (IPO). The IPO is also known as the Decadal Pacific Oscillation, or DPO.

In Sect. 2.1, we investigate the nature of the climate of Asia and the Pacific in more detail, and identify some evidence of longer-term variability and change in these large-scale features of the climate. A particular question is whether identified changes can be attributed to human activities, and there is continuing research to clarify this issue.

1.2.2 Trends in the Mean Climate Across Asia and the Pacific

Analysis of the surface climate records from around the world has clearly established that the world is warming, with similar trends in the sea-surface temperature and in the land temperature. Identifying trends in precipitation is much more difficult, partly because its natural variability is high and its spatial and temporal coherence is low. The basic measurement of precipitation over the Indian and Pacific Oceans is limited by the lack of *in situ* observations, and so there is great dependence on indirect satellite-based instruments.

The Himalayas and Tibetan Plateau (HTP) are of particular interest because the region includes a large number of glaciers. The mass balance of a glacier is determined by the combined impacts of precipitation and temperature, which can vary locally. Nonetheless, it is found that in Asia (and globally) glaciers have been retreating for some decades. The remoteness of the HTP region together with the large number of small glaciers means that comprehensive monitoring is very difficult (Sect. 2.1).

1.2.3 *Trends in Extreme Climate Events Across Asia and the Pacific*

While there is interest in identifying trends in the mean climate, many natural ecosystems (as well as humans) respond dramatically to extremes in temperature and rainfall. For example, heat waves where the over-night minimum temperature does not fall below about 24 °C can lead to substantial increases in human mortality in Melbourne, Australia (Nicholls et al. 2008b). Several studies have been carried out in the Asia-Pacific region to prepare systematic analyses of current trends in climate extremes (for example, Choi et al. 2009). Given the range in orography across the region from the HTP to small islands in the Pacific, it is useful to consider whether trends are different in the high mountain areas, but analysis of temperatures across South Asia suggests that trends at high altitudes are often affected by local features (Revadekar et al. 2012).

As with mean climate trends, we expect trends in precipitation extremes to be more difficult to detect. In Sect. 2.1, we report on the findings from recent studies of temperature and precipitation extremes across Asia and the Pacific. In general, the indicators are based on percentiles so that meaningful comparisons can be made across different climate regimes.

1.2.4 *Challenges and Opportunities for Modeling the Climate of Asia and the Pacific*

Climate modeling involves the use of computers to solve the complex equations that describe the physical basis of variations in the atmosphere and ocean; these models also take into account the interactions between the atmosphere and the land surface. Indeed, modeling provides an effective means to assimilate observations to improve our understanding of climate variations on a range of time and space scales. Together modeling and monitoring of climate provide the foundation for analysis of the interactions between climate, natural ecosystems and human socio-economic systems.

In recent decades there has been much use of global climate models (GCMs) to provide projections of future climate under the effects of enhanced emissions of greenhouse gases (GHGs). Similar models are also used routinely for weather forecasting and for seasonal outlooks of climate variability. While global climate models generally provide information on scales of 100 km or so, regional climate models (RCMs) are used to give details at much finer scales. Statistical methods can also be used to relate the output of climate models to local-scale features, such as the temperature and rainfall at a specific location.

Climate modeling for the Asia and the Pacific has particular challenges because of the complexities in the topography. Over the Pacific there are many small islands with extensive coastlines that generate small-scale weather features such as sea breezes. Over Asia the steep and rugged orography of the HTP region is difficult to resolve in most numerical climate models. Progress in modeling for Asia and the Pacific is discussed in Sect. 2.2.

1.2.5 Current Projections for the Future Climate Across Asia and the Pacific Under Climate Change

Projections of future climate in Asia are dependent upon the ability of climate models to represent the Asia monsoon, which dominates the seasonal variability of the region. Tropical cyclones, which are relatively small-scale features to be represented in climate models, also have significant impacts on the climate of Asia and the Pacific. Nonetheless, a number of studies have been carried out to assess the likely variations in the monsoon under climate change scenarios developed through the international Climate Model Intercomparison Project Phase 3 (CMIP3). There are also some early results from the more recent CMIP5, which uses updated emission scenarios and climate models.

In Sect. 2.2, we also consider the results of studies carried out using regional climate models, focused on specific subregions of Asia and the Pacific. The Regional Model Intercomparison Project (RMIP) is a collaborative study by about ten groups from the region aimed at providing climate projections to support the impact and adaptation community (Fu et al. 2005) across Asia. The Pacific Climate Change Science Program (PCCSP) provides detailed projections for 15 island states in the Pacific (Power et al. 2011) to support impact and adaptation studies. The World Climate Research Programme (WCRP) has established a new programme, the Coordinated Regional Downscaling Experiment (CORDEX), to produce climate information at regional scales across the globe (Giorgi et al. 2009), and there are CORDEX projects focused on South Asia, East Asia and Southeast Asia. These projects are collaborating to generate consistent and useful projections across monsoon Asia.

1.2.6 Climate: Society, Security and Sustainability

The climate of the Asia-Pacific region clearly plays an essential role in the functioning of the environments and societies across the region. Chapters 3, 4, 5 and 6 of the book consider the complex interactions between these features of the region, with an emphasis on the impacts of climate variability and change and on the responses of human and environmental systems to climate. Chapter 3 considers the interactions between urbanization and climate, while Chap. 4 discusses the relationships between climate and the security of societies for food, water and energy, as well as the need for the management of climate-related natural disasters. In Chap. 5 we consider societal issues of governance, remote communities, and human health in relation to climate variability and change. The interactions between climate and sustainability are discussed in Chap. 6, with a focus on integrated assessments and the management of natural ecosystems.

1.2.7 Trends in Urbanization Across Asia and the Pacific

The move of rural populations to towns and cities has been the basis of economic progress for thousands of years around the world. In Sect. 3.2, trends in urbanization since 1950 across Asia and the Pacific are documented and the expected trends to 2050 are discussed. The fraction of urban population in Asia increased from about 15 % in 1950 to 30 % by 1990, and to 40 % by 2010. The rapid increase in urbanization has been accompanied with significant economic growth across the region.

A noticeable trend in Asia has been the growth of mega-cities, which have populations greater than ten million. Urbanization is expected to continue in the coming decades with about three billion people in urban areas by 2050. However, while the number of mega-cities will continue to increase, the majority of urban residents (about 60 % of the total urban population) will live in cities of less than one million people. The increase in urbanization has profound implications on the interactions between climate, land and energy use on scales varying from global to local. The expected concentration of people in smaller urban areas means that there should be greater focus on mitigation and adaptation planning for these areas.

1.2.8 Interactions Between Urban Areas and Climate

The urban heat island effect is a well-known feature of a significant impact of cities on climate (Oke 1973), and the effect has been documented in the cities of Asia. Once a heat island effect is established, there is evidence that the urban trends in temperature become similar to those in neighboring rural areas, so that the effect is not seen in hemispheric or global temperature trends (Peterson 2003).

In Sect. 3.3 we also consider the impacts of urban areas on precipitation and air quality. While it is clear that the land-use change associated with urban areas affects local precipitation patterns, the details can vary from place to place and from season to season. On the other hand urban activity, especially those associated with fossil fuels, leads to reductions in air quality in urban areas. Aerosols emitted from the burning of fossil fuel are a major problem for human health and for their impacts on local and regional climate across Asia.

Urban areas are the main source of greenhouse gases (GHGs) associated with global warming. Indeed, the provision of food, water and energy for cities is the main driver of the increasing emissions of GHGs in Asia and the world. The handling of waste from urban areas is also a significant source of GHGs. Recognizing that some of these functions (especially the production of energy and food) tend to take place in the areas around rather than within cities, it is important to include the peri-urban areas when considering the total impact of cities on GHG emissions.

While the largest cities are the highest emitters of GHGs across Asia, the per capita emission rate depends upon a range of factors, such as local climate, community wealth and population density. Higher emission rates per capita tend to be in lower density cities in colder climates and with greater wealth. There is evidence that low density urban areas (and rural areas) are less energy efficient than areas with higher population densities (Marcotullio et al. 2012). Such observations imply that mitigation policies should not unintentionally promote anti-urban outcomes.

1.2.9 Climate Hazards and Vulnerabilities for Urban Areas

Climate change is manifested through changes in local features such as temperature, sea level (storm surges), air quality, precipitation and hydrology. All these features are impacting on urban communities across Asia, often resulting in weather-related disasters. In Sect. 3.4, it is noted that about 40 % of the reported flooding events across the globe from 2000 to 2009 occurred in Asia, with cities of China, India and Thailand seen as most vulnerable to future coastal flooding (Nicholls et al. 2008a).

Urban infrastructure is vulnerable to climate variability and change in Asia and the Pacific. Vulnerabilities are seen in transportation, water supply and sanitation, food production and distribution, energy production and distribution, and manufacturing industries. Within communities, there are sections with higher vulnerability to the hazards of climate because of their lower capacity for adaptation: these groups include the poor, the elderly and the very young.

1.2.10 Climate Change Mitigation and Adaptation Strategies for Urban Areas

The significance of climate variability and change is recognized in most regions of Asia and the Pacific and Sect. 3.5 describes a range of mitigation and adaptation strategies that are being developed and implemented. The wide variation in the nature of urban areas across the region means that different strategies are being applied. Cities are generally encouraged to use their local knowledge in developing optimal strategies, taking into account their local biophysical and socio-economic conditions.

The policy emphasis in Asia tends to be on mitigation rather than adaptation strategies at present (Satterthwaite et al. 2007). However, some mitigation policies, such as those to reduce GHG emissions from transportation, also assist the process of adaptation to climate change through, for example, improved air quality. Urban design is a key area for innovative strategies to both reduce GHG emissions and adapt to climate variability and change. For example, the reclaimed Cheonggyecheon

River in the middle of Seoul has provided benefits for tourism, recreation and reductions in heat island effects (Cho 2010). The development of acceptable and effective policies for mitigation and adaptation will require enhanced understanding of urban governance, particularly of the power relationships that influence outcomes at the local level.

1.2.11 Meeting Future Needs for Rice and Wheat Across Asia

It is anticipated that global food production will need to double by 2050 to feed a world population of 9.2 billion. There are now 450 million small-farm holders largely across Asia whose livelihoods are vulnerable to climate variability and change. Increasingly, food producers are also susceptible to market forces at regional or even global scales, as the interactions between markets become more complex. Section 4.1 considers the interactions between the range of hazards and vulnerabilities for farmers in Asia and the Pacific.

Rice is the dominant staple food across most of Asia. Indeed, it feeds about half the world population, and about 750 million of the poorest people depend on rice. Rice is therefore the focus of Sect. 4.1. The production of a kilogram of rice requires 2,500 l of water. The production of rice is clearly susceptible to the availability of water and land and to the impacts of climate change. The impacts of climate change are compounded with the continuing impacts of inter-annual climate variability, generally driven by the El Niño phenomenon. In parts of Asia there are additional pressures on food production from the competitive generation of biofuels, based on uncertain attempts at the mitigation of GHG emissions.

Wheat is also an important crop in parts of Asia, especially in India, Pakistan and China. The issues associated with the production of wheat are similar to those of rice. Strategies are being developed in Asia for both the mitigation of climate change and the adaptation to its impacts. Because of the continuing challenges of seasonal to inter-annual climate variability, many strategies are based on the management of these issues.

1.2.12 Meeting Future Needs for Food from Fisheries Across Asia and the Pacific

The fisheries of the Pacific and Indian Oceans are an essential source of food for Asia and the Pacific. These fisheries are under stress not only from climate change and variability, but also from direct human activities such as pollution from industries along coastal areas and increased fishing due to population growth. In Sect. 4.1 we consider the impacts of these pressures, which have led to significant changes in the diversity and biomass of marine ecosystems (Fig. 1.2).

Fig. 1.2 Local fisheries in Ha Long Bay, Viet Nam (Source: APN)

While the concentration of GHGs has increased in the atmosphere leading to global warming, the oceans have absorbed about 40 % of the carbon dioxide emitted from anthropogenic activities and 80 % of the heat associated with global warming. These additional loadings have caused the acidity of thc ocean to increase by about 30 % and the temperature of the upper ocean to increase by a fraction of a degree. Changes in evaporation and precipitation lead to changes in salinity near the surface. Marine ecosystems are susceptible to all these changes; for example, (Takasuka et al. 2004) suggests that shifts in the 'warm' and 'cool' anchovy regimes in the North Pacific Ocean are due to temperature variations in the ocean.

Fisheries management for the future needs to be built on an ecosystems approach, where marine, climate and human influences are taken into account. The life span of marine species varies considerably, and so management measures should be developed and implemented early in order to account for the time lags in the growth of stocks to sustainable levels.

1.2.13 Future Prospects for Water Security Across Asia and the Pacific

For many countries of Asia and the Pacific, water security is a critical issue due to growing populations and economies matched against the finite availability of water in any one nation. Climate change intensifies the issue with its enhanced uncertainties in the annual cycle of rainfall in many regions. These issues are discussed in Sect. 4.2.

Fig. 1.3 Himalayas from Kathmandu Valley (Source: APN)

The "water tower" for Asia is the HTP region, which covers about seven million square kilometers and includes the largest and highest glaciers in the world (see Fig. 1.3). This region provides water for about 20 % of the world population, principally through the annual melting of snow and ice into the major rivers that then flow through several countries to the sea. Trans-border issues are therefore special challenges that require multi-national agreements to ensure water security for all countries along each river.

The HTP region is especially sensitive to climate change, with higher temperatures promoting increased melting of glaciers. The rate of glacier retreat is, however, a complex function of local conditions (Fujita and Nuimura 2011). The relationship between glacier melt and river stream flow is further complicated by the varying hydrology along each river. Glacier melt not only provides vital water resources for downstream communities and natural ecosystems, but they also bring the threat of natural disasters such as glacier lake outburst floods (GLOF).

The rivers of the HTP provide hydro-electric power at an increasing number of locations. There are plans to build a further 100 dams to generate about 150 gigawatts of power across the region. The local and regional impacts of the infrastructure associated with these generators are not fully understood at this time. Their construction does emphasize the need for multi-national agreements on flows along the relevant river systems.

Water security for Pacific Islands tends to be a national issue. The supply of fresh water and the disposal of waste are critical for many island communities, owing to both natural and human factors. Population growth and climate change exacerbate the challenges. The variations in climate, topography and culture across the 30,000 islands of the region mean that a range of strategies will need to be developed to ensure water security into the future.

1.2.14 Balancing Energy Demands Across Asia and the Pacific with Reduced GHG Emissions

As the economies of Asia and the Pacific grow to support and enhance the wellbeing of communities, the demand for energy increases. Fossil fuels have been the main source of the required energy. The challenge for the region and the world is to maintain the increase in community wellbeing while limiting the emission of carbon dioxide from the combustion of fossil fuels. The factors associated with this balance are considered in Sect. 4.3.

The demand for energy can be reduced by improving energy efficiency, so that there is no longer a linear relationship between energy use and national economic indicators. The emission of GHGs can be reduced by increasing the effectiveness of energy generation, through the use of non-fossil fuels or increasing the efficiency of traditional generators. In Sect. 4.3, the policy options being taken by the major economies of Asia (namely China, India and Japan) are analyzed and compared.

1.2.15 Institutional Arrangements in Asia and the Pacific to Manage and Mitigate the Impacts of Climate Change

The development and implementation of strategies to secure food, water and energy for the communities of Asia and the Pacific are essentially dependent on the governance processes in operation across the region. Section 5.1 considers the governance processes that allow the societies of the region to function effectively. National and regional institutional arrangements have been developed, sometimes over centuries, to ensure safe and stable environments for each community. Climate change is imposing new stresses on these institutional arrangements.

Mitigation of climate change and adaptation to climate change impacts require distinct governance mechanisms, with mitigation tending to be a longer term issue compared with adaptation. Indeed, human societies have a propensity to develop *ad hoc* solutions in adapting to environmental changes. There has been concerted, if not altogether successful, efforts at the international level for the development of policies to mitigate the emission of GHGs, and this is complemented by work at national and regional levels. Adaptation policies are generally taken up at local or national levels, but for parts of Asia and the Pacific international cooperation is likely to play a significant role.

The acceptance and implementation of policy strategies by communities require public education and communication. These efforts are needed especially for strategies that involve demand management for resources such as energy or water.

Fig. 1.4 Community village in the Himalayas (Source: APN)

1.2.16 Future Prospects for Remote Communities Under Climate Change

There is a particular challenge for governance in ensuring that the remote communities of Asia and the Pacific are not disadvantaged by climate change. In Sect. 5.2 we consider the vulnerabilities of these communities. More than 30 million people live in the HTP region. They are seen as one of the groups most vulnerable to climate change impacts, as their natural environment is very harsh and they are always at risk from a range of natural disasters. There is increasing evidence of climate change affecting their environment, with loss of biodiversity and tree-line shifts as temperatures rise. As noted earlier, there is also an increased risk of natural disasters, such as GLOFs. Changes in the seasonal stream flow of rivers, due to changes in the times of snow and glacier melt, are impacting on their water security for agriculture and domestic uses (Fig. 1.4).

As with remote mountain communities, the small island communities of the Pacific and Indian Oceans are recognized as being especially vulnerable to climate change impacts. There are thousands of small islands across the Pacific supporting more than two million people in small island developing states (SIDS). There is great cultural and social diversity between these communities, which have adapted to their natural environments over many years. The impacts of climate change, together with the modern social and economic pressures associated with growth, increase their risks. The environmental hazards come from sea level arise, as well as from the direct effects of increasing temperatures and changes in extreme climate events. A particular challenge is associated with the maintenance of water resources.

SIDS make small contributions to the global emission of GHGs, but are particularly susceptible to the impacts of those emissions. Thus, they have been very active in the international negotiations on mitigation. There has also been some work on the development of adaptation strategies for SIDS, but migration remains a possible option for many of these states in the future.

1.2.17 Mitigating the Effects of Climate Change on Human Health Across Asia and the Pacific

Climate change poses a major threat to human health for many societies of Asia and the Pacific. In Sect. 5.3, we discuss these threats and strategies to alleviate them. For many communities of the region, the natural variability of climate already poses risks from natural hazards such as typhoons and heat waves. Poor design in urbanization and industrialization can lead to additional stresses from air pollution and the urban heat island effect. These hazards are being increased by climate change, especially by increases in temperatures and their associated extremes.

There is evidence of vector-borne diseases, such as malaria and dengue, spreading to new areas as the climate changes. A range of public health programs are being developed at local, national and international levels to mitigate the impacts of climate change on human health in Asia and the Pacific.

1.2.18 Adoption of Low Carbon Development Pathways Across Asia and the Pacific

The management of the societal risks from climate change has become a key aspect of the broader goal of sustainable development at national, regional and global levels. An essential component of sustainable development is the move towards low carbon societies, which aim to reduce carbon dioxide emissions while promoting economic development and community wellbeing. The issues associated with low carbon development are discussed in Sect. 6.1.

Countries across Asia and the Pacific have adopted GHG emission targets in order to reduce their dependence on carbon-based economies. National, regional and international mechanisms are being developed to support these voluntary objectives. The best means of investigating low carbon strategies and their implementation is through integrated assessment models (IAMs). Low carbon investigations using IAMs (for example, Nordhaus 1979) now incorporate all the relevant aspects of the economy to allow studies of the effectiveness of policy strategies, and they have been applied to a range of case studies across Asia and the Pacific. Both national and regional approaches are being investigated.

1.2.19 Ecosystem Services to Support Mitigation of GHG Emissions and to Enhance the Adaptive Capacity of Communities Across the Asia and the Pacific

The 1987 World Commission on Environment and Development provides a milestone in attempting to reconcile economic development and protection of the environment through sustainable development. Thus, natural ecosystems are seen as a part of sustainable development through the delivery of ecosystem services. These concepts, which are discussed in Sect. 6.2, can be applied in Asia and the Pacific to assess the sustainability of natural ecosystems (such as forest and fresh water resources) that support a range of human industries. Coastal communities are dependent upon the sustainability of fisheries and marine ecosystems.

Climate change is imposing additional challenges for management policies by increasing the environmental stresses on natural ecosystems. However, careful analysis shows that the natural ecosystems of Asia and the Pacific have an essential role in the mitigation of climate change. For example, forests in the Philippines (Lasco and Pulhin 2001) and other countries of the region have the potential to contribute to climate change mitigation. Such analyses need to take into account the interactions between climate, natural ecosystems and human communities. They can lead to strategies to better manage the risks associated with climate variability and change.

1.3 Conclusions

The climate of Asia and the Pacific varies greatly, and these variations have played a major role in the development of natural ecosystems and of human societies. Human activities, principally the growth of populations and economies across the region, are placing additional stresses on both natural and societal systems. Anthropogenic climate change is a significant factor in these increases in stress. In the chapters of this book, we document the climate of the region and the interactions of the climate with the environment and the societies of Asia and the Pacific. There is a particular focus on the impacts of climate change and on strategies aimed at both mitigating and adapting to those impacts.

References

Cho, M. R. (2010). The politics of urban nature restoration: The case of Cheonggyecheon restoration in Seoul, Korea. *International Development Planning Review, 32*, 145–165.

Choi, G., Collins, D., Ren, G., Trewin, B., Baldi, M., Fukuda, Y., … & Zhou, Y. (2009). Changes in means and extreme events of temperature and precipitation in the Asia-Pacific Network region, 1955–2007. *International Journal of Climatology, 29*, 1906–1925. doi:10.1002/joc.1979.

Fu, C., Wang, S., Xiong, Z., Gutowski, W. J., Lee, D. -K., McGregor, J. L., … & Suh, M. -S. (2005). Regional climate model intercomparison project for Asia. *Bulletin of the American Meteorological Society, 86*, 257–266.

Fujita, K., & Nuimura, T. (2011). Spatially heterogeneous wastage of Himalayan glaciers. *PNAS, 108*, 14011–14014.

Giorgi, F., Jones, C., & Asrar, G. R. (2009). Addressing climate information needs at the regional level: The CORDEX framework. *WMO Bulletin, 58*(3), 175–183.

Goswami, B. N., Wu, G., & Yasunari, T. (2006). The annual cycle, intraseasonal oscillations, and roadblock to seasonal predictability of the Asian summer monsoon. *Journal of Climate, 19*, 5078–5099. doi:10.1175/JCLI3901.1.

Lasco, R. D., & Pulhin, F. B. (2001). Forestry mitigation options in the Philippines: Application of the COMAP model. *Mitigation and Adaptation Strategies for Global Change, 6*, 313–334.

Manton, M. J., Heath, L., Salinger, J., & Stevenson, L. A. (2011). *Climate in Asia and the Pacific: A synthesis of APN activities*. Kobe: Asia Pacific Network for Global Change Research.

Marcotullio, P. J., Sarzynski, A., Albrecht, J., & Schulz, N. (2012). The geography of urban greenhouse gas emissions in Asia: A regional analysis. *Global Environmental Change, 22*(4), 944–958.

Nicholls, R. J., Hanson, S., Herweijer, C., Patmore, N., Hallegatte, S., Corfee-Morlot, J., Chateau, J., & Muir-Wood, R. (2008a). *Ranking port cities with high exposure and vulnerability to climate extremes: Exposure estimates* (OECD environment working papers, Vol. 1). Paris: OECD.

Nicholls, N., Skinner, C., Loughman, M., & Tapper, N. (2008b). A simple heat alert system for Melbourne, Australia. *International Journal of Biometeorology, 52*, 375–384.

Nordhaus, W. D. (1979). *The efficient use of energy resources*. New Haven: Yale University Press.

Oke, T. R. (1973). City size and the urban heat island. *Atmospheric Environment, 7*, 769–799.

Peterson, T. C. (2003). Assessment of urban versus rural in situ surface temperatures in the contiguous United States: No difference found. *Journal of Climate, 16*, 2941–2959.

Power, S. B., Schiller, A., Cambers, G., Jones, D., & Hennessy, K. (2011). The Pacific climate change science program. *Bulletin of the American Meteorological Society, 92*, 1409–1411.

Revadekar, J. V., Hameed, S., Collins, D., Manton, M., Sheikh, M., Borgaonkar, H. P., … & Shrestha, M. L. (2012). Impact of altitude and latitude on changes in temperature extremes over South Asia during 1971–2000. *International Journal of Climatology, 33*(1), 199–209. doi:10.1002/joc.3418.

Satterthwaite, D., Huq, S., Pelling, M., Rei, H., & Lankao, P. R. (2007). *Adapting to climate change in urban areas, the possibilities and constraints in low- and middle-income nations* (Cited in: Human settlements discussion paper series, Theme: Climate change and cities, Vol. 1). London: International Institute for Environment and Development.

Takasuka, A., Oozeki, Y., Kimura, R., Kubota, H., & Aoki, I. (2004). Growth-selective predation hypothesis revisited for lraval anchovy in offshore waters: Cannibalism by juveniles versus predation by skipjack tunas. *Marine Ecology Progress Series, 278*, 297–302.

Walker, G. T. (1924). *Correlation in seasonal variations of weather. IX. A further study of world weather. Memoirs of the India Meteorological Department, 24*, 275–333.

Chapter 2
Climate in Asia and the Pacific: Climate Variability and Change

Michael James Salinger, Madan Lall Shrestha, Ailikun, Wenjie Dong, John L. McGregor, and Shuyu Wang

Abstract The geographic extent of Asia and the Pacific leads to great variation in the climate of the region. Major influences on global climate arise from the scale and elevation of the Himalayan Tibetan Plateau (HTP), and from the air-sea interactions in the Pacific associated with the El Niño–Southern Oscillation (ENSO). The monsoon has a profound effect on the climate of Asia, with its strong seasonal cycle. Variability is also caused by ENSO, the Indian Ocean Dipole (IOD) and

M.J. Salinger (✉)
University of Auckland, 3/7 Mattson Road, Pakuranga, 2010 Auckland, New Zealand
e-mail: salinger@orcon.net.nz

M.L. Shrestha (✉)
Nepal Academy of Science and Technology, Khumaltar, Lalitpur, Nepal
e-mail: madanls1949@gmail.com

Ailikun
Institute of Atmospheric Physics, Monsoon Asia Integrated Regional Study (MAIRS) IPO, Chinese Academy of Sciences, 40# Hua Yan Li, Qi JIa Huo Zi, Chao Yang District, P.O. Box 9804, 100029 Beijing, China
e-mail: aili@mairs-essp.org; aili@tea.ac.cn

W. Dong
State Key Laboratory of Earth Surface Processes and Resource Ecology, College of Global Change and Earth System Science, Beijing Normal University, 19 Xinjiekou Wai Street, 100875 Beijing, China
e-mail: dongwj@bnu.edu.cn

J.L. McGregor
CSIRO Marine and Atmospheric Research, 107-121 Station Street, 3195 Aspendale, VIC, Australia
e-mail: John.McGregor@csiro.au

S. Wang
Institute of Atmospheric Physics, Chinese Academy of Sciences, Building 40, Huayan Li, Chaoyang District, 100029 Beijing, China
e-mail: wsy@tea.ac.cn

M.J. Manton and L.A. Stevenson (eds.), *Climate in Asia and the Pacific: Security, Society and Sustainability*, Advances in Global Change Research 56, DOI 10.1007/978-94-007-7338-7_2,

Pacific Decadal Oscillation (PDO). ENSO is the principal source of inter-annual global climate variability. ENSO has significant climate and societal impacts on both regional and global scales. The climate effects of ENSO are modulated on decadal time scales by the PDO. IOD affects the climate in the Indian Ocean and Australasia. Climate observations show significant warming trends in temperature across Asia and the Pacific; not only is there an increase in mean temperature, but there is more warming in North Asia and less in the Pacific during the twentieth century. Observed trends in precipitation are more variable, with some evidence of increasing intensity of storms. Glacier mass balance studies show dramatic decline in ice mass in the Himalayas and New Zealand, with monitored ice mass losses of 0.3–0.5 km^2/year, in the last three decades. Temperature extremes have changed region wide: cool nights, cold days have very significantly decreased universally, and the frequency of hot days has increased. Projections of future climate change for the region suggest longer summer heat waves in South and East Asia and Australia, and increases in precipitation in several areas. Regional downscaling techniques are used to project future climate: warming is projected to be largest in high latitude (Northern Asia, Central Asia) and high altitude (Tibetan Plateau) regions; with a suppression of the south Asian summer monsoon, along with a delay of monsoon onset and increase of monsoon break periods. Monsoon precipitation is projected to increase over South Asia.

Keywords Monsoon Asia • Pacific • Climate variability trends • Climate change trends

2.1 Observed Climate Variability and Trends

2.1.1 Introduction

Asia-Pacific surface climates span a wide range of latitudes from sub-polar regimes in the Russian Federation, monsoon Asia, which dominates a large proportion of Asia, into the Pacific Ocean, which extends from temperate to equatorial latitudes. Altitude range is extreme, spanning from sea-level to more than 8,500 m in the Himalayas, with the average altitude of the Himalayan Tibetan Plateau (HTP) being 4,500 m. This provides the most continental climates on the planet. In contrast, the Pacific Basin is the largest oceanic basin extending from the Arctic in the north to the Southern Ocean (or, depending on definition, to Antarctica) in the south, bounded by Asia and Australia in the west, and the Americas in the east. The climates of small Pacific Islands are the most oceanic in the world. The Asian highland massif contains numerous glaciers. The Himalayan range alone has a total snow and ice cover of 35,110 km^2 containing 3,735 km^3 of permanent snow and ice (Qin et al. 2006).

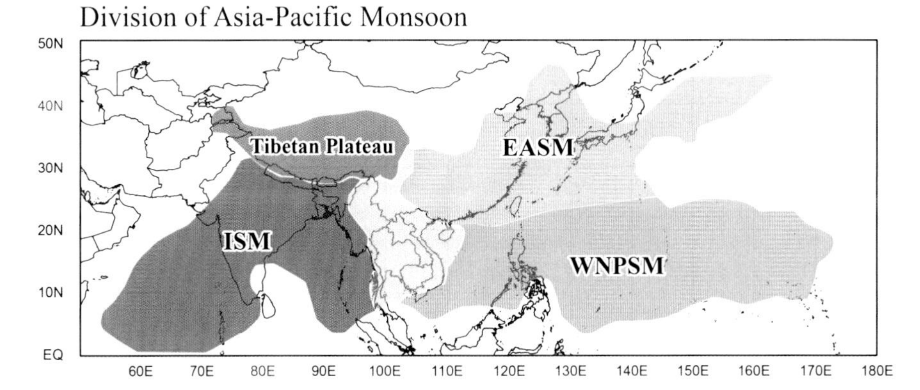

Fig. 2.1 Schematic diagram of the Asian monsoon region: *blue* – the east Asian summer monsoon (*EASM*), *purple* – the Indian summer monsoon (*ISM*), *green* – northwest pacific summer monsoon (*WNPSM*), *yellow* – monsoon buffer, *red* – Qinghai-Tibet Plateau (Reprinted with permission from "Rainy Season of the Asian–Pacific Summer Monsoon," by B. Wang and H. Lin, 2002, Journal of Climate, 15, p. 392. Copyright 2002 by American Meteorological Society. Modified from source)

2.1.2 Large Scale Circulation and the Monsoon System

2.1.2.1 Asia

Monsoon circulation is described as seasonal reversing of wind along with a change of precipitation caused by the asymmetric heating between land and ocean. The Asian monsoon is a most significant component of the global climate system, influencing lifestyles and livelihoods, and providing water resources to nearly 60 % of the world's population. Usually, the summer monsoon season in Asia starts from mid or late May and ends in late September. The Asian monsoon can be classified into some sub-systems, such as south Asian monsoon, east Asian monsoon, south-east Asian monsoon and western north Pacific monsoon (Wang and Lin 2002), but South Asia (India) and East Asia are the two main monsoon areas that have been studied extensively (Fig. 2.1). Besides the strong annual cycle, the Asian monsoon has a wide range of variability from the intra-seasonal, inter-annual to inter-decadal time scales. The intra-seasonal oscillation, with time scales from weeks to months, determines the "active raining period" and "break dry period" of the monsoon, which highlights the importance of intra-seasonal monsoon variability to annual prediction (Webster et al. 1998).

The Asian monsoon is characterized by seasonal migration (northward and withdrawal) of a precipitation belt or the inter-tropical convergence zone (ITCZ) in summer (Gadgil 2003). As shown by Goswami et al. (2006a), the onset of the Indian monsoon starts with a rapid transition of the high precipitation zone from near the equator to about 15°N toward the end of May or beginning of June. Another characteristic is the splitting of the ITCZ into a primary branch over the continental region between 20°N and 25°N, and a secondary branch between the equator and

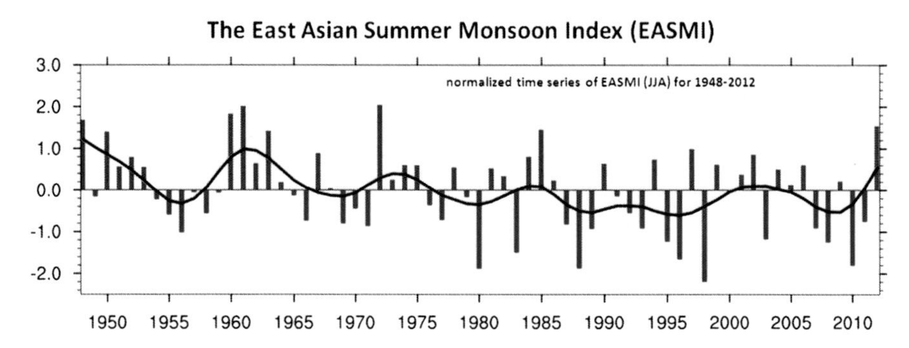

Fig. 2.2 The east Asian summer monsoon index (*EASMI*). Normalized time series (*JJA*) for 1948–2012 (Source: After Li et al. 2010)

10°S. The monsoon tends to withdraw in late September. The northward propagation of the Indian monsoon onset has slowed during the past decade compared with prior decades, and this is consistent with the observed weakening of the May–June mean easterly vertical shear in the region and weakening of the north–south gradient of low-level humidity across the equator. There is evidence to support the hypothesis that the weakening of the easterly shear is due to an eastward shift of the Walker circulation, associated with strengthening of the El Niño phenomenon (Goswami et al. 2010).

The long term trend of the Indian monsoon can be monitored from the observed historical daily precipitation record over India from 1951 to 2001. Goswami et al. (2006a) found rising trends in the frequency and the magnitude of extreme rain events and a significant decreasing trend in the frequency of moderate events over central India during the monsoon seasons. Moreover, a substantial increase in hazards related to heavy rain is expected over central India in the future.

The seasonal migration of the east Asian summer monsoon (EASM) has two stages. The first onset happens in mid-May with the rapid establishment of the Meiyu-Baiu-Changma front. The second onset takes place in early June with the southern ITCZ moving rapidly to about 20°N and establishing dry conditions south of the equator (Goswami et al. 2006a). Many studies have shown weakening of the EASM in recent decades, especially precipitation decreasing from the 1980s in northern China. At the same time, Li et al. (2010) noted a southward shift of the main components of the EASM (the subtropical westerly jet stream, the western Pacific ocean subtropical high, the subtropical Meiyu-Baiu-Changma front, and the tropical monsoon trough) from 1958 to 2008 (Fig. 2.2). Such a southward shift may be due to meridional asymmetries in the regional response to global warming.

The inter-annual variability of the south Asian monsoon and southeast Asian monsoon is strongly related to the El Niño Southern Oscillation (ENSO) phenomenon, with heavy monsoon rainfall in La Niña years and a weak monsoon in El Niño years (for example, Parthasarathy et al. 1994). This interaction is primarily through changes in the equatorial Walker circulation influencing the regional Hadley Circulation (HC) associated with the Asian monsoon (Lau and Nath 2000). Feng et al. (2011a) show a significant weakening of the northern part of the summer HC and a reverse see-saw relationship of the zonal-mean updraft over 10°N to 20°N and around the equator. This transition is accompanied by the southward retreat of the HC core and is well correlated with the weakening of tropical summer Asian monsoons.

2.1.2.2 Pacific

Surface climates of the tropical Pacific islands are dominated by the vast surrounding ocean and the large-scale atmospheric and oceanic circulations (Streten and Zillman 1984; Terada and Hanzawa 1984). The major atmospheric circulation features include the northeast and southeast trade wind regimes, which originate in the subtropical high pressure belts of each hemisphere where air sinks and dries. These tropical easterly flows are characterized by their constancy in speed and direction, although they tend to be strongest in the respective hemispheric winter season and extend further poleward in the respective hemispheric summer season. The trade winds from the two hemispheres converge in the ITCZ and South Pacific Convergence Zone (SPCZ), where rising air forms the ascending branch of the Hadley Circulation. The HC represents the main north to south component of the Pacific atmospheric circulation. In addition, the Walker Circulation operates in the east to west plane of the tropical Pacific with normally rising air over Indonesia and sinking air in the southeast tropical Pacific. This circulation is intimately linked with the major source of inter-annual tropical climate variability, ENSO.

The SPCZ is one of the most significant features of subtropical southern hemisphere climate (Kiladis et al. 1989; Vincent 1994). It is characterized by low-level convergence of air flow leading to uplift and a band of cloudiness and rainfall stretching from the 'Warm Pool' in the western Pacific south-eastwards towards French Polynesia (Streten and Troup 1973; Kiladis et al. 1989; Vincent 1994). It shares some characteristics with the ITCZ, which lies just north of the Equator, but is more extra tropical in nature, especially east of the Date Line (Trenberth 1976). To the west, it is linked to the ITCZ over the Warm Pool. To the east, it is maintained by the interaction of the trade winds and transient disturbances in the mid-latitude westerly winds propagating from the Australasian region. It tends to lie over a region of large sea surface temperature (SST) gradient, rather than the maximum of SST, and is most active in the Austral (southern hemisphere) summer period (November-April). The location of the convergence maximum of the SPCZ shows considerable variability between seasons, varying by 10–15° of latitude. This causes large variability in rainfall throughout the southwest Pacific.

The Pacific North American (PNA) oscillation (Wallace and Gutzler 1981) describes large-scale features over the North Pacific Ocean and the North American continent, and it has a significant influence on the weather of the Pacific. In its positive phase, the PNA is associated with enhanced ridging of the pressure over western North America and deeper troughs over the central north Pacific and southeastern USA. Variations in the PNA oscillation on time scales of days to months can be predicted with some skill (Johansson 2007), and there is some evidence of links between the PNA oscillation pattern and variations in the large-scale ENSO (Straus and Shukla 2002).

2.1.3 Drivers of Climate Variability and Trends Across Asia-Pacific

Superimposed on the average seasonal cycles of surface climate and observed trends in Asia-Pacific surface climate are various sources of natural climate variability that modulate atmospheric and oceanic climate on time scales from weeks to decades.

2.1.3.1 El Niño-Southern Oscillation (ENSO)

The ENSO phenomenon is the principle source of inter-annual global climate variability. This highly coupled ocean–atmosphere phenomenon is centered in the tropical Pacific. ENSO has significant climate and societal impacts both within the region and, through teleconnections (Fig. 2.3), to many distant parts of the world (Troup 1965; Trenberth 1991, 1997; McPhaden et al. 2006; Trenberth et al. 2007). ENSO fluctuates between two phases, which change the normal Asia-Pacific atmospheric and oceanic circulations. During El Niño events, the easterly trade winds weaken along the equatorial Pacific and a large part of the equatorial Pacific experiences unusually warm SSTs, and warmer than normal SSTs in a large part of the Indian Ocean and warmer than normal surface temperatures over south and south east Asia. At the same time, SSTs are cooler than normal in the subtropical southwest and northwest Pacific and Papua-New Guinea. This is associated with a weakening of the horizontal Walker Circulation and strengthening of the meridional HC. The centre of intense tropical convection shifts eastward towards the Date Line and the ITCZ and SPCZ move closer to the equator. As a result, precipitation is higher than normal in the equatorial Pacific, but lower than normal over South and Southeast Asia, Indonesia, eastern Australia, and the Southwest Pacific with some regions experiencing drought conditions. There are also shifts in the preferred location of tropical cyclone activity. The slope of the thermocline (separating warmer surface and cooler deeper waters) flattens across the Pacific Ocean, and the Warm Pool shifts eastwards.

Fig. 2.3 (continued) for 1958–2004, and GPCP precipitation for 1979–2003 (*bottom left*), updated from Trenberth and Caron (2000). The Darwin-based SOI, in normalized units of standard deviation, from 1866 to 2005 (Können et al. 1988; *lower right*) features monthly values with an 11-point low-pass filter, which effectively removes fluctuations with periods of less than 8 months (Trenberth 1984). The smooth *black* curve shows decadal variations. *Red* values indicate positive sea level pressure anomalies at Darwin and thus El Niño conditions (Source: Trenberth et al. 2007) (Reprinted with permission from "Climate Change 2007: The Physical Science Basis. Working Group I Contribution to the Fourth Assessment Report of the Intergovernmental Panel on Climate Change," Cambridge University Press, Figure 3.27, p. 288)

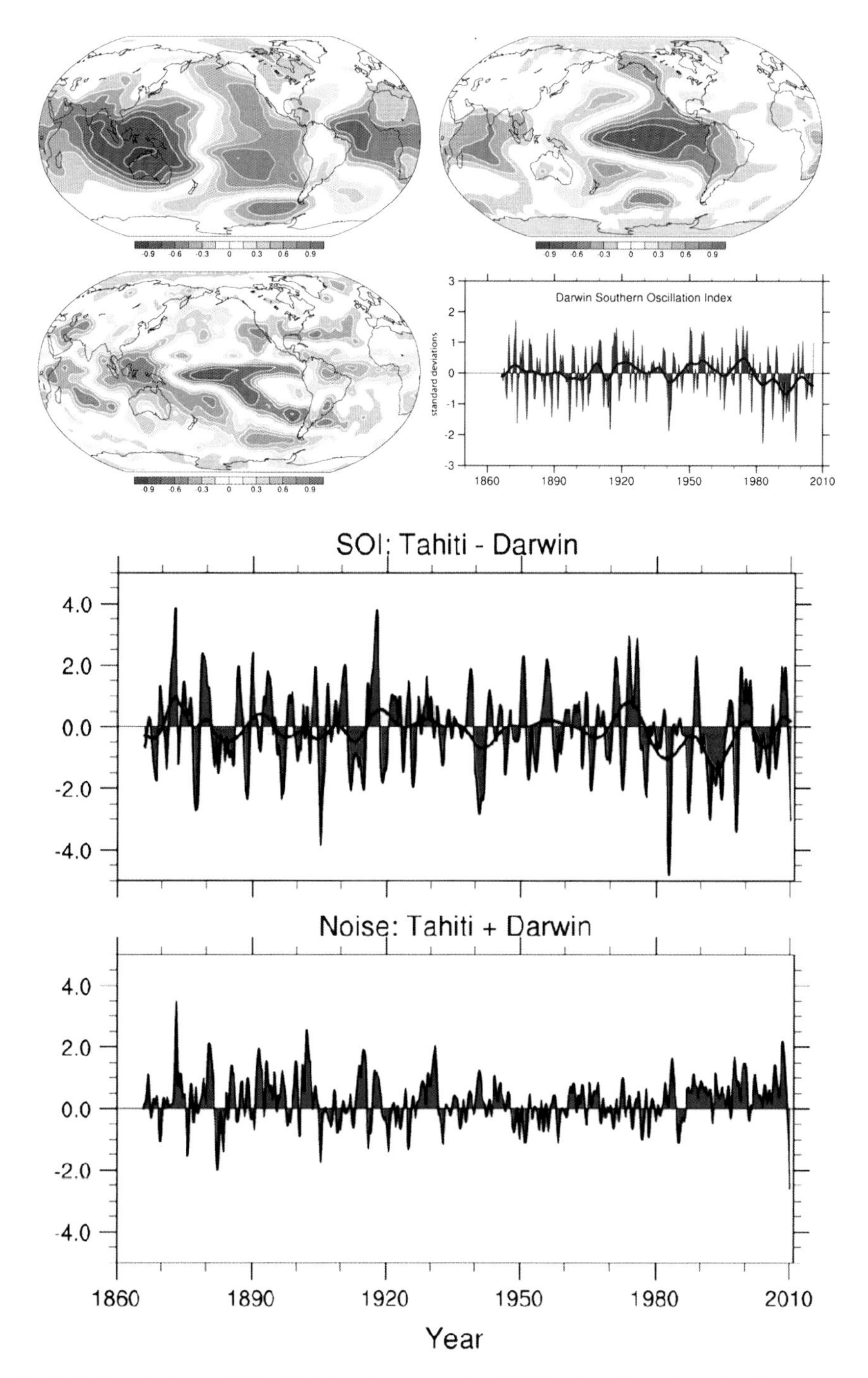

Fig. 2.3 Correlations with the SOI, based on normalised Tahiti minus Darwin sea level pressures, for annual (May to April) means for sea level pressure (*top left*) and surface temperature (*top right*)

Climate anomalies during La Niña events are typically opposite to those of El Niño events, with stronger trade winds and large parts of the Pacific, as well as the Indian Ocean and south Asia, experiencing cooler than normal SSTs. Higher than normal temperatures and SSTs occur in the southwest and northwest Pacific. There are changes in the usual locations of tropical cyclones and a shift of the heaviest rainfall zone to the far western tropical Pacific, East Australia, Indonesia and South and Southeast Asia. The depth of the thermocline also increases from east to west across the Pacific for La Niña; this difference is reduced during El Niño.

Both phases of ENSO typically evolve over a period of 12–18 months and have some predictability once they have started to develop. Seasonal outlooks of ENSO conditions have improved significantly in reliability and are based on being able to successfully observe and model the development of SST anomalies in the tropical Pacific up to a year in advance of an event (McPhaden 2004). Two commonly used indices of ENSO activity are (1) the Southern Oscillation Index (SOI), which measures the atmospheric component and represents the anomalous sea level pressure difference between Tahiti in the southwest Pacific and Darwin in northern Australia, and (2) the Niño 3.4 region (5°N to 5°S, 170°W to 120°W) average SST anomaly, which captures the oceanic component of ENSO. These indices are very similar, indicating the highly-coupled ocean–atmosphere nature of ENSO but also show differences in the timing and magnitude of individual events, which typically recur every 3–7 years.

Although each ENSO event evolves slightly differently, there are common features to these different 'flavours' (Trenberth and Stepaniak 2001) and features typical of the two phases can be determined by averaging the surface climate anomalies across several events. The traditional El Niño, also called Eastern Pacific (EP) El Niño, involves temperature anomalies in the eastern Pacific. However, in the last two decades of the twentieth century non-traditional El Niño patterns are observed, in which the usual place of the temperature anomaly (Niño 1 and 2) is not affected, but an anomaly arises in the central Pacific (Niño 3.4). The phenomenon is called Central Pacific (CP) El Niño, "Date Line" El Niño or El Niño "Modoki" (Larkin and Harrison 2005). Depending on the season, the impacts over regions such as East Asia, New Zealand, and the western coast of USA can be different from those of the traditional ENSO.

ENSO events also affect the spatial occurrence of tropical cyclone activity in the southwest Pacific. During El Niño episodes, the overall number of tropical cyclones tends to be lower, with highest occurrences between Vanuatu and Fiji, and chances of occurrence higher further east in Samoa, southern Cook Islands and French Polynesia. During La Niña events, tropical cyclones are more frequent in the Coral Sea, with highest occurrence around New Caledonia, and higher occurrence between the coast of Queensland and Vanuatu. During these seasons, there is an absence of tropical cyclones from the Cook Islands eastwards. The location of the SPCZ also varies systematically with ENSO-related expansion and contraction of the Warm Pool (Folland et al. 2002). Such movements can result in very large precipitation anomalies on either side of the mean location of the SPCZ (Salinger et al. 1995), as it moves northeast during El Niño events and southwest during La Niña events.

A recent modeling study simulated the effects of climate change on ENSO over the twenty-first century. The study found no significant changes in its extent or frequency, but the warmer and moister atmosphere of the future could make ENSO events more extreme (Stevenson et al. 2011). However, there remains considerable uncertainty in the future behavior of ENSO under climate change conditions.

2.1.3.2 Indian Ocean Dipole

ENSO is found to be associated with inter-annual variability of the Indian summer monsoon (for example, Webster et al. 1998), but this relationship may be weakening in recent decades (Kinter et al. 2002). The linkage between the Indian monsoon and Indian Ocean Dipole (IOD) has been the subject of many investigations, such as Saji et al. (1999) and Webster et al. (1999). The IOD manifests through an east–west gradient of tropical SST, which in one extreme phase in boreal autumn shows cooling and drying off Sumatra and warming off Somalia in the west, combined with anomalous easterlies along the Equator. In a negative dipole year, the reverse occurs, with Indonesia much warmer and wetter. Several recent IOD events have occurred simultaneously with ENSO events. The strongest IOD episode ever observed occurred in 1997–1998. Trenberth et al. (2002) showed that Indian Ocean SSTs tend to rise about 5 months after the peak of ENSO in the Pacific. Monsoon variability (Lau and Nath 2004) is also likely to play a role in triggering or intensifying IOD events. Decadal variability in correlations between SST-based indices of the IOD and ENSO has been documented (Clark et al. 2003). At interdecadal time scales, the SST patterns associated with the variability of ENSO indices are very similar to the SST patterns associated with the Indian monsoon rainfall (Krishnamurthy and Goswami 2000) and with the North Pacific inter-decadal variability (Deser et al. 2004), raising the issue of coupled mechanisms modulating both ENSO-monsoon system and IOD variability (Terray et al. 2005). It has been shown that a positive IOD index can reduce the effect of ENSO, resulting in increased monsoon rains in some ENSO years such as 1983, 1994 and 1997. Further, it has been shown that the two poles of the IOD – the eastern pole (around Indonesia) and the western pole (off the African coast) – can independently and cumulatively affect the rainfall for the monsoon in the Indian subcontinent.

2.1.3.3 Pacific Decadal Oscillation (PDO)/Interdecadal Pacific Oscillation (IPO)

The inter-annual variability of ENSO and the strength of its climate teleconnections are modulated on decadal time scales by a long-lived pattern of Pacific climate variability known as the Pacific Decadal Oscillation (PDO) (Mantua et al. 1997; Zhang et al. 1997) or the Interdecadal Pacific Oscillation (IPO) (Power et al. 1999). The PDO is the North Pacific part of a Pacific basin-wide pattern encompassed by the IPO and is described by an "El Niño-like" pattern of Pacific SST anomalies and appears to persist in either a warm or cool phase for several decades (Fig. 2.4). Warm phases

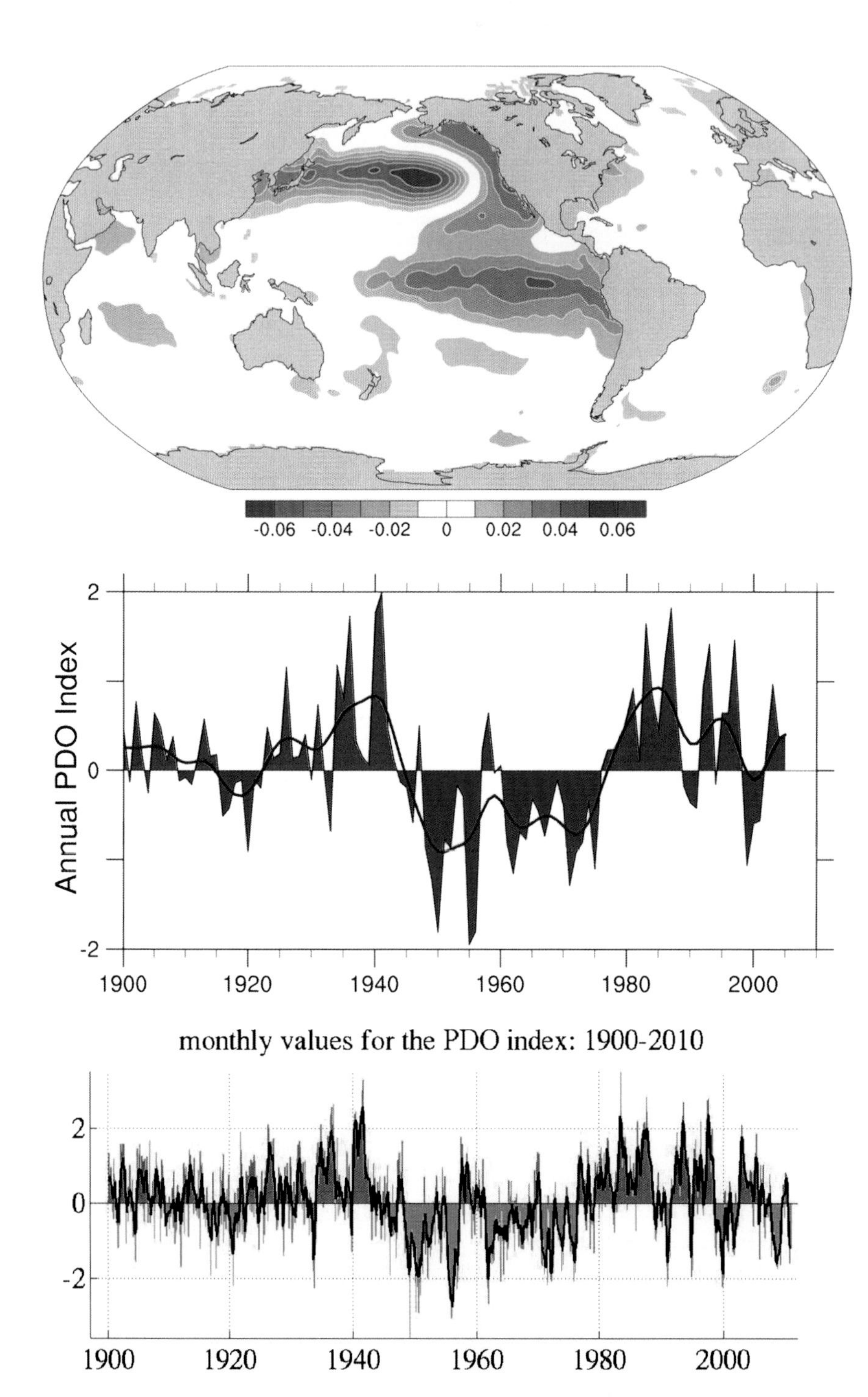

Fig. 2.4 Pacific Decadal Oscillation: (*top*) SST based on the leading EOF SST pattern for the Pacific basin north of 20°N for 1901–2004 (updated; see Mantua et al. 1997; Power et al. 1999) and projected for the global ocean (units are non-dimensional); and (*bottom*) annual time series (Updated from Mantua et al. 1997) (Reprinted with permission from "Climate Change 2007: The Physical Science Basis. Working Group I Contribution to the Fourth Assessment Report of the Intergovernmental Panel on Climate Change," Cambridge University Press, Figure 3.28, p. 289)

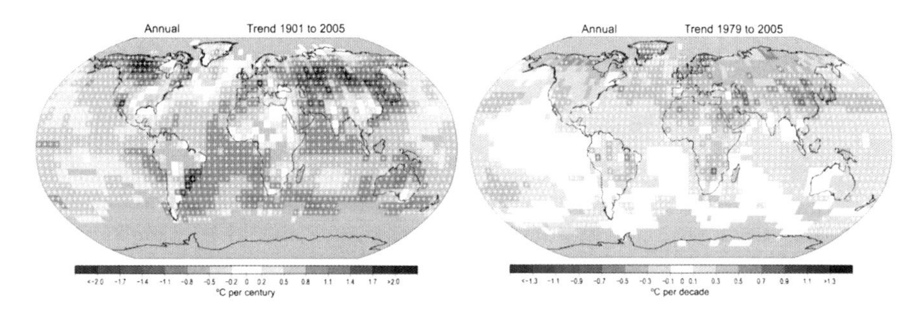

Fig. 2.5 Linear trend of annual temperatures for 1901–2005 (*left*; °C per century) and 1979–2005 (*right*; °C per decade). Areas in *grey* have insufficient data to produce reliable trends. The minimum number of years needed to calculate a trend value is 66 years for 1901–2005 and 18 years for 1979–2005. An annual value is available if there are 10 valid monthly temperature anomaly values. The data set used was produced by NCDC from Smith and Reynolds (2005). Trends significant at the 5 % level are indicated by *white + marks* (Reprinted with permission from "Climate Change 2007: The Physical Science Basis. Working Group I Contribution to the Fourth Assessment Report of the Intergovernmental Panel on Climate Change," Cambridge University Press, Figure 3.9, p. 250)

characterized the 1920s–1940s and from the mid-1970s to the late 1990s. In these periods, ENSO was a weaker source of inter-annual climate variability. The warm phases were preceded and separated by IPO and PDO cool phases from the 1900s to 1920s and 1940s to 1970s, when ENSO was a major source of inter-annual climate variability (Deser et al. 2004). Decadal variability in the SST field of the Pacific is associated with decadal variability in atmospheric variables, such as sea-level pressure, winds and precipitation (Deser et al. 2004; Burgman et al. 2008). The nature of ENSO has varied considerably over time. The 1976–1977 PDO/IPO climate shift (Trenberth 1990) was associated with marked changes in El Niño evolution (Trenberth and Stepaniak 2001), a shift to generally above-normal SSTs in the eastern and central equatorial Pacific and a tendency towards more prolonged and stronger El Niños. This tendency reversed in 1998/1999 with the latest negative phase of the PDO.

2.1.4 *Trends and Extremes*

2.1.4.1 Surface Temperature and Precipitation

Warming in mean surface air temperature over the period 1901–2005 (Fig. 2.5) range from as high as 1.5 °C over Siberia, and at least 1.0 °C/century over northern Asia (Trenberth et al. 2007). Over the remainder of the Asian continent the rate has been 0.8–1.0 °C/century as well as in northern Australia. This compares with a warming rate of 0.5–0.8 °C/century for the Pacific and remainder of Oceania. Higher latitude regions in Asia are experiencing a faster rate of warming than temperate regions. For example, winter temperatures in Mongolia have increased on average by 3.6 °C over the past 60 years (Bohannon 2008). A number of Asian countries have found that winter temperatures are changing faster than summer, and that heat waves are lasting longer (Cruz et al. 2007).

For the period 1979–2005 the warming trend per decade reflects a similar pattern with rates in excess of 0.5 °C/decade over much of the entire Asian continent. In South Asia and Oceania warming rates are between 0.1 °C and 0.3 °C/decade, with the least warming in the tropical and eastern Pacific Ocean (0.1 °C/decade).

Land-based records of accurate trends in precipitation for the Asia-Pacific region can only be gleaned from a much sparser network of observing stations. For the period 1901–2005 (Fig. 2.6) most stations over Asia and Australia record a precipitation increase of between 20 % and 40 % per century. Increases from 10 % to 30 % per decade have occurred over northwest Asia and northwest Australia, western New Zealand and parts of the western tropical Pacific. Over India, eastern precipitation has decreased from 5 % to 15 % per decade (Trenberth et al. 2007).

Over much of Asia the regional climate has been affected not only by the effects of enhanced levels of greenhouse gases (GHGs), but also by high levels of aerosols produced mainly from biomass burning and the consumption of fossil fuels (Ramanathan et al. 2007). The resulting atmospheric brown cloud (ABC) affects the radiation budget of the atmosphere (through both the scattering and absorption of radiation) and the properties of clouds. There is both observational and modeling evidence of ABC impacts on both temperature and precipitation in Asia (Nakajima et al. 2007). There is considerable evidence (Wild 2009a) that the scattering effect of aerosols reduced solar radiation reaching the surface ('global dimming') for some decades in the last half of the twentieth century, and hence masked some of the warming effects of enhanced levels of GHGs. While this dimming effect has been reduced in some parts of the world over the last decade or so, there has been a continuing decline in surface solar radiation in China and India (Wild et al. 2009).

2.1.4.2 Tropical Cyclones

At the global level there are no clear trends in the intensity and frequency of tropical cyclones over the last half of the twentieth century (Knutson et al. 2010). However, there are some indications of trends at the regional level. For Asia and the Pacific there is evidence of increases in the intensity and frequency of tropical cyclones as well as intense rainfall, tornadoes and thunderstorms (Cruz et al. 2007). Analysis of Pacific Ocean extra-tropical cyclones over the past 50 years suggests that, while the overall frequency of storms has not changed, the frequency of intense storms has increased. Moreover, both observational studies and modeling projections indicate that the intensity of cyclones in the North Pacific is increasing (Lambert and Fyfe 2006).

2.1.4.3 Glaciers

(a) *Glacier mass balance studies*

There have been few field studies on glaciers in Asia, and there is not a clear picture of how the region is expected to change in the coming decades. The main

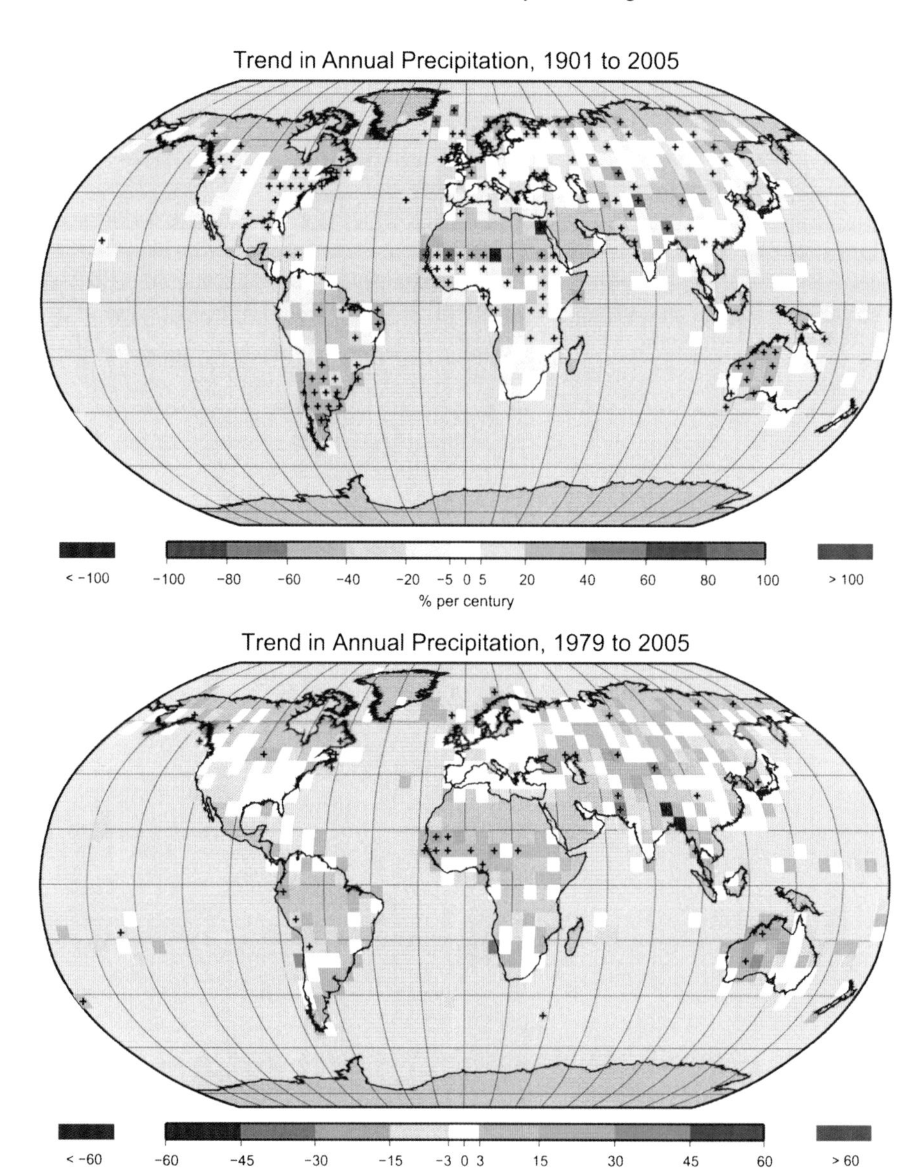

Fig. 2.6 Trend of annual land precipitation amounts for 1901–2005 (*top*, % per century) and 1979–2005 (*bottom*, % per decade), using the GHCN precipitation data set from NCDC. The percentage is based on the means for the 1961–1990 period. Areas in *grey* have insufficient data to produce reliable trends. The minimum number of years required to calculate a trend value is 66 for 1901–2005 and 18 for 1979–2005. An annual value is complete for a given year if all 12 monthly percentage anomaly values are present. Note the different colour bars and units in each plot. Trends significant at the 5 % level are indicated by *black + marks* (Reprinted with permission from "Climate Change 2007: The Physical Science Basis. Working Group I Contribution to the Fourth Assessment Report of the Intergovernmental Panel on Climate Change," Cambridge University Press, Figure 3.13, p. 256)

difficulty is the accessibility of glaciers because of the steep topography with rugged terrain. Moreover, large costs are incurred in carrying out studies in such hard-to-reach places. Assessment of mass balance of the glaciers of the region has been carried out by evaluating each succeeding year's glacier surface vis-à-vis its position in the previous year, or by evaluating the respective glacier surface at the end of the accumulation and ablation season. From these trends the glacier mass balance can be evaluated.

Since the glacier inventory work carried out by Muller (1970) in the Mount Everest region, Glaciological Expedition to Nepal (GEN), a Nepal-Japan joint venture, undertook the task of making a glacier inventory of Nepal early in the 1970s (Higuchi et al. 1976; Watanabe 1976). Most of the works were carried out with the help of aircraft observations and ground field surveys. The advancement of remote sensing technology facilitated this survey work of making an inventory of glaciers in other parts of the Himalaya-Tibetan Plateau (HTP) region. The Hindu Kush-Himalaya (HKH) is an extensive area, which alone contains more than 15,000 glaciers and snow fields (Ives et al. 2010).

(b) *Field studies*

In order to understand the dynamics of glaciers, field studies become an important activity and some examples are presented from the Asia and Pacific region.

Nepal

The Nepal Himalayas contain around 3,252 glaciers. Field studies are constrained by the rugged terrain and remoteness. However, some studies have been carried out in the central and eastern Himalaya regions of Nepal. The mass balance studies of the glaciers were carried out in Hidden Valley located in the northern side of the central Himalaya during the monsoon season of 1974 (Nakawo et al. 1976; Fujii et al. 1976) and such activities are beginning in Nepal. There have been few studies of the glaciers on the northern side of the Nepal Himalayas; Watanabe et al. (1967) is an exception. The Hidden Valley work therefore has special significance. Among the studied glaciers in the Hidden Valley, Rikha Samba glacier is the largest with an area of 4.81 km^2 and an altitude span from 5,245 to 5,985 m. For the glaciers of Nepal, accumulation and ablation occur mainly in the boreal summer. Although the Hidden Valley is situated on the northern side of the Himalayas with low precipitation totals, it is still influenced by the Asia monsoon (Shrestha et al. 1976). Glaciers in the Hidden Valley have shown considerable retreat during the period 1974–1994 (Fujii et al. 1996).

Figure 2.7 shows the evolution of the Rikha Samba Glacier from 1974 to 2010. It clearly shows the depletion of the snout of the glacier. The rate at which Himalayan glaciers are shrinking generally remains poorly constrained because ground-based observations are limited by the high altitude and remoteness of the region.

Fujita and Nuimura (2011) looked at the mass wastage of the Himalayan glaciers and they selected three benchmark glaciers in the Nepal Himalayas on which observations have been made since the 1970s. Based on in situ measurements,

Evolution of Rikha Samba Glacier, Hidden Valley

1974 1994 1998 2010

Fig. 2.7 Time sequence of Rikha Samba glacier, Hidden Valley (Source: Madan L. Shrestha, Koji Fujita, Glaciological Expedition of Nepal (*GEN*), and Department of Hydrology and Meteorology (*DHM*) Nepal)

the wastage rate of the glaciers is equivalent to the global mean during the recent decade (2000–2010), but is higher than the global mean during the previous two decades. This study also indicates that some glaciers at lower altitudes may disappear sooner than those at higher altitudes (Fig. 2.8). However, the heterogeneous distribution of the Equilibrium Line Altitudes (ELAs) trends suggest that it is unwarranted to draw conclusions regarding the fate of the Himalayan glaciers based on such a small number of examples, especially when the benchmark glaciers are chosen partly because of their small size, small elevation range, and simple geometry. In addition, other variables like nature, intensity and quantity of precipitation along with temperature, cloud cover, wind and radiation play an important role and respond in a complex manner through their impact on glacier fluctuations.

India

In the India Himalayas, all available sources illustrate that almost all the glaciers are in a state of depletion. As an example, the Gangotri glacier has been in a continuous state of retreat and fragmentation during the past century. The length of the glacier has been computed for different years based on available data, showing that the length of the glacier has reduced by about 0.59 km in 33 years, from 1976 to 2009, with an average retreat rate of 17.59 m/year. The analysis shows that the glacier is not only receding in length but also in terms of

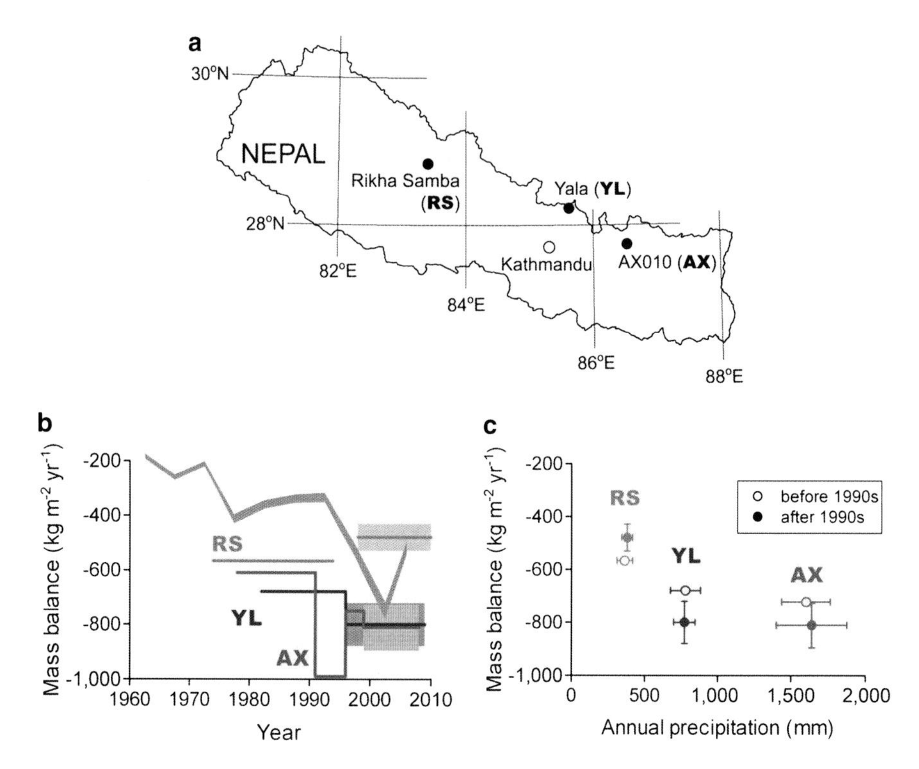

Fig. 2.8 Location of the three benchmark glaciers in Nepal (**a**), temporal changes in the area-averaged mass balances of the glaciers compared with the pentadal global mean (*grey line*) (Cogley et al. 2010) (**b**), and mass balances compared with annual precipitation (**c**). *Colour* shading in **b** and *vertical* bars in **c** denote measurement errors of mass balance. *Horizontal* bars in **c** denote variability of annual precipitation (Source: Fujita and Nuimura 2011) (Reprinted from "Spatially heterogeneous wastage of Himalayan glaciers," by K. Fujita and T. Nuimura, 2011, PNAS, 108(34), Figure 1, p. 2)

glaciated area from all sides. Analysis shows that between 1976 and 2006, the glacier area has reduced by 15.5 km^2, with an average loss of 0.51 km^2 per year. With a reduction in the area and length of the Gangotri glacier, there has also been a retreat in the snout position. Similar reductions have also been observed in most of the other glaciers (Kumar et al. 2009) indicating a significant change in the glaciers in the India Himalayas.

New Zealand

There are few places in the Pacific region with glaciers. New Zealand has glaciers with a long, continuous record of annual end-of-summer-snowline measurements for a set of Southern Alps 'index glaciers' from 1977 to present. Chinn et al. (2012) used these index glaciers to estimate annual mass balance and volume water equivalent changes to the over 3,000 glaciers on the Southern Alps. Results show that estimated ice volume in water equivalents for the Southern Alps has decreased from 54.5 km^3 in 1976 to 46.1 km^3 by 2008. This equates to a loss rate of 0.3 km^3a^{-1} over the last three decades, but this is considerably less than the rate of ice volume loss estimated for the previous 100 years. Results show that there are significant correlations with an index of southwest/northeast circulation over New Zealand in the ablation season. El Niño years are associated with mass balance gains and La Niña years with mass balance losses. There are also significant correlations with the Southern Annular Mode, also known as the Antarctic Oscillation (Kidson 1999).

2.1.4.4 Extreme Events

Manton et al. (2001) and Kwon (2007) analyzed trends in climate extremes in the Asia-Pacific region using 20 extreme temperature indices and 10 extreme precipitation indices classified into three groups: percentile-based indices, fixed-threshold-based indices, and others. Percentile-based extreme temperature indices include cool/warm nights/days (upper and lower 10th percentiles). Similarly, percentile-based extreme precipitation indices include very wet days (95th percentile). Other indices included the number of consecutive dry/wet days. Thirty-year (1971–2000) average values were used to calculate each extreme index from daily maximum/minimum or precipitation data at individual weather stations. A linear regression was fitted to the time series of extreme climate indices for each weather station over the period 1955–2007, and significant levels of the slope values are calculated using the RClimDex software (Zhang and Yang 2004).

Figures 2.9, 2.10 and 2.11 show spatial patterns of linear trends of extreme climate indices over the period 1955–2007 across the region. Among the fifth upper or lower percentile-based indices including cool/warm days/nights, the magnitude of changes in cool nights are greatest at both low- and mid-latitude regions. As shown in Fig. 2.9, cool nights have decreased most in Southeast Asia at the rate of 20 days/decade or more. In the mid-latitude regions above 30°N, cool nights have decreased at the rate of 0–10 days/decade. In central Australia, the trends are not statistically significant and also show reversed signs.

The magnitude of trends in cool days lies within the range of 0–10 days/decade in most of the Asia-Pacific land areas (Fig. 2.10). In India (Kothawale et al. 2009) the frequency of hot days and warm nights shows a widespread increase from 1970 to 2005, especially over southern India. Cool days and cold nights have a widespread decrease. Trends in Nepal (Baidya et al. 2008) show an increase in hot days and warm nights. Most of the temperature extreme indices show a consistently different pattern in the mountainous and southern plains of Nepal (Terai belt). The trend has relatively higher magnitude in mountainous regions where permanent snow and glaciers occur.

Frost days do not occur in the tropical regions between 30°N and 30°S. Regions where frost days do not occur extend more southward in the Southern Hemisphere compared with the northern limit of no frost days. For instance, the southern limit of no frost days occurs in New Zealand (mid-latitude), while the northern limit is located in southern China (subtropical region). In many regions in the mid-latitude

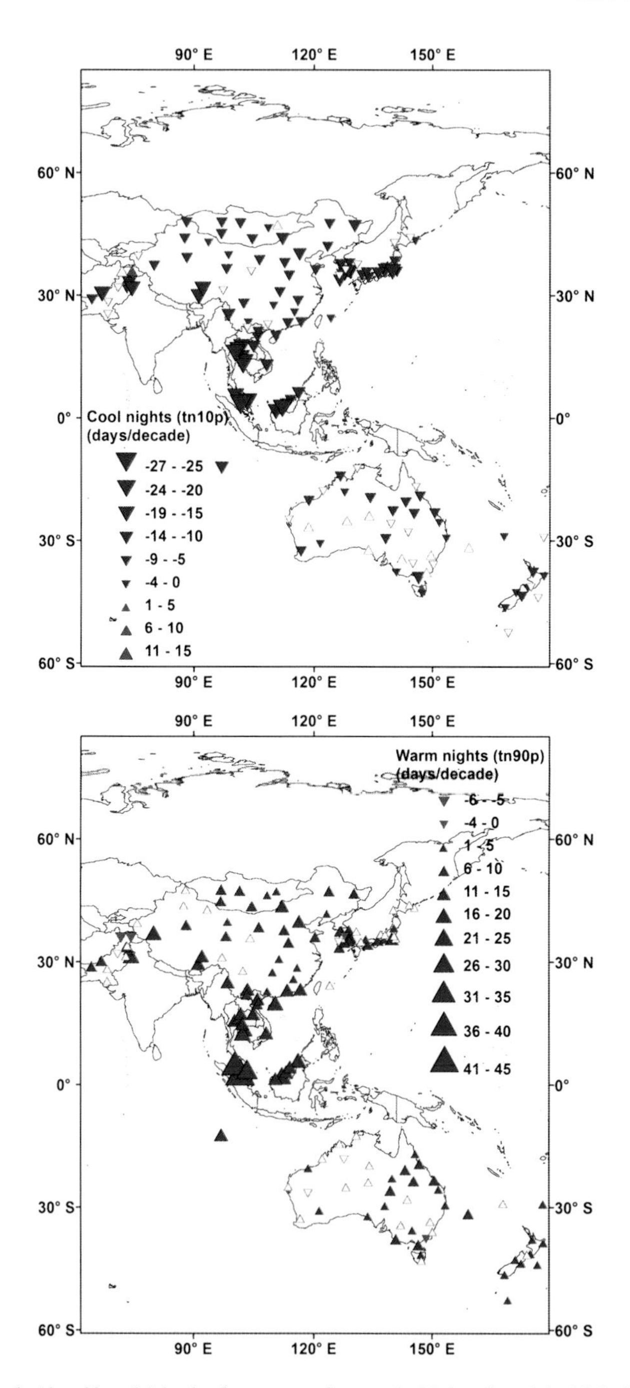

Fig. 2.9 Trends (days/decade) in the frequency of warm (*tn90p*) and cool (*tn10p*) nights over the period 1955–2007 across ten Asia-Pacific countries; *colour-filled symbols* indicate trend is significant at 95 % level; the frequency of cool nights is decreasing across the region (Source: Kwon 2007)

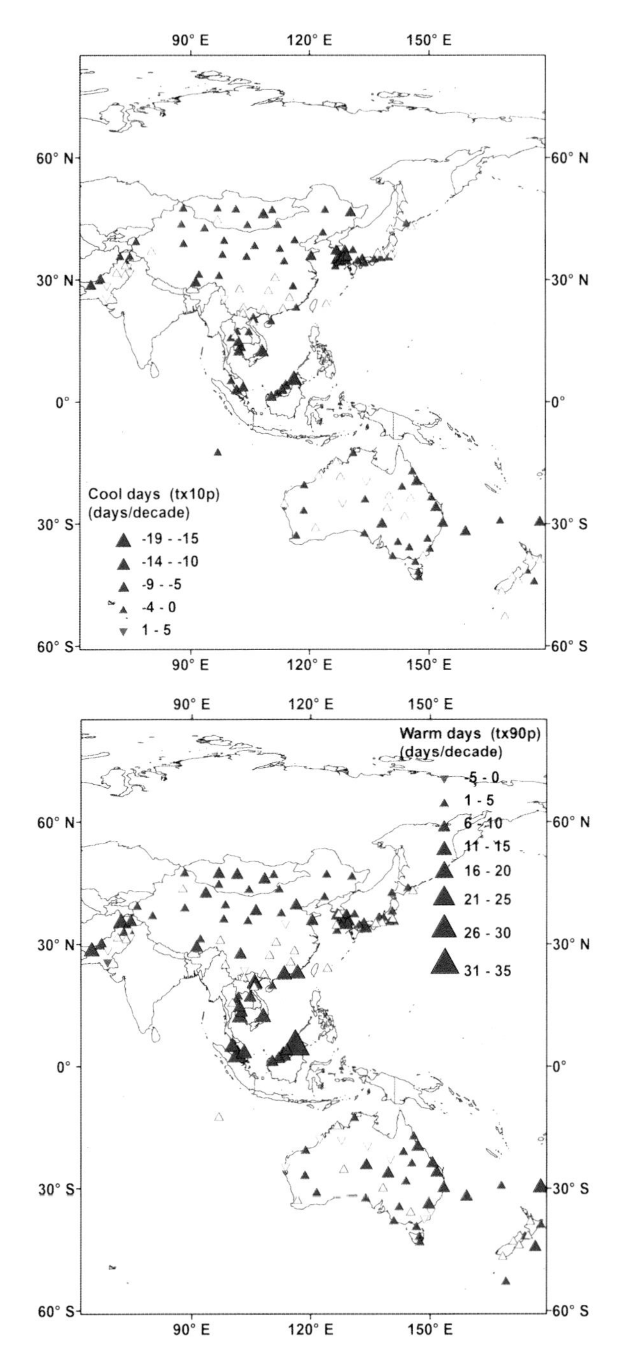

Fig. 2.10 Trends (days/decade) in the frequency of cool (*tx10p*) and warm (*tx90p*) days over the period 1955–2007 across ten Asia-Pacific countries; *colour-filled symbols* indicate trend is significant at 95 % level; the frequency of cool nights is decreasing across the region (Source: Kwon 2007)

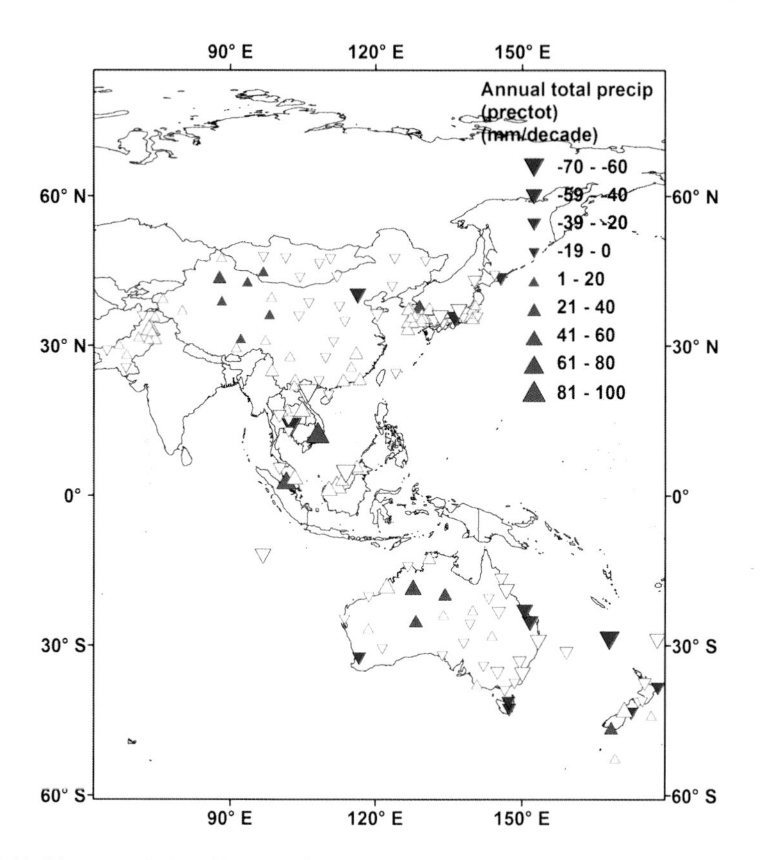

Fig. 2.11 Linear trends (mm/decade) of annual total precipitation day amount (*prcptot*) over the 1955–2007 period across ten across ten Asia-Pacific countries. *Colour-filled symbols* indicate that the linear trend is significance at the 95 % level (Source: Kwon 2007)

of the Southern Hemisphere, frost days are relatively rare. In contrast, over the continents including China and Mongolia, the number of frost days has decreased in the range of 0–8 days/decade. Compared with frost days, significant trends of summer days are observed in both 50°N and 40°S.

Spatial patterns of linear trends of annual total precipitation amount (prcptot) are illustrated in Fig. 2.11. Overall, the trends vary from one location to another. It is difficult to identify regionally-coherent significant trends. However, if insignificant trends are also considered, the overall decreasing trends of annual total precipitation amount are observed in northern China and south eastern Australia, while the increasing trends are found in the Tibetan plateau, Southeast Asia, Republic of Korea and north-western Australia, implying that the summer monsoon system may be intensified in these regions.

As shown in Fig. 2.12, similar patterns are observed for trends in very wet days. For India (Joshi and Rajeevan 2006), for the period 1901–2000, positive trends are seen over the west coast and north western parts of the Indian peninsula. In contrast

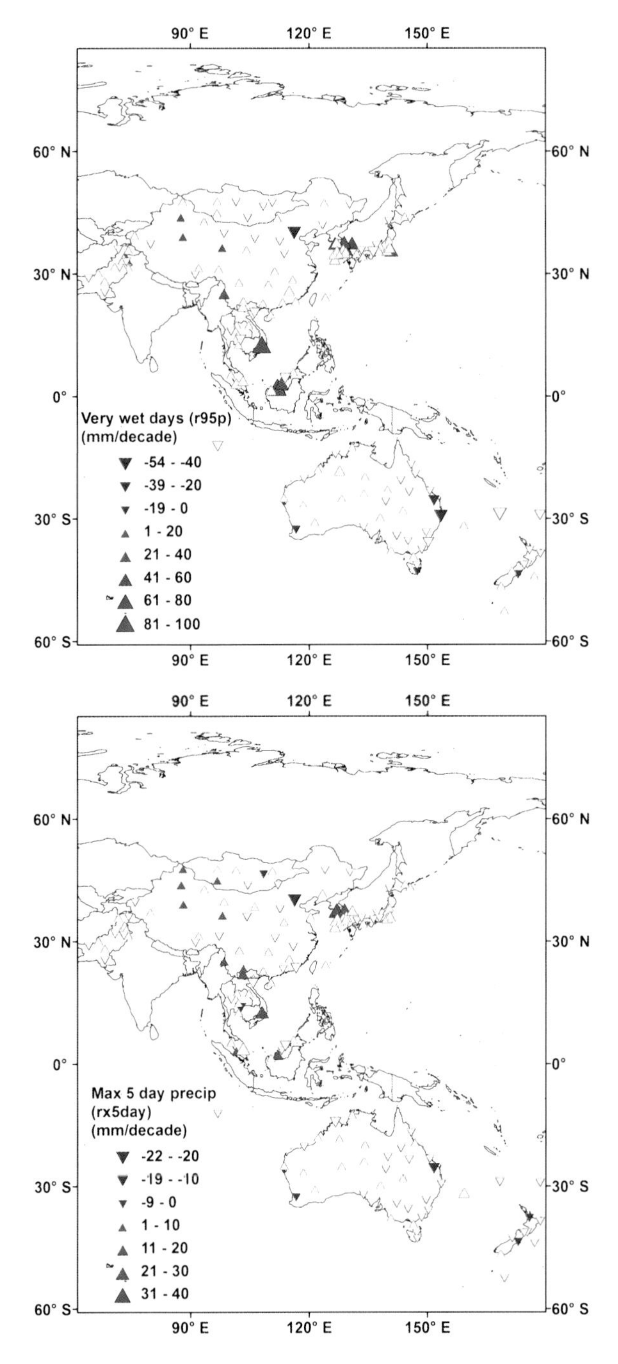

Fig. 2.12 Trends for very wet days (*r95p*) and consecutive dry days (*cdd*) over the 1955–2007 period across ten Asia-Pacific countries. *Colour-filled symbols* indicate that the linear trend is significant at the 95 % level (Source: Kwon 2007)

to extreme rainfall events, a decrease is seen in the frequency of moderate rainfall events over central India during the monsoon season from 1950 to 2000 (Goswami et al. 2006b). Strong increasing trends of the very wet day index are observed in Southeast Asia and Republic of Korea, where heavy rainfall events occur during the monsoon period. In contrast, there are decreasing trends in northern China and south-eastern Australia. There is no noticeable regionally-coherent pattern in trends in consecutive dry days. The magnitude and sign of the trends vary substantially from place to place.

Recently, Liu et al. (2009) obtained some interesting results through a series of analyses on how the intensity of precipitation reacts to global warming. Starting with observations from Taiwan, they found that the 90th percentile of precipitation intensity increases by about 95 % for each degree C increase in global mean temperature; the global average precipitation intensity increases by about 23 % per degree C. Their study clearly showed the linkage between increasing extreme events and global warming.

2.1.5 *Attribution of Change*

IPCC (2007b) notes that anthropogenic change has been detected in global surface temperature with very high significance levels (at least the 1 % level). This conclusion is strengthened by the detection of changes in the upper ocean with high significance level. Upper ocean warming argues against the surface warming being due to natural internal processes. The observed change is very large relative to climate model-simulated internal variability. Surface temperature variability simulated by models is consistent with the variability estimated from the instrumental record. These conclusions apply to observed temperature change over Asia and the Pacific.

It can also be concluded that anthropogenic forcing has contributed to widespread glacier retreat during the twentieth century in Central Asia, the Himalayan plateau and the Southern Alps of New Zealand.

A range of observational evidence indicates that temperature extremes are changing. An anthropogenic influence on warm and cold nights is consistent with that expected from global warming, as is the increase in high intensity rainfall events. The detection of changes in temperature extremes is supported by other comparisons between models and observations. Model uncertainties in changes in temperature extremes are greater than for mean temperatures and there is limited observational coverage and substantial observational uncertainty (IPCC 2012).

The current understanding of climate change in the monsoon regions remains considerably uncertain with respect to circulation and precipitation. IPCC (2007a) records that the Asian monsoon circulation is likely to decrease by 15 % by the late twenty-first century under the SRES A1B scenario (Tanaka et al. 2005). These results are consistent with simulations (Ramanathan et al. 2005; Tanaka et al. 2005) of weakening monsoons due to anthropogenic factors, but further model and empirical studies are required to confirm this.

Observations indicate that the HC has widened by about 2–5° since 1979 (Johanson and Fu 2009). This widening and the concomitant poleward displacement of the subtropical dry zones may be accompanied by large-scale drying near 30°N and 30°S. Simple and comprehensive global climate models (GCMs) indicate that the HC may widen in response to global warming, warming of the western Pacific, or polar stratospheric cooling. Observational and modeling evidence has been found of changes in the HC affecting the rainfall of south eastern Australia (Lucas et al. 2012; Timbal and Drosdowsky 2012). The observed HC widening cannot be explained by natural variability alone, and it is also significantly larger than in simulations of the twentieth and twenty-first centuries (Kent et al. 2011). These results illustrate the need for further investigation into the nature and causes of the widening of the HC.

2.2 Modeling Projections and Regional Downscaling

2.2.1 Introduction

Climate modeling over the Asian region involves some significant complexities. The topography of the region includes many small islands with extensive coastlines, producing many complicated sea-breeze and land-breeze effects. There is also much significant orography, including the Himalayas and Tibetan plateau where there are extra effects from snow and snow melt. Monsoon behavior in the region is still difficult to capture well in global climate models (GCMs); this is partly due to a lack of horizontal resolution in GCMs, and so various topographic interactions with the atmospheric flow are not adequately captured. The complex topography also affects the ability to properly represent relevant physical processes within the models. A particular problem occurs with handling moist convective processes, which require complex diurnally-varying triggering of the convection, and an associated detailed treatment of the atmospheric boundary layer, in order to adequately represent the sea- and land-breeze effects.

As a result, there is a requirement for quite fine resolution over the Asian region to resolve the complex topographic features. One approach, called dynamical downscaling, is to use high-resolution models formulated with the same dynamical equations of motion as the GCMs, driven in some manner by the GCM simulations, which typically have a horizontal resolution of 100–200 km. The dynamical downscaling technique has been available for about 20 years, and the models have come to be called regional climate models (RCMs). Some early studies were those of Giorgi and Bates (1989) over northern America, McGregor and Walsh (1993) over Australia, and Kida et al. (1991) over Japan. The resolution of RCMs is often around 50 km, but in recent years some studies have been performed at 10 km resolution or finer.

Another method for obtaining finer resolution climate change information is to use some form of statistical downscaling from dynamical climate models, typically from coupled atmosphere–ocean GCMs, though the technique may also be applied to further downscaling RCM simulations. A feature of this type of methodology is that some error-correction (for present-day climate) may be incorporated. Both dynamical and statistical downscaling methods are discussed in the following subsections, and examples of their use over the Asian region are presented.

2.2.2 Regional Model Downscaling

2.2.2.1 Dynamical Downscaling

As mentioned, dynamical downscaling methodologies were first developed about 20 years ago. The original technique involved the use of limited-area models "nested" within outputs from a GCM. Over the last decade, an increasing number of researchers are also using variable-resolution global atmospheric models for dynamical downscaling purposes. The dynamical downscaling techniques may be broadly arranged into four groups, and all may be referred to as RCMs.

(a) *Limited-area models with lateral boundary forcing*
Climate simulations with this technique commenced about 20 years ago, and the technique continues to be widely used. The technique is a natural evolution from limited-area numerical weather prediction (NWP) models, which have been used for operational forecasting for some decades. Although NWP models are usually only run for forecasts of a few days, they are well-suited to longer simulations if there is a reasonable flow through the domain, and the domain is not so large as to allow undesired internal circulations to develop. Where there is some incompatibility between the internal flow, and the prescribed lateral boundary conditions, some unwanted reflections may occur, with associated spurious rainfall effects. It is common in such models to include a lateral boundary zone, with flow progressively damped near the boundaries, to suppress artificial reflections. Reviews of this type of model are provided by Giorgi and Mearns (1991) and McGregor (1997).

(b) *Limited-area model with internal forcing (or nudging)*
A development for limited-area models was proposed by Kida et al. (1991), to provide internal forcing (within the limited-area domain) of larger-length scale features, such that they remain consistent with flow features of similar scale in the host GCM. Lateral boundary forcing is usually also included in such models. This type of dynamical downscaling permits rather larger downscaled domains than (a), because the flow will be broadly consistent with that of the host model over the whole domain.

(c) *Stand-alone variable-resolution global models, run in time slice mode*
Application of variable-resolution global atmospheric models was first described by Déqué and Piedlievre (1995), using ARPEGE, an adaptation of the ECMWF IFS NWP model to regional climate modeling. A time-slice approach is often used, where initial conditions are taken from analyses or a host GCM, and

the simulation is "free-standing", forced only by sea surface temperatures (SSTs) and sea-ice from the host GCM. By correcting the monthly biases of the SSTs of the host GCM (as compared to present-day observed SSTs) and applying these same monthly SST bias corrections for the duration of the climate simulation, it is possible in principle to avoid replicating the circulation biases of the host GCM. This type of simulation is usually only modestly stretched, in order to avoid significant refraction effects of waves passing through regions of varying resolution (Caian and Geleyn 1997). An intercomparison by Fox-Rabinovitz et al. (2006) of four such models over northern America at 50 km resolution demonstrated the efficacy of the methodology.

Under this category can be included high-uniform-resolution atmospheric GCMs run in time-slice mode. A notable example is provided by the 20 km climate simulations of the MRI-AGCM3.2 (Mizuki et al. 2012). The Meteorological Research Institute of Japan (MRI) have also performed 60 km global time-slice simulations, as have CSIRO with conformal-cubic atmospheric model, or CCAM, (Nguyen et al. 2011a, b); the latter for an ensemble driven by IPCC AR4 bias-corrected SSTs.

(d) *Strongly-stretched variable-resolution global models*
It is possible to downscale the time-slice variable-resolution simulations of (c) to even finer resolution. The CSIRO group running CCAM have adopted this dynamical downscaling methodology. A long time-slice simulation is first performed as in (c). The larger-scale atmospheric variables of this simulation may then be applied periodically (for example, every 6 h) to drive a more strongly-stretched simulation. The forcing by the larger-scale variables may be conveniently performed by means of a digital filter (Thatcher and McGregor 2009). The technique may be repeated successively to obtain even finer-resolution dynamically-downscaled simulation.

Dynamical downscaling over Asia and the Pacific Islands is being carried out by many groups, using one or more of the above methods. Table 2.1 lists a number of the groups, arranged according to the four downscaling methodologies. The various models are listed, and the typical resolution being used.

2.2.2.2 Areas Needing Improvements in RCMs

The impacts and adaptation communities require estimates of the uncertainty of regional climate simulations. Even for a given GHG emissions scenario, the 23 models of the IPCC Fourth Assessment produced a wide range of regional climate outcomes. One may be able to reduce this range by considering the "most credible" of the coupled GCMs, for example by the quality of their present-day climate simulation, or their climatological performance for indices such as ENSO. This desirability to produce ensembles of simulations also applies to dynamical downscaling. It may variously be addressed by a multi-RCM approach (for example model intercomparisons such as RMIP and CORDEX), by using a single RCM to downscale a variety of host GCMs, or by using a variety of RCMs driven by a variety of host GCMs.

Table 2.1 Some examples of dynamical downscaling methods used in the Asia-Pacific region

Downscaling method	Institute	Typical resolution and duration	Modeled regions
Limited-area – boundary forced			
RIEMS, WRF	CAS	50 km	China
WRF	Nat. Univ. Sing.	50, 12.5 km, 150y	Singapore
RegCM3	CMA	20 km; 2×30y	China
	Kyungpook Nat. U.	50 km 2×75y	CORDEX-EA
PRECIS	IMHEN	25 km; 20y, 2×10y	Vietnam
	Univ. Keb. Malaysia	25 km 2×30y	Malaysia
HadGEM3-RA	KMA/Met Office	50 km	CORDEX-EA
Limited-area – internally forced			
MM5/SNURCM, WRF	Seoul Nat.Univ	50 km 2×75y	CORDEX-EA
RSM	Yonsei Univ.	50 km 2×75y	CORDEX-EA
NHRCM	MRI	5 km; 3×20y	Japan
Global variable res. – time slice			
CCAM	CSIRO	60 km 140y	Global
LMDZ	IIMT (Pune)	35 km	India
	CSIR-CMMACS	35 km	India
MRI-AGCM3.2	MRI	60 km, 20 km 2×25y	Global
HadGEM2A	KMA	60 km, 2×30y	Global
Global highly-variable res.			
CCAM	CSIRO	60, 50, 14, 8 km 140y	RMIP/CORDEX, Indonesia, Pacific Is.

Within individual RCMs, there is a need to improve the representation and parameterization of physical processes. The following lists some of the more important issues for the Asian region:

- Land-cover change (LCC) needs improved treatments and methodologies are needed for estimating the future evolution of LCC;
- Improvements of the land surface to consider urban effects, river routing, and other related processes;
- Improvements to deep and shallow convection, to improve the diurnal convective behavior, and improve the distribution of convective heating throughout the depth of the atmosphere;
- Possible use of convective super-parameterizations (Grabowski 2001) though as model resolution gets very fine (around 2 km), there should then be no need for any convective parameterization, as the processes are explicitly resolved;
- Improved aerosol schemes for both direct and indirect effects;
- Simulation of the carbon cycle, source and sinks of nutrients for both land and ocean;
- Refined resolution, to better represent small islands and the Maritime Continent; and
- Coupling to a mixed-layer ocean or, ideally, a regional ocean model.

2.2.2.3 Statistical Downscaling

There are broadly three common statistical downscaling methodologies:

(a) *Perfect Prognosis approach*
In this methodology, statistical relationships are sought from observations for variables in which there is high confidence, whilst ignoring those in which there is low confidence. The assumption is that the model simulation of the large-scale predictors is "perfect." Some of these models include a noise component to help capture variability and extremes.
(b) *Model Output Statistics approach*
Model Output Statistics (MOS) methods develop statistical relationships between simulated predictors and observed predictands. They are most often applied to climate-model simulated fields of the same variable being predicted. At their simplest, MOS methods provide a bias correction of the present-day simulated field to match the observations.
(c) *Weather Generator approach*
Weather generators are statistical models that produce random sequences of climate variables with statistical properties that match those of the observed variables.

The above categorization and the various approaches are described in some detail by Evans et al. (2012). When comparing outputs of dynamical and statistical downscaling for present-day climate, the statistical methods tend to be closer to observed station data and present-day climate (for example, Frost et al. 2011; Iizumi et al. 2011). This is to be expected from the nature of the statistical methods, which are derived by strongly taking into account the observed values. A major difficulty with statistical downscaling is that the future validity is not known of the present-day correlations upon which the techniques are based. A recommended downscaling approach is to use dynamical downscaling to reach as fine a resolution as can be achieved with available computing resources, then proceed to still-finer scales (for example, catchment or station scale) by means of statistical downscaling. Note that the final statistical downscaling step is able to remove the various biases of the dynamically-downscaled simulations (for example, Corney et al. 2010).

2.2.3 *Asian Climate Change Projections*

2.2.3.1 Projection of Asian Climate by GCMs

Generally, for the Asia-Pacific region, the IPCC Fourth Assessment Report (AR4) finds that future warming may lead to summer heat waves of longer duration and greater intensity frequency in East Asia with fewer very cold days in East Asia and

South Asia. It is also possible that precipitation will increase in northern areas and the Tibetan Plateau, as well as in eastern Asia and the southern parts of Southeast Asia (Christensen et al. 2007).

However, the climate across Asia is different from that in other areas because it is dominated by the most significant monsoon in the world. Traditionally, the Asian monsoon was separated into the south Asian monsoon, east Asian monsoon, western north Pacific monsoon and southeast Asian monsoon, depending on the different physical monsoon processes. Besides the heating contrast between land and ocean, ENSO is one of the key processes affecting the intensity of the southeast Asian monsoon and south Asian monsoon. At the same time, tropical cyclones bring moisture to Asian monsoon areas. The east Asian monsoon is thought to be influenced by the heat conditions over the Tibetan Plateau and the temperature gradient along the east coast, which means the western Pacific Ocean is important to the east Asian monsoon.

The east Asian summer monsoon shows typical characteristics of a rainy season in June and July, which is called Meiyu in China, Baiu in Japan and Changma in Korea. This Meiyu-Baiu-Changma front can be well represented in high resolution GCMs. Kitoh and Uchiyama (2006) found that the withdrawal of the Meiyu-Baiu-Changma rainy season may be delayed under climate change conditions; they used 15 GCM outputs under the IPCC A1B emissions scenario. By using a 20-km resolution atmospheric GCM, Kusunoki and Mizuta (2008) found increasing rainfall over the Yangtze River valley and western Japan may occur in the future. By checking CMIP3 modeling ensemble products under the A1B scenario, Ninomiya (2011) showed the northward shift of both Meiyu and Baiu fronts, and that the rainfall decreased slightly in twenty-first century projections.

It is noted in Sect. 2.1.4 that aerosols have a significant impact on climate across Asia owing to emissions from biomass burning and fossil fuel consumption. Wild's (2009b) study of Climate Model Intercomparison Project Phase 3 (CMIP3) models suggests that, while the models capture many features of the climate of the last 50 years, their decadal-scale variability is less than observed partly because they do not account for the variability of aerosols in the atmosphere.

CMIP3 remains an important source of model output to support a wide range of applications. However, recognizing that there is now an increased emphasis both on long-term projections to support mitigation scenarios and on short-term climate change a couple of decades ahead, the World Climate Research Programme (WCRP) has commenced a much broader set of model experiments as Phase 5 of CMIP (CMIP5). Taylor et al. (2012) provide a summary of the aims and plans for CMIP5, which will involve research institutes around the world including the Asia-Pacific region.

By analyzing outputs from eight CMIP5 models under the new scenarios (RCP 2.6, 4.5, 8.5 indicating low, middle, high emission scenarios), a study in China has found that the increase in global temperature over the present century (2006–2099) under the new scenarios is lower and the uncertainty range is less than for the corresponding AR4 scenarios. For example, the increase in global temperature is found to be around 1.4–2.2 °C over this century under RCP4.5 in CMIP5, while it is 1.7–4.4 °C in A1B scenario in CMIP4.

2.2.3.2 Projection of Asia-Pacific Climate by RCMs

Much information on climate projections for Asia can be found in the IPCC AR4 (2007a). In Chap. 11 (regional climate projection) of the Working Group 1 report, Asia is separated into six sub-areas, which are northern Asia (50–70°N, 40–180°E), Central Asia (30–50°N, 50–100°E), Tibetan Plateau (30–75°N, 50–100°E), East Asia (20–50°N, 100–145°E), South Asia (5–50°N, 64–100°E) and Southeast Asia (11°S–20°N, 95–115°E). Regarding the twenty-first century Asian climate projection for the A1B scenario in IPCC AR4, warming is found to be largest in high latitude (northern Asia, central Asia) and high altitude (Tibetan Plateau) regions; East Asia and South Asia are higher than the global mean, and southeast Asia is similar to the global mean. In most of the sub-areas of Asia, winter (December-February (DJF)) warming contributes more than other seasons; an exception is summer (June-August (JJA)) in Central Asia.

For precipitation changes due to enhanced GHG effects, a suppression of the south Asian summer monsoon, along with a delay of monsoon onset and increase of monsoon break periods are indicated by Ashfaq et al. (2009). In the research of Kumar et al. (2011) by analyzing the outputs from the PRECIS model, the model projections indicate significant warming over India towards the end of the twenty-first century. The summer monsoon precipitation over India is expected to be 9–16 % more in the 2080s compared to the baseline 1970s (1961–1990) under global warming conditions. Also, rainy days are projected to be less frequent and more intense over central India.

By using the ensemble results of five GCMs, Immerzeel et al. (2010) analyzed the impact of global warming on the water resources of the Himalayan river basins in the mid twenty-first century (2046–2065) under the A1B scenario. The Indus and Brahmaputra basins are found to be more influenced by climate warming than the Ganges, Yangtze and Yellow rivers, because the large population and large-scale irrigation heavily rely on melting water in these two areas. By using the CMIP3 and CMIP4 outputs, Wang and Zhang (2010) found that if there is extensive ice-melt in the Arctic under global warming then it is likely to result in stronger east Asian summer monsoons and more rainfall in northern China, north west China and the Maritime continent of Indonesia.

A recent study included running CCAM in time-slice mode at 60 km global resolution for the A2 emissions scenario, forced by the bias-corrected SSTs from an ensemble of six coupled GCMs from AR4 (Nguyen and McGregor 2009; Nguyen et al. 2011a, b). The pattern of warming of surface air temperature from the simulations is shown in Fig. 2.13 with largest oceanic increases over the equatorial region, showing similarities over the Pacific to the observed increases in recent decades as shown in Fig. 2.14. An analysis of the six CCAM ensemble members and their host GCMs suggests that the horizontal gradient of the surface sea-level pressure across the tropical Pacific will weaken under the warm future climate condition. This is consistent with a weakening of large-scale vertical circulations. The model simulations project indicates that the Pacific annual mean rainfall will significantly increase along the ITCZ, with weak changes elsewhere. For the southern

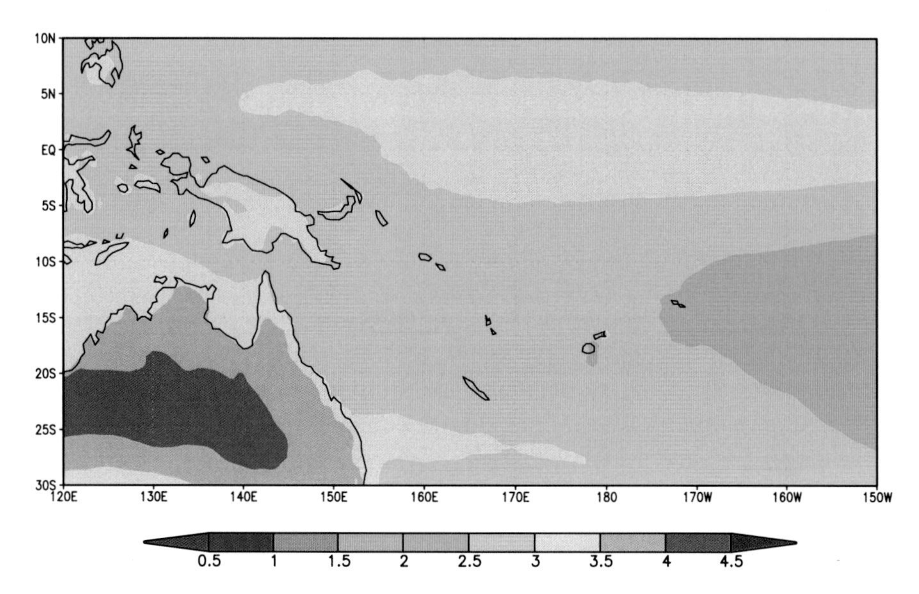

Fig. 2.13 Change in annual maximum near-surface air temperature (°C) 2080–2099 and 1980–1999 for the CCAM 60 km multi-model simulations using the A2 emission scenario (Created by J.M. McGregor with data from Nguyen et al. 2011a, b)

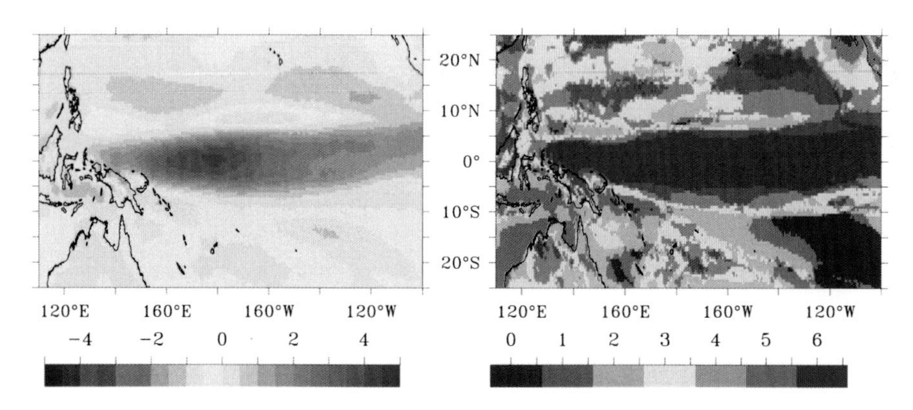

Fig. 2.14 Simulated annual change in rainfall (mm/day) between 2080 and 2099 and 1980–1999 (*left*) and amount of agreement of increase (*right*, *0* means all show decrease, *6* indicates all 6 model runs show increase). Results are for CCAM 60 km simulations for the A2 scenario, driven by SST changes of 6 AR4 coupled GCMs (Created by J.M. McGregor with data from Nguyen et al. 2011a, b)

hemisphere, rainfall is projected to decrease, with the largest reduction seen over the subtropical high. On the other hand, an increase in annual mean rainfall is projected for the northern hemisphere Pacific and over most land areas.

The situation appears somewhat different over the maritime continent region, with its complicated convective rainfall behavior, related to its many islands.

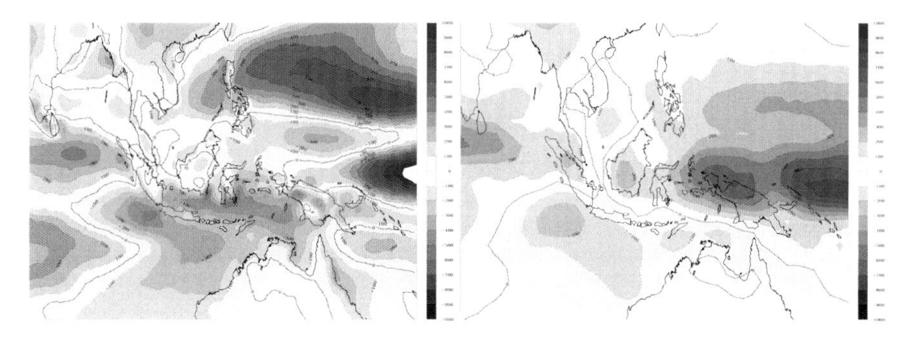

Fig. 2.15 Simulated average annual rainfall changes (changed mm/year amounts) in (2080–2100) compared to (1961–1990) from an ensemble of 60 km CCAM simulations over Indonesia (*left*, McGregor et al. 2009), and the six host GCMs (*right*) (Created by J.M. McGregor with data from Nguyen et al. 2011a, b)

In a downscaling study over the region using an ensemble of 60 km CCAM simulations driven by six of the AR4 coupled GCMs (McGregor et al. 2009), the ensemble of downscaled simulations, and also the host GCM ensemble, indicate a future reduction in annual rainfall (Fig. 2.15) for most of the larger islands of the maritime continent.

2.2.4 *Coordinated Projects on Regional Modeling*

Many research groups now run either global or regional climate models for climate research or for the support of climate impact studies. Through regional collaboration such groups have developed inter-comparison programs that compare the results of an individual model with those of all other models as well as with region climate observations (Wang 2012). These coordinated projects allow the uncertainties in model results to be quantified and hence develop confidence in the application of models to climate impact studies.

2.2.4.1 Regional Model Intercomparison Project (RMIP)

The Regional Climate Model Intercomparison Project (RMIP) involves long-term regional coordination on Asian climate change studies (Fu et al. 2005). It has three phases, each focusing on different scientific objectives. Phases I and II were designed to assess the ability of models to simulate the seasonal cycle, climate extremes, and general climatology across Asia (Fu et al. 2005; Lee et al. 2007; Feng et al. 2011b; Feng and Fu 2006).

In 2009, RMIP III on "Building Asian Climate Change Scenarios by Multi-Regional Climate Models Ensemble" was established and, at the time of writing, is continuing with financial support from the APN (Wang and Zhang 2010). The objective of

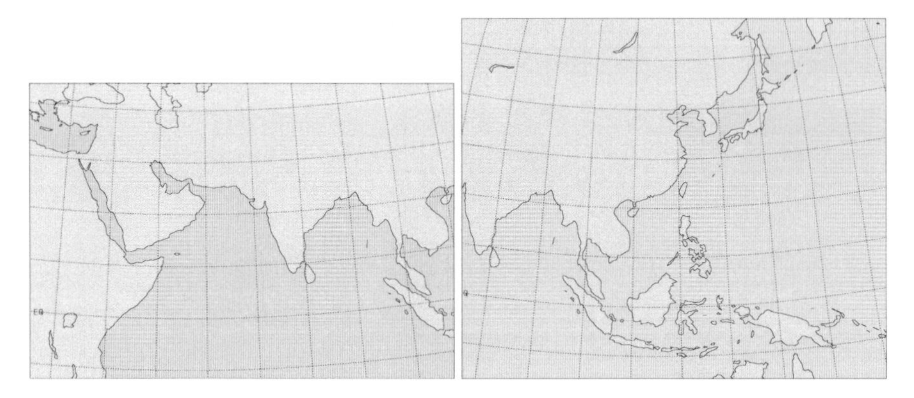

Fig. 2.16 CORDEX domains for West Asia and East Asia; each have about 50 km grid resolution (Plotted using the domain specifications available at http://wcrp-cordex.ipsl.jussieu.fr)

RMIP III is to generate robust climate change scenarios for the impact and adaptation research communities, including detailed evaluation and assessment of the source and magnitude of uncertainties. The project also aims to increase the understanding of variability of the east Asia monsoon system in future climate and its impact on regional climate.

Ten RCMs and two GCMs from seven regional climate modeling groups across the Asia-Pacific region are participating in RMIP III. The CORDEX East Asia domain is adopted for the study, and two time slices are composed of present-day conditions (1900–2000) and future climate (2000–2070).

2.2.4.2 Coordinated Regional Downscaling Experiment (CORDEX)

A WCRP initiative in partnership with START and other key institutions, CORDEX aims to produce climate information at regional to local scales to help and support local decision makers respond to potential climate change. Even though there exist several successful intercomparison programs for each continent (for example, ENSEMBLE over Europe, PIRCS and NARCAPP over North America, RMIP over Asia), the strength of CORDEX compared to previous projects is to provide a common framework in a global-wide perspective for regional climate projections in order to understand their uncertainties as well as provide model evaluation (Giorgi et al. 2009). The domain size for East Asia (Fig. 2.16) is the largest among the 12 domains of CORDEX. This large domain may lead to problems both scientifically and practically. Nevertheless, the benefit of this domain is that the CORDEX-East Asia domain covers the maritime continent in the western Pacific, which is a somewhat vulnerable area with respect to climate change. South Asia has its own large CORDEX domain, with coordination of the simulations and analyses being undertaken at the Indian Institute of Tropical Meteorology (IITM). IITM has a particular interest in long-term monsoon change and modeling of the south Asian monsoon.

The Korea Meteorological Administration (KMA) is participating in CORDEX, in collaboration with the UK Met Office Hadley Centre. Four dynamical regional climate models and one statistical model are involved in this collaborative research. Phase I (2010–2011) focuses on the CORDEX-East Asia domain with 50 km resolution and Phase II (2012~) focuses on the area of the Korean peninsula but with resolution 12.5 km or finer. The strategy of CORDEX includes current climate simulations and regional projections forced by CMIP5 global projections based on Representative Concentration Pathways (RCP) emission scenarios.

2.2.5 Applications and Case Studies

This sub-section introduces some successful cases of regional climate projection.

2.2.5.1 RMIP

Based on the evaluation of the ability of RCMs to simulate Asian climate on different time scales in RMIP I and II (Fu et al. 2005; Feng and Fu 2006; Lee, et al. 2007; Feng et al. 2011c), RMIP III considers projections of Asian climate change for the period 2040–2070. Compared with the present-day climate (1980–2000), initial results from nine RCMs suggest that annual Asian surface air temperature will increase by 2.0 °C around 2050, with more warming in the winter season (DJF) of 2.1 °C. Precipitation is expected to increase by about 2.4 % annually, with a greater increase over land.

2.2.5.2 Adapting Climate Change in China (ACCC) project

The Adapting Climate Change in China (ACCC) project is an international project focused on understanding the impact of climate change on Chinese agriculture, such as the risk of droughts, extreme weather events, increasing temperatures and changes in water availability. The ACCC project expects to promote and enhance climate change adaptation capacity of local policy makers and the general public. These aims are achieved through improving climate science and impact assessments, and incorporating the views of local communities and decision-makers on potential adaptation responses to climate impacts. The UK Meteorological Office's use of their PRECIS modeling system has produced high-resolution downscaled future scenarios in a dryland area of China (Ningxia Province, China) to be used in an assessment study of agricultural production and water management. Vulnerability indicators for dryland, water and agriculture sectors were developed and the vulnerability assessment methodology was used by scientists and local policy makers in the Adapting to Climate Change in China Project (see http://www.ccadaptation.org.cn/en/index.aspx).

2.2.5.3 Pacific Climate Change Science Program

The Pacific Climate Change Science Program (PCCSP) was initiated and funded by the Government of Australia. Under this program, the Australian Bureau of Meteorology and Commonwealth Scientific and Industrial Research Organisation (2011) conducted a comprehensive modeling and impacts study for 15 of the Pacific's island countries. Projects under the PCCSP included both dynamical and statistical downscaling, and produced impacts and adaptation advice for these countries to facilitate effective adaptation planning and implementation (Power et al. 2011).

2.3 Conclusions

The geographic extent of Asia and the Pacific leads to great variation in the climate of the region. Major influences on global climate arise from the scale and elevation of the Himalayas Tibetan Plateau (HTP), and from the air-sea interactions in the Pacific associated with the El Niño–Southern Oscillation (ENSO). The monsoon has a profound effect on the climate of Asia, with its strong seasonal cycle driving a range of human activities, especially agriculture, as well as the life-cycles of natural ecosystems.

Climate observations show significant trends in temperature across Asia and the Pacific; not only is there an increase in mean temperature but also in extremes such as the frequency of hot days. Observed trends in precipitation are more variable, but there is some evidence of increasing intensity of storms. Projections of future climate change for the region suggest longer summer heat waves in East Asia and increases in precipitation in several areas.

Our understanding of the climate of the region is critically dependent on the availability and quality of observations of the atmosphere, ocean and land surface, and so it is essential for such data to be collected and analyzed routinely and carefully. Because of the complex interactions of the climate system, these data cannot be usefully studies in isolation and so there needs to be sharing of climate data and associated expertise across the whole region.

Even collecting climate data in some parts of the region is challenging. For example, the isolation of some islands of the Pacific makes routine observation difficult, while the terrain and climate of the Himalayas give rise to an unfriendly environment for routine measurement. Cooperation at regional and international levels needs to be promoted to ensure that an effective record of the climate of the region is maintained.

Numerical modeling provides a means for assimilating observations in a dynamically consistent manner and for then making predictions of future weather and climate. A number of major centers in Asia and the Pacific maintain complex climate modeling systems, and these groups generally collaborate with similar agencies around the world through projects such as the Coupled Model Intercomparison Project (CMIP). Such projects are important in ensuring the quality of model output, and in promoting the development of model improvements.

However, modeling as a tool for the support of climate impact and adaptation studies is of interest to many groups across the region. Regional climate models can be used for this purpose, and these models can be used by a wide range of groups to 'downscale' the output of global models to local scales. Downscaling can also be achieved through statistical methods, which are readily accessible to the broad science community. Projects like the Coordinated Regional Downscaling Experiment (CORDEX) provide a means to promote cooperation across the modeling and climate applications communities.

References

Ashfaq, M., Shi, Y., Tung, W. W., Trapp, R. J., Gao, X., Pal, J. S., & Diffenbaugh, N. S. (2009). Suppression of south Asian summer monsoon precipitation in the 21st century. *Geophysical Research Letters, 36*, L01704. doi:10.1029/2008GL036500.

Australian Bureau of Meteorology & CSIRO. (2011). *Climate change in the Pacific: Scientific assessment and new research.* Volume 1: Regional overview. Volume 2: Country reports.

Baidya, S. K., Shrestha, M. L., & Sheikh, M. M. (2008). Trend in daily climatic extremes of temperature and precipitation in Nepal. *Journal of Hydrology and Meteorology, 5*(1), 38–51. SOHAM-Nepal.

Bohannon, J. (2008). The big thaw reaches Mongolia's pristine north. *Science, 319*, 567–568.

Burgman, R. J., Clement, A. C., Mitas, C. M., Chen, J., & Esslinger, K. (2008). Evidence for atmospheric variability over the Pacific on decadal timescales. *Geophysical Research Letters, 35*, L01704. doi:10.1029/2007GL031830.

Caian, M., & Geleyn, J. -F. (1997). Some limits to the variable-mesh solution and comparison with the nested-LAM solution. *Quarterly Journal of the Royal Meteorological Society, 123*, 743–766.

Chinn, T., Fitzharris, B. B., Willsman, A., & Salinger, J. (2012). Annual ice volume changes 1976–2008 for the New Zealand Southern Alps. *Global and Planetary Change, 92*. doi:10.1016/j.gloplacha.2012.03.002.

Christensen, J. L., Hewitson, B., Busuioc, A., Chen, A., Gao, X., Held, I., … & Kumar K. (2007). Regional climate projections. Climate change 2007: The scientific basis. Contribution of working group I to the fourth assessment report of the intergovernmental panel on climate change. In S. Solomon, D. Qin, M. Manning, Z. Chen, K. Averyt, M. Marquis, K. B. M. Tignor & H. L. Miller (Eds.) (pp. 847–940). Cambridge: Cambridge University Press.

Clark, C. O., Webster, P. J., & Cole, J. E. (2003). Interdecadal variability of the relationship between the Indian Ocean zonal mode and East African coastal rainfall anomalies. *Journal of Climate, 16*, 548–554.

Cogley, J. G., Kargel, J. S., Kaser, G., & Van der Veen, C. J. (2010). Tracking the source of glacier misinformation. *Science, 327*, 522.

Corney, S. P., Katzfey, J. J., McGregor, J. L., Grose, M. R., Bennett, J., White, C. J., … & Bindoff, N. L. (2010). *Climate futures for Tasmania: Modelling technical report.* Hobart: Antarctic Climate and Ecosystems Cooperative Research Centre.

Cruz, R. V., Harasawa, H., Lal, M., Wu, S., Anokhin, Y., Punsalmaa, B., … & Huu Ninh, N. (2007). Asia. Climate change 2007: Impacts, adaptation and vulnerability. Contribution of working group II to the Fourth Assessment Report of the Intergovernmental Panel on Climate Change. In M.L. Parry, O.F. Canziani, J.P. Palutikof, P.J. van der Linden, & C.E. Hanson (Eds.) (pp. 469–506). Cambridge: Cambridge University Press.

Déqué, M., & Piedlievre, J. P. (1995). High resolution climate simulation over Europe. *Climate Dynamics, 11*, 321–339.

Deser, C., Phillips, A. S., & Hurrell, J. W. (2004). Pacific interdecadal climate variability: Linkages between the tropics and the north Pacific during boreal winter since 1900. *Journal of Climate, 17*, 3109–3124.

Evans, J., McGregor, J. L., & McGuffie, K. (2012). Future regional climates. In H. Henderson-Sellers, & K. McGuffie (Eds.), *Future of the World's climate*. Elsevier, 223–252.

Feng, J., & Fu, C. (2006). Inter-comparison of 10-year precipitation simulated by several RCMs for Asia. *Advances in Atmospheric Sciences, 23*(4), 531–542.

Feng, R., et al. (2011a). Regime change of the boreal summer Hadley circulation and its connection with the tropical SST. *Journal of Climate, 24*, 3867–3877. doi:10.1175/2011JCLI3959.1.

Feng, J. M., Wang, Y. L., & Fu, C. (2011b). Simulation of extreme climate events over China with different regional climate models. *Atmospheric and Oceanic Science Letters, 4*(1), 47–56.

Feng, J., Lee, D. K., Fu, C., Tang, J., Sato, Y., Kato, H., … & Mabuchi, K. (2011c). Comparison of four ensemble methods combining regional climate simulations over Asia. *Meteorology and Atmospheric Physics, 11*(1–2), 41–53. doi:10.1007/s00703-010-0115-7.

Folland, C. K., Renwick, J. A., Salinger, M. J., & Mullan, A. B. (2002). Relative influences of Zone. *Geophysical Research Letters, 29*, 1643. doi:10.1029/2001GL014201.

Fox-Rabinovitz, M., Côté, J., Dugas, B., Déqué, M., & McGregor, J. L. (2006). Variable resolution general circulation models: Stretched-grid model intercomparison project (SGMIP). *Journal of Geophysical Research, 111*, D16104. doi:10.1029/2005JD006520.

Frost, A. J., Charles, S. P., Timbal, B., Chiew, F. H. S., Mehrotra, R., Nguyen, K. C., … & Kent, D. M. (2011). A comparison of multi-site daily rainfall downscaling techniques under Australian conditions. *Journal of Hydrology, 408*(1–2), 1–18. doi:10.1016/j.jhydrol.2011.06.021.

Fu, C., Wang, S., Xiong, Z., Gutowski, W. J., Lee, D. K., McGregor, J. L., … & Suh, M.S. (2005). Regional climate model intercomparison project for Asia. *Bulletin of the American Meteorological Society, 86*, 257–266.

Fujii, Y., Nakawo, M., & Shrestha, M. L. (1976). Mass balance studies of the glaciers in hidden valley, Mukut Himal. *Journal of the Japanese Society of Snow and Ice, 38*, 17–21.

Fujii, Y., Fujita, K., & Paudyal, P. (1996). Glaciological research in hidden valley, Mukut Himal in 1994. *Bulletin of Glacier Research, 14*, 7–11.

Fujita, K., & Nuimura, T. (2011). Spatially heterogeneous wastage of Himalayan glaciers. *PNAS, 108*(34), 14011–14014.

Gadgil, S. (2003). The Indian monsoon and its variability. *Annual Review of Earth and Planetary Sciences, 31*, 429–467.

Giorgi, F., & Bates, G. T. (1989). On the climatological skill of a regional model over complex terrain. *Monthly Weather Review, 117*, 2325–2347.

Giorgi, F., & Mearns, L. O. (1991). Approaches to the simulation of regional climate change: A review. *Reviews of Geophysics, 29*, 191–216.

Giorgi, F., Jones, C., & Asrar, G. R. (2009). Addressing climate information needs at the regional level: The CORDEX framework. *WMO Bulletin, 58*(3), 175–183.

Goswami, B. N., Wu, G., & Yasunari, T. (2006a). The annual cycle, intraseasonal oscillations, and roadblock to seasonal predictability of the Asian summer monsoon. *Journal of Climate, 19*, 5078–5099. doi:10.1175/JCLI3901.1.

Goswami, B. N., Venugopal, V., Sengupta, D., Madhusoodanan, M. S., & Xavier, P. K. (2006b). Increasing trend of extreme rain events over India in a warming environment. *Science, 314*(5804), 1442–1445. doi:10.1126/science.1132027.

Goswami, B. N., Kulkarni, J. R., Mujumdar, V. R., & Chattopadhyay, R. (2010). On factors responsible for recent secular trend in the onset phase of monsoon intraseasonal oscillations. *International Journal of Climatology, 30*, 2240–2246. doi:10.1002/joc.2041.

Grabowski, W. W. (2001). Coupling cloud processes with the large-scale dynamics using the Cloud-Resolving Convection Parameterization (CRCP). *Journal of the Atmospheric Sciences, 58*, 978–997.

Higuchi, K., Iozawa, T., & Higuchi, H. (1976). Flight observations for the inventory of glaciers in the Nepal Himalaya. *Journal of the Japanese Society of Snow and Ice, 38*, 6–9.

Iizumi, T., Nishimori, M., Dairaku, K., Adachi, S. A., & Yokozawa, M. (2011). Evaluation and intercomparison of downscaled daily precipitation indices over Japan in present-day climate: Strengths and weaknesses of dynamical and bias correction-type statistical downscaling methods. *Journal of Geophysical Research, 116*, D01111. doi:10.1029/2010JD014513.

Immerzeel, W. W., van Beek, L. P. H., & Bierkens, M. F. P. (2010). Climate change will affect the Asian water towers. *Science, 328*, 1382–1385.

IPCC. (2007a). In S. Solomon, D. Qin, M. Manning, Z. Chen, M. Marquis, K. B. Averyt, M. Tignor, & H. L. Miller (Eds.), *Climate change 2007: The physical science basis, Contribution of working group I to the Fourth Assessment Report of the Intergovernmental Panel on Climate Change*. Cambridge, UK/New York: Cambridge University Press.

IPCC. (2007b). Summary for policymakers. In S. Solomon, D. Qin, M. Manning, Z. Chen, M. Marquis, K. B. Averyt, M. Tignor, & H. L. Miller (Eds.), *Climate change 2007: The physical science basis. Contribution of working group I to the Fourth Assessment Report of the Intergovernmental Panel on Climate Change*. Cambridge, UK/New York: Cambridge University Press.

IPCC. (2012). In C. B. Field, V. Barros, T. F. Stocker, D. Qin, D. J. Dokken, K. L. Ebi, M. D. Mastrandrea, K. J. Mach, G.-K. Plattner, S. K. Allen, M. Tignor, & P. M. Midgley (Eds.), *Managing the risks of extreme events and disasters to advance climate change adaptation. A Special report of working groups I and II of the Intergovernmental Panel on Climate Change*. Cambridge, UK/New York: Cambridge University Press.

Ives, J. D., Shrestha, R. B., & Mool, P. K. (2010). *Formation of glacial lakes in the Hindu Kush-Himalayas and GLOF risk assessment*. Kathmandu: ICIMOD.

Johanson, C. M., & Fu, Q. (2009). Hadley cell widening: Model simulations versus observations. *Journal of Climate, 2*, 2713–2725.

Johansson, A. (2007). Prediction skill of the NAO and PNA from daily to seasonal time scales. *Journal of Climate, 20*, 1957–1975.

Joshi, U. R., & Rajeevan, M. (2006). *Trends in precipitation extremes over India* (Research report, Vol. 3/2006). Pune: Indian Meteorological Department.

Kent, D. M., Kirono, D. G. C., Timbal, B., & Chiew, F. H. S. (2011). Representation of the Australian sub-tropical ridge in the CMIP3 models. *International Journal of Climatology, 33*(1), 48–57. doi:10.1002/joc.3406.

Kida, H., Koide, T., Sasaki, H., & Chiba, M. (1991). A new approach for coupling a limited area model to a GCM for regional climate simulations. *Journal of the Meteorological Society of Japan, 69*, 723–728.

Kidson, J. W. (1999). Principal modes of Southern Hemisphere low frequency variability obtained from NCEP-NCAR reanalyses. *Journal of Climate, 12*, 2808–2830.

Kiladis, G. N., von Storch, H., & van Loon, H. (1989). Origin of the South Pacific convergence zone. *Journal of Climate, 2*, 1185–1195.

Kinter, J. L., Miyakoda, K., & Yang, S. (2002). Recent change in the connection from the Asian monsoon to ENSO. *Journal of Climate, 15*, 1203–1215.

Kitoh, A., & Uchiyama, T. (2006). Changes in onset and withdrawal of the East Asian summer rainy season by multi-model global warming experiments. *Journal of the Meteorological Society of Japan, 84*, 247–258.

Knutson, T. R., McBride, J. L., Chan, J., Emanuel, K., Holland, G., Landsea, C., … & Sugi, M. (2010). Tropical cyclones and climate change. *Nature Geoscience, 3*, 157–163.

Können, G. P., Jones, P. D., Kaltofen, M. H., & Allan, R. J. (1998). Pre-1866 extensions of the Southern Oscillation index using early Indonesian and Tahitian meteorological readings. *Journal of Climate, 11*, 2325–2339.

Kothawale, J. V., Revasdekar, J. V., & Kumar, K. R. (2009). Recent trends in pre-monsoon daily temperature extremes over India. *Journal of Earth System Science, 119*(1), 51–65.

Krishnamurthy, V., & Goswami, B. N. (2000). Indian monsoon-ENSO relationship on interdecadal timescale. *Journal of Climate, 13*, 579–595.

Kumar, R., Areendran, G., & Rao, P. (2009). Witnessing change: Glaciers in the Indian Himalayas. Pilani, WWF-India and Birla Institute of Technology.

Kumar, K. K., Patwardhan, S. K., Kulkarni, A., Kamala, K., Rao, K. K., & Jones, R. (2011). Simulated projections for summer monsoon climate over India by a high-resolution regional climate model (PRECIS). *Current Science, 101*, 312–326.

Kusunoki, S., & Mizuta, R. (2008). Future changes in the Baiu rain band projected by a 20-km mesh global atmosphere model: Sea surface temperature dependence. *SOLA, 4*, 85–88.

Kwon, W. T. (2007) Development of indices and indicators for monitoring trends in climate extremes and its application to climate change projection. (APN Project Report for APN project: ARCP2007-20NSG). Retrieved from http://www.apn-gcr.org/resources/items/show/1537

Lambert, S.J., & Fyfe, J.C. (2006). Changes in winter cyclone frequencies and strengths simulated in enhanced greenhouse warming experiments: results from the models participating in the IPCC diagnostic exercise. *Climate Dynamics, 26*, 713–728.

Larkin, N. K., & Harrison, D. E. (2005). On the definition of El Niño and associated seasonal average U.S. weather anomalies. *Geophysical Research Letters, 32*, L13705. doi:10.1029/2005GL022738.

Lau, K. M., & Nath, M. J. (2000). Impact of ENSO on the variability of the Asian–Australian monsoons as simulated in GCM experiments. *Journal of Climate, 13*, 4287–4309.

Lau, N. C., & Nath, M. (2004). Coupled GCM simulation of atmosphere–ocean variability associated with zonally asymmetric SST changes in the tropical Indian Ocean. *Journal of Climate, 17*, 245–265.

Lee, D.-K., Gutowski, W., Kang, H. S., & Kim, C. J. (2007). Intercomparison of precipitation simulated by regional climate models over East Asia in 1997 and 1998. *Advances in Atmospheric Sciences, 24*(4), 539–554.

Li, J. P., Zhu, Z. W., Jiang, Z. H., & He, J. H. (2010). Can global warming strengthen the East Asian summer monsoon? *Journal of Climate, 23*, 6696–6705. doi:10.1175/2010JCLI3434.1.

Liu, S. C., Fu, C., Shiu, C.-J., Chen, J.-P., & Wu, F. (2009). Temperature dependence of global precipitation extremes. *Geophysical Research Letters, 36*, L17702.

Lucas, C., Nguyen, H., & Timbal, B. (2012). An observational analysis of Southern Hemisphere tropical expansion. *Journal of Geophysical Research, 117*, D17112. doi:10.1029/2011JD017033.

Manton, M. J., Della-Marta, P. M., Haylock, M. R., Hennessy, K. J., Nicholls, N., Chambers, L. E., … & Yee, D. (2001). Trends in extreme daily rainfall and temperature in Southeast Asia and the south Pacific: 1961–1998. *International Journal of Climatology, 21*, 269–284.

Mantua, N. J., Hare, S. R., Zhang, Y., Wallace, J. M., & Francis, R. C. (1997). A Pacific interdecadal climate oscillation with impacts on salmon production. *Bulletin of the American Meteorological Society, 78*, 1069–1079.

McGregor, J. L. (1997). Regional climate modelling. *Meteorology and Atmospheric Physics, 63*, 105–117.

McGregor, J. L., & Walsh, K. (1993). Nested simulations of perpetual January climate over the Australian region. *Journal of Geophysical Research, 98*, 23283–23290.

McGregor, J. L., Nguyen, K., Katzfey, J. J., & Thatcher, M. (2009). Regional climate modelling over island countries. Extended abstracts, international symposium on equatorial monsoon system, Kuta Paradiso Hotel, Bali, 16–18 July 2009.

McPhaden, M. J. (2004). Evolution of the 2002/03 El Niño. *Bulletin of the American Meteorological Society, 85*, 677–695.

McPhaden, M. J., Zebiak, S. E., & Glantz, M. H. (2006). ENSO as an integrating concept in earth science. *Science, 314*, 1740–1745.

Mizuta, R., Yoshimura, H., Murakami, H., Matsueda, M., Endo, H., Ose, T., Kamiguchi, K., Hosaka, M., Sugi, M., Yukimoto, S. Kusunoki, S. & Kitoh, A. (2012). Climate simulations using MRI-AGCM3.2 with 20-km grid. *Journal of the Meteorological Society of Japan, 90A*, 233–258.

Muller, F. (1970). Inventory of glaciers in the Mount Everest region. In *Perennial ice and snow masses* (Technical papers in hydrology, Vol. 1, pp. 47–59). Paris: UNESCO/IAHS.

Nakajima, T., Yoon, S. C., Ramanathan, V., Shi, G. Y., Takemura, T., Higurashi, A., … & Schutgens, N. (2007). Overview of the Atmospheric Brown Cloud East Asian Regional Experiment 2005 and a study of the aerosol direct radiative forcing in East Asia. *Journal of Geophysical Research, 112*(D24), D24S91.

Nakawo, M., Fujii, Y., & Shrestha, M. L. (1976). Flow of glaciers in hidden valley, Mukut Himal. *Seppyo, 38*, 39–43.

Nguyen, K. C., & McGregor, J. L. (2009). Modelling the Asian summer monsoon using CCAM. *Climate Dynamics, 32*, 219–236.

Nguyen, K., Katzfey, J. J., & McGregor, J. (2011a). Climate change projections based on downscaling. In G. Cambers. (Ed). *Climate change in the Pacific*. Pacific Climate Change Science Program Technical Report.

Nguyen, K. C., Katzfey, J. J., & McGregor, J. L. (2011b). Global 60 km simulations with CCAM: Evaluation over the tropics. *Climate Dynamics*. doi:10.1007/s00382-011-1197-8.

Ninomiya, K. (2011). Characteristics of the Meiyu and Baiu frontal precipitation zone in the CMIP3 20th century simulation and 21st century projection. *Journal of the Meteorological Society of Japan, 89*, 151–159.

Parthasarathy, B., Munot, A. A., & Kothawale, D. R. (1994). All India monthly and seasonal rainfall series: 1871–1993. *Theoretical and Applied Climatology, 49*, 217–224.

Power, S., Casey, T., Folland, C., Colman, A., & Mehta, V. (1999). Interdecadal modulation of the impact of ENSO on Australia. *Climate Dynamics, 15*, 319–324.

Power, S. B., Schiller, A., Cambers, G., Jones, D., & Hennessy, K. (2011). The Pacific climate change science program. *Bulletin of the American Meteorological Society, 92*, 1409–1411.

Qin, D., Liu, S., & Li, P. (2006). Snow cover distribution, variability, and response to climate change in Western China. *Journal of Climate, 19*, 1820–1833.

Ramanathan, V., Chung, C., Kim, D., Bettge, T., Buja, L., Kiehl, J. T., Washington, W. M., … & Wild, M. (2005). Atmospheric brown clouds: Impacts on South Asian climate and hydrological cycle. *Proceedings of the National Academy of Sciences of the United States of America, 102*, 5326–5333.

Ramanathan, V., Ramana, M. V., Roberts, G., Kim, D., Corrigan, C., Chung, C., & Winker, D. (2007). Warming trends in Asia amplified by brown cloud solar absorption. *Nature, 448*, 575–578.

Saji, N. H., Goswami, B. N., Vinayachandran, P. N., & Yamagata, T. (1999). A dipole mode in the tropical Indian Ocean. *Nature, 401*, 360–363.

Salinger, M. J., Basher, R. E., Fitzharris, B. B., Hay, J. E., Jones, P. D., MacVeigh, J. P., & Leleu, I. (1995). Climate trends in the south-west Pacific. *International Journal of Climatology, 15*, 285–302.

Shrestha, M. L., Fujii, Y., & Nakawo, M. (1976). Climate of hidden valley, Mukut Himal during the monsoon in 1974. *Journal of the Japanese Society of Snow and Ice, 38*, 105–108.

Smith, T. M., & Reynolds, R. W. (2005). A global merged land and sea surface temperature reconstruction based on historical observations (1880–1997). *Journal of Climate, 18*, 2021–2036.

Stevenson, S., Fox-Kemper, B., Jochum, M., Neale, R., Deser, C., & Meehl, G. (2011). Will there be a significant change to El Niño in the 21st century? *Journal of Climate, 25*, 2129–2145. doi:10.1175/JCLI-D-11-00252.1.

Straus, D. M., & Shukla, J. (2002). Does ENSO force the PNA? *Journal of Climate, 15*, 2340–2358.

Streten, N. A., & Troup, A. J. (1973). A synoptic climatology of satellite observed cloud vortices over the Southern Hemisphere. *Quarterly Journal of the Royal Meteorological Society, 99*, 56–72.

Streten, N. A., & Zillman, J. W. (1984). Climate of the South Pacific Ocean. In H. van Loon (Ed.), *Climates of the oceans, World survey of climatology volume 15* (pp. 263–429). Amsterdam: Elsevier.

Tanaka, H. L., Ishizaki, N., & Nohara, N. (2005). Intercomparison of the intensities and trends of Hadley, Walker and Monsoon Circulations in the global warming predictions. *Scientific Online Letters on the Atmosphere, 1*, 77–80.

Taylor, K. E., Stouffer, R. J., & Meehl, G. A. (2012). An overview of CMIP5 and the experimental design. *Bulletin of the American Meteorological Society, 9*, 485–498. doi:10.1175/BAMS-D-11-00094.1.

Terada, K., & Hanzawa, M. (1984). Climate of the North Pacific Ocean. In H. van Loon (Ed.), *Climates of the oceans, World survey of climatology volume 15* (pp. 431–503). Amsterdam: Elsevier.

Terray, P., Dominiak, S., & Delecluse, P. (2005). Role of the southern Indian Ocean in the transition of the monsoon-ENSO system during recent decades. *Climate Dynamics, 24*, 169–195.

Thatcher, M., & McGregor, J. L. (2009). Using a scale-selective filter for dynamical downscaling with the conformal cubic atmospheric model. *Monthly Weather Review, 137*, 1742–1752.

Timbal, B. & Drosdowsky, W. (2012). The relationship between the decline of Southeastern Australian rainfall and the strengthening of the subtropical ridge. doi:10.1002/joc.3492

Trenberth, K. E. (1976). Spatial and temporal variations of the Southern Oscillation. *Quarterly Journal of the Royal Meteorological Society, 102*, 639–653.

Trenberth, K. E. (1984). Signal versus noise in the Southern Oscillation. *Monthly Weather Review, 112*, 326–332.

Trenberth, K. E. (1990). Recent observed interdecadal climate changes in the Northern Hemisphere. *Bulletin of the American Meteorological Society, 71*, 988–993.

Trenberth, K. E. (1991). General characteristics of El Niño-Southern Oscillation. In M. H. Glantz, R. W. Katz, & N. Nicholls (Eds.), *Teleconnections linking worldwide climate anomalies. Scientific basis and societal impact* (pp. 13–42). Cambridge, UK: Cambridge University Press.

Trenberth, K. E. (1997). The definition of El Niño. *Bulletin of the American Meteorological Society, 78*, 2771–2777.

Trenberth, K. E., & Caron, J. M. (2000). The Southern Oscillation revisited: Sea level pressures, surface temperatures and precipitation. *Journal of Climate, 13*, 4358–4365.

Trenberth, K. E., & Stepaniak, D. P. (2001). Indices of El Niño evolution. *Journal of Climate, 14*, 1697–1701.

Trenberth, K. E., Caron, J. M., Stepaniak, D. P., & Worley, S. (2002). The evolution of ENSO and global atmospheric surface temperatures. *Journal of Geophysical Research, 107*, D8. doi:10.1029/2000JD000298. http://www.cgd.ucar.edu/cas/papers/2000JD000298.pdf.

Trenberth, K. E., Jones, P. D., Ambenje, P., Bojariu, R., Easterling, D., Tank, A. K., … & Zhai, P. (2007). Observations: Surface and atmospheric climate change. In S. Solomon, D. Qin, M. Manning, Z. Chen, M. Marquis, K. B. Averyt, M. Tignor, & H. L. Miller. (Eds.), *Climate change 2007: The physical science basis. Contribution of working group I to the Fourth Assessment Report of the Intergovernmental Panel on Climate Change* (pp. 235–336). Cambridge, UK/New York: Cambridge University Press.

Troup, A. J. (1965). The Southern Oscillation. *Quarterly Journal of the Royal Meteorological Society, 91*, 490–506.

Vincent, D. G. (1994). The South Pacific Convergence Zone (SPCZ): A review. *Monthly Weather Review, 122*, 1949–1970.

Wallace, J. M., & Gutzler, D. S. (1981). Teleconnections in the geopotential height field during the northern hemisphere winter. *Monthly Weather Review, 109*, 784–812.

Wang, S. (2012.) Building Asian climate change scenarios by regional climate models ensemble. (APN Project Report: ARCP2011-01CMY-Wang). Retrieved from http://www.apn-gcr.org/resources/items/show/1567

Wang, B., & Lin, H. (2002). Rainy season of the Asian–Pacific summer monsoon. *Journal of Climate, 15*, 386–398.

Wang, H., & Zhang, Y. (2010). Model projections of east Asian summer monsoon climate under "free Arctic" scenario. *Atmospheric and Oceanic Science Letters, 3*, 176–180.

Watanabe, O. (1976). On the types of glaciers in the Nepal Himalayas and their characteristics. *Journal of the Japanese Society of Snow and Ice, 38*, 10–16.

Watanabe, O., Endo, Y., & Ishida, T. (1967). Glaciers and glaciations in the Nepal Himalaya, mainly on the results of field research on two glaciers in Nepal Himalaya. *Low Temperature Science, Series A, 25*, 197–218.

Webster, P. J., Magana, V. O., Palmer, T. N., Shukla, J., Tomas, R. A., Yanai, M., & Yasunari, T. (1998). Monsoons: Processes, predictability, and prospects of prediction. *Journal of Geophysical Research, 103*, 14451–14510.

Webster, P. J., Andrew, W. M., Loschnigg, J. P., & Leben, R. R. (1999). Coupled ocean-temperature dynamics in the Indian Ocean during 1997–98. *Nature, 40*, 356–360.

Wild, M. (2009a). Global dimming and brightening: A review. *Journal of Geophysical Research, 114*, D00D16. doi:10.1029/2008JD011470.

Wild, M. (2009b). How well do IPCC-AR4/CMIP3 climate models simulate global dimming/brightening and twentieth-century daytime and night-time warming? *Journal of Geophysical Research Atmospheres, 114*, D00D11. doi:10.1029/2008JD011372.

Wild, M., Trussel, B., Ohmura, A., Long, C. N., Konig-Langlo, G., Dutton, E. G., & Tsvetkov, A. (2009). Global dimming and brightening: An update beyond 2000. *Journal of Geophysical Research, 114*, D00D13. doi:10.1029/2008JD011382.

Zhang, X., & Yang, F. (2004). *RClimDex (1.0) user manual*. Ontario: Environment Canada.

Zhang, Y., Wallace, J. M., & Battisti, D. S. (1997). ENSO-like variability: 1900–93. *Journal of Climate, 10*, 1004–1020.

Chapter 3
Climate and Urbanization

Peter Marcotullio, Richard Cooper, and Louis Lebel

Abstract Urbanization is a major factor across Asia and the Pacific, and so the scope of this chapter is somewhat restricted. There is a focus on larger urban areas, as the small communities of rural areas are discussed in other chapters. The breadth of the topic of urbanization also means that reports by government agencies and NGOs (grey literature) are cited, as well as the formal academic literature. The six sections of the present Chapter systematically review literature in the field. In the first section we overview urbanization trends in the region. In the second section we review the history of urbanization in the region. The third section examines urbanization and climate in Asia and the Pacific. The fourth section describes the risks in urban areas due to climate change-related hazards. The fifth section overviews mitigation and adaptation measures in the region. The final section concludes with the needs for resilient cities and addresses uncertainties, research gaps and policy measures.

Future predictions suggest that large cities will not hold most of the region's total urban population. In 1990, cities of larger than one million held almost 35.1 % of the total urban population and by 2025 the UN predicts that cities of one million or more will hold 41.2 % of the total urban population. The share of those living in

P. Marcotullio (✉)
Department of Geography, Hunter College, 695 Park Ave.,
New York, NY 10065, USA
e-mail: peter.marcotullio@hunter.cuny.edu; pjm@columbia.edu

R. Cooper
Southeast Asia START Regional Centre (SEA START RC),
IW LEARN, Chulalongkorn University, 5th Floor Chulawich Building,
Bangkok 10330, Thailand
e-mail: richard@iwlearn.org

L. Lebel
Unit for Social and Environmental Research (USER),
Faculty of Social Sciences, Chiang Mai University, Chiang Mai 50200, Thailand
e-mail: louis@sea-user.org

M.J. Manton and L.A. Stevenson (eds.), *Climate in Asia and the Pacific: Security, Society and Sustainability*, Advances in Global Change Research 56,
DOI 10.1007/978-94-007-7338-7_3,

mega-cities is expected to increase from 8.9 % of the total urban population to 12.8 % during the same period. While expectations are for an increase in share, there still remains 58.2 % of the urban population living in settlements smaller than one million. This fact is all the more impressive given the large population that is expected to move into cities in the region between 2010 and 2025 (over 549 million in 15 years, or over 36 million people a year).

Climate resilient cities are those that can withstand climate effects and not change dramatically. They include biophysical and socio-economic sub-systems that can withstand various climate impacts and continue to develop in a fairly predictable manner. Cities that are not resilient change dramatically to new states with new relationships emerging both within the socio-economic sub-system and between the socio-economic and biophysical sub-systems. Resilient cities are sustainable cities. Resilience can be achieved when urban areas move along a more sustainable pathway. The goal of policy makers and stakeholders for their individual urban centres, urban regions as well as nations in the face of climate change is to enhance resilience. In the present review of cities in the Asia-Pacific region, we identify some important aspects that impinge on this goal. Addressing uncertainties, research gaps and policy needs related to climate change and urbanization will help make cities in the region more resilient.

Keywords GHG emissions • Climate hazards • Mega-cities • Urbanization • Climate vulnerabilities

3.1 Introduction

The present chapter reviews literature on trends of urban development and their relationships between this development and climate at the local, regional and global levels in Asia and the Pacific. It includes an examination of the vulnerabilities of urban cities and its residents to climate-related hazards. We also briefly outline and sample urban mitigation and adaptation strategies currently being formulated or implemented.

Chapter 3 has six sections that systematically review literature in the field. In the first section we overview urbanization trends in the region. In the second section we review the history of urbanization in the region. The third section examines urbanization and climate in Asia and the Pacific. The fourth section describes the risks in urban areas due to climate change-related hazards. The fifth section overviews mitigation and adaptation measures in the region. The final section concludes with the needs for resilient cities and addresses uncertainties, research gaps and policy measures.

Urbanization is a major factor across Asia and the Pacific, and so the scope of this chapter must be somewhat restricted; for example, information on cities of small island developing states on the Pacific is covered elsewhere in the present volume. Moreover, there is a focus on larger urban areas, as the small communities of rural areas are discussed in other chapters. The breadth of the topic of urbanization also means that reports by government agencies and NGOs (grey literature) are cited, as well as the formal academic literature.

3.2 Urbanization Trends in Asia and the Pacific

The story of contemporary Asia and the Pacific urbanization is underpinned by the growing strength and complexity of globalization processes operating in the region. During the second half of the twentieth century, Japanese-led growth gave way to an increasing development of the international division of labour, trade, foreign direct investment (FDI) flows, and movements of people, information and resources into and between the four Tigers, the ASEAN-4, and most recently in China, India and Viet Nam. The "flying geese" model of development with Japan as the lead goose (Akamatsu 1962; Bernard and Ravenhill 1995; Hatch and Yamamura 1996; Kojima 2000) complemented by institutional and regulatory policies (Amsden 1989; Rowan 1998) that influence economic forces (Dicken 1992) set the context for changes that are now manifested across the region.

Given the strength and importance of regional urban integration processes operating across Asia and the Pacific, some have predicted convergence in form and function (Cohen 1996; Dick and Rimmer 1998; Hack 2000). This notion has not proven true as cities across the region continue diverging in important characteristics. For example, different transportation-related development trajectories have been identified for Asian cities. This is also true of the relationship between cities and climate. For example, current estimates of greenhouse gas (GHG) emissions from urban areas across the region suggest that they vary with population size, density, wealth and climate, implicating diversity in mitigation and adaptation strategies (Marcotullio et al. 2012). Diversity among uniqueness is an important aspect of urbanization within the region.

Another notable aspect of urbanization in the region is the emergence of mega-cities (populations over ten million). Much research has focused on these cities (Douglass 2000; McGee and Robinson 1995; Laquian 2005; Stubbs and Clarke 1996; UN-Habitat 2011a). The emphasis on these cities (Tokyo, Shanghai, Jabotabek, Manila, Bangkok, Jakarta, Mumbai, etc.) is understandable given their rise in number and economic importance.

Future predictions suggest that large cities, while increasing in importance, will not hold most of the region's total urban population. In 1990, cities of larger than one million held almost 35.1 % of the total urban population and by 2025 the UN predicts that cities of one million or more with hold 41.2 % of the total urban population. The share of those living in mega-cities is expected to increase from 8.9 % of the total urban population to 12.8 % during the same period. While expectations are for an increase in share, there still remains 58.2 % of the urban population living in settlements smaller than one million. This fact is all the more impressive given the large population that is expected to move into cities in the region between 2010 and 2025 (over 549 million in 15 years, or over 36 million people a year).

3.2.1 Contemporary History and Future Predictions

The trends in urbanization in Asia and the Pacific follow uneven growth over both time and space and, in the present section, we divide the history of urbanization into three eras: 1950–1990, 1990–2010 and 2010–2050.

3.2.1.1 1950–1990 Trends

The early post-War period finds much of the region largely rural, but urbanizing rapidly. This apparent contradiction was due to rapidly increasing rural and urban populations (Drakakis-Smith 1992). Underpinning this dynamic was a changing age structure of the population. During 1960–2007, the proportion of Asia's population in the 15–24 age bracket increased from 17 % in 1960 to 21 % in 1985, before beginning to decline (18 % in 2007) (UN-Habitat 2011a).

Besides the city-states and developed countries in Oceania, most nations had shares of urban populations of a third or less of the total populations (Table 3.1). The urban share for all of Asia increased from approximately 16 % in 1950 to 29.5 % in 1990. The economies that urbanized the most rapidly include the four Tigers of East Asia (Hong Kong, Singapore, Taiwan and Republic of Korea) and the Association of Southeast Asian Nations (ASEAN)-4 (Malaysia, Indonesia, Philippines and Thailand). For example, during this 40-year period, the Republic of Korea urbanization level increased from 21.4 % to 74 %.

Rapid urbanization led to dense settlements in the period 1950–1990. In the region the population increases ranged from 211 million to over 870 million, averaging 3.6 % per year. The four Tigers experienced an average growth rate of 4.6 % while ASEAN-4 urban population grew by 4.4 %. Other rapid urbanization in this period occurred in Mongolia, Bhutan, Bangladesh, Nepal, Lao People's Democratic Republic and some of the small island states in the Pacific, but these were related to population growth and not economic growth. The populous economies of South Asia also retained strong population growth, particularly in India and Pakistan, each growing by 3.2 % and 4.1 %, respectively (Table 3.2).

For many developing economies, the 1980s was an era of economic slow-down, particularly, as the world economy responded to a global recession. During the 1980s, urbanization in the region continued and, in fact, urban growth rates increased across much of the region. In some countries, such as India and the Philippines, growth slowed during the decade, but urbanization and urban growth continued.

During the 1980s, China began to enter a phase of rapid urban growth, particularly around coastal areas. At this time, Chinese national development policy prioritized the implementation of six special economic zones, which were later expanded to include another 14 cities all of which were along the eastern coast. This shift to a more open market-based policy changed the country. In South Asia, urban population growth was strong (2.4 % per year), but the sub-region remained largely rural. By 1990, most developing nations in the Asia-Pacific region were less than 30 % urban (Table 3.2).

Table 3.1 Asia and the Pacific urbanization levels, 1950–1990 (percent)

	1950	1970	1990	Absolute Percent Change (1950–1990)	Average Annual Percent Change (1950–1990)
World	**28.83**	**36.08**	**42.62**	**13.78**	**0.981**
Asia and the Pacific	**15.99**	**21.58**	**29.47**	**13.48**	**1.540**
Eastern Asia	**15.51**	**22.88**	**32.21**	**16.70**	**1.843**
China	11.80	17.40	26.44	14.64	2.037
China, Hong Kong SAR	85.20	87.73	99.52	14.32	0.389
China, Macao SAR	96.89	97.03	99.76	2.87	0.073
Dem. People's Republic of Korea	31.00	54.20	58.38	27.38	1.595
Japan	34.85	53.20	63.09	28.24	1.495
Mongolia	20.00	45.05	57.03	37.03	2.654
Republic of Korea	21.35	40.70	73.84	52.49	3.150
Taiwan[a]	–	60.19	75.76	15.57	1.545
Southern Asia	15.94	18.88	25.14	6.26	1.929
Bangladesh	4.28	7.59	19.81	15.53	3.904
Bhutan	2.10	6.09	16.39	14.29	5.271
India	17.04	19.76	25.55	8.51	1.017
Maldives	10.61	11.89	25.84	15.23	2.250
Nepal	2.68	3.96	8.85	6.18	3.036
Pakistan	17.52	24.82	30.58	13.06	1.402
Sri Lanka	15.33	21.89	18.61	3.28	0.486
South-Eastern Asia	**15.48**	**21.50**	**31.62**	**16.14**	**1.802**
Brunei Darussalam	26.76	61.68	65.83	39.08	2.276
Cambodia	10.20	15.97	12.60	2.40	0.530
Indonesia	12.40	17.07	30.58	18.18	2.283
Lao People's Democratic Republic	7.24	9.63	15.44	8.20	1.911
Malaysia	20.36	33.45	49.79	29.43	2.261
Myanmar	16.16	22.83	24.71	8.55	1.068
Philippines	27.14	32.98	48.59	21.46	1.467
Singapore	99.45	100.00	100.00	0.56	0.014
Thailand	16.48	20.89	29.42	12.95	1.460
Timor-Leste	9.89	12.89	20.84	10.95	1.881
Viet Nam	11.64	18.30	20.26	8.61	1.394
Oceania	**62.00**	**70.80**	**70.70**	**8.70**	**0.329**
Australia/New Zealand	**76.16**	**84.51**	**85.29**	**9.13**	**0.284**
Australia	77.00	85.27	85.40	8.40	0.259
New Zealand	72.52	81.11	84.74	12.22	0.390
Melanesia	**5.44**	**14.88**	**19.92**	**14.48**	**3.299**
Fiji	24.35	34.76	41.61	17.26	1.348
New Caledonia	24.59	51.23	59.54	34.95	2.235
Papua New Guinea	1.70	9.80	14.99	13.29	5.593
Solomon Islands	3.80	8.92	13.68	9.88	3.254
Vanuatu	8.75	12.33	18.72	9.96	1.918

(continued)

Table 3.1 (continued)

	1950	1970	1990	Absolute Percent Change (1950–1990)	Average Annual Percent Change (1950–1990)
Micronesia	**31.62**	**46.28**	**62.60**	**30.98**	**1.722**
Guam	41.30	61.92	90.80	49.50	1.989
Kiribati	11.00	24.09	34.99	23.99	2.935
Marshall Islands	23.34	53.49	65.05	41.72	2.596
Micronesia (Fed. States of)	20.00	24.81	25.82	5.82	0.640
Nauru	100.00	100.00	100.00	–	0.000
Northern Mariana Islands	42.00	70.06	89.73	47.73	1.916
Palau	53.88	59.72	69.59	15.71	0.642
Polynesia	**23.24**	**33.89**	**40.07**	**16.83**	**1.372**
American Samoa	61.77	70.38	80.95	19.18	0.678
Cook Islands	38.03	53.29	57.72	19.69	1.048
French Polynesia	34.09	55.25	55.86	21.78	1.243
Niue	21.50	21.10	30.90	9.40	0.910
Pitcairn	–	–	–	–	–
Samoa	12.89	20.35	21.20	8.31	1.252
Tokelau	–	–	–	–	–
Tonga	12.89	20.19	22.70	9.82	1.426
Tuvalu	11.19	22.08	40.66	29.47	3.278
Wallis and Futuna Islands	–	–	–	–	–

Source: Data from UN DESA 2009 World Urbanization Prospects: The 2009 Revision, File 2 Percentage of Population Residing in Urban Areas by Major Area, Region and Country, 1950–2050, POP/DB/WUP/Rev.2009/1/F2

[a]Taiwan: CEPD, 1975–2011, Urban and Regional Development Statistics, Taiwan, 1970 data are for 1975

One striking feature of Asian urbanization is the concentration of the urban population in large cities, in particular, mega-cities. In 1950, there was one mega-city in Asia (Tokyo), compared to five mega-cities in 1990 (Tokyo, Mumbai, Osaka-Kobe, Kolkata and Seoul) in the region (Table 3.2). Moreover, during this period, the share of the urban population residing in cities of over one million increased from 27.3 % to 35.1 %. The number of those living in mega-cities increased from 11.2 million to 77.3 million. By 1990, 8.9 % of the total urban population resided in mega-cities.

3.2.1.2 1990–2010 Trends

During the first half of this era, economic expansion and urbanization continued. Rapid and prolonged wealth creation for nations prompted some to call Asian development a "miracle" (World Bank 1993). Economic indicators alone, however, did not capture the social transformations experienced in the region. It was

Table 3.2 Asia and the Pacific distribution of urban population and urban agglomeration by size, 1950–1990 (population in thousands)

Asia and the Pacific distribution of urban population and urban agglomeration by size, 1950–1990 (population in thousands)

Urban agglomeration size	1950	1970	1990	Percent Total 1990	Absolute Change (1950–1990)	Percent Total Change (1950–1990)
Ten million and larger						
Number	1	1	5	5.2	4	5.7
Population	11,275	23,298	77,306	8.9	66,032	10.0
Five million to less than ten million						
Number	0	5	10	10.3	10	14.3
Population	0	33,493	71,156	8.2	71,156	10.8
One million to less than five million						
Number	26	43	82	84.5	56	80.0
Population	46,453	87,954	157,827	18.1	111,374	16.9
Less than one million						
Population	153,711	285,479	565,900	64.9	412,189	62.4
Total number urban agglomeration over one million	27	49	97		70	
Total urban population	211,439	430,224	872,189		660,750	

Source: Data from World Urbanization Prospects: The 2009 Revision, File 12
Population of Urban Agglomerations with 750,000 Inhabitants or More in 2009, by Country, 1950–2025, POP/DB/WUP/Rev.2009/2/F12

not only wealth that increased, but also the quality of life dramatically improved with reductions in poverty levels, longer life expectancy, reductions in birth mortality, increases in access to basic services and greater literacy (Deolalikar et al. 2002; UN-Habitat 2011a).

During the later 1990s, as economic development spread to more locations, global patterns shifted and the numbers of those in poverty dropped worldwide. The world total number of those in poverty (less than US$ 1 a day) decreased from 1.248 billion in 1990 to 969 million in 2004. Asia and the Pacific's share of the global poor, however, shrunk faster than that for the world. In 1990, there were approximately 955.4 million in poverty throughout the region accounting for 77 % of the global poverty. By 2004, this number shrank to 615.3 million, which accounted for 64 % of the total. Most of those escaping poverty were from East and Southeast Asia, but numbers in South Asia also decreased. Moreover, Asia and the Pacific's share of the ultra-poor (living on less than US$0.5 a day) dropped even more dramatically. In 1990, of the 1.248 billion that were poor, 193 million were ultra-poor. During this year Asia and the Pacific housed 93.4 million (48 % of the world's ultra-poor population). By 2005, the number of ultra-poor throughout the world dropped to 162 million. In the Asia-Pacific region, however, the number dropped to 28.5 million, i.e. 17 % of the world's ultra-poor (Ahmed et al. 2007). One of the main features of Asia's success was the social transformation and poverty reduction that accompanied the region's rapid economic growth and urbanization (Deolalikar et al. 2002; UN-Habitat 2011c).

During this period, the region has undergone rapid demographic transition. The United Nations Development Programme's (UNDP) "human development index" trends confirmed that there have been significant social and economic advances over the last three to four decades. While these figures portray the aggregate human development experienced in some countries within the region, there is great diversity in, among and within nations. Initial conditions, before rapid growth, may be primarily responsible for the more egalitarian experience of Northeast Asia, and there is less clear evidence that growth has been directed to more equitable income distributions elsewhere (Jomo 1998).

From late 1997 and most of 1998, several countries in Asia suffered severe economic contractions and the region as a whole suffered dwarfing what was experienced in the 1980s. The speed and intensity with which the crisis mounted within country after country surprised the world. Some commented that during this period, globalization took a step backwards (Asian Development Bank 1999). However, at the end of 1998, the crises were contained in most countries.

The reprieve of rapid growth facilitated the examination of its costs. The crisis itself intensified and re-focused attention on social tensions (Daniere 1996; Schmidt 1998). Concerns also emerged over environmental conditions including pollution levels that have increased faster than GDP even during the most rapid growth periods (Asian Development Bank 1997; Dua and Esty 1997; Brandon 1994; Setchell 1995). These trends facilitated a questioning of the future viability of globalization-driven growth strategies. By the end of the era, sustainable urban development, pro-poor growth and green buildings, economies and cities had become buzzwords.

Over the 20 years between 1990 and 2010, the urban residential population in Asia and the Pacific expanded by 670 million growing from 872 million to 1.54 billion people compared to the pervious era's increase of 449 million over 40 years (Table 3.3). Asian urbanization continued to include the growth of large cities. During this era large cities become the predominant form of urbanization throughout the region. The number of mega-cities increased from five to ten, including Tokyo, Delhi, Mumbai, Shanghai, Kolkata, Dhaka, Karachi, Beijing, Manila and Osaka-Kobe. Moreover the share of the urban population living in cities over one million increased from 35.1 % to 40.6 %. The population in mega-cities reached approximately 174 million or 11.3 % of the total dwellers in urban areas.

At the end of this era, there remained places with low urbanization levels; Sri Lanka, Bangladesh, Bhutan, Nepal, Papua New Guinea and some of the Small Island States. Most economies, however, reached levels of 30 % or more.

Table 3.3 Asia and the Pacific urbanization levels, 1990–2010 (percent)

Major area, region, economy	1990	2000	2010	Absolute Percent Change (1990–2010)	Average Annual Percent Change (1990–2010)
World	**42.62**	**46.40**	**50.46**	**7.85**	**0.849**
Asia and the Pacific	**29.47**	**34.81**	**40.56**	**11.09**	**1.610**
East Asia	**32.21**	**40.39**	**50.17**	**17.96**	**2.241**
China	26.44	35.76	46.96	20.51	2.913
China, Hong Kong SAR	99.52	100.00	100.00	0.48	0.024
China, Macao SAR	99.76	100.00	100.00	0.24	0.012
Dem. People's Republic of Korea	58.38	59.41	60.22	1.83	0.155
Japan	63.09	65.22	66.83	3.74	0.288
Mongolia	57.03	56.86	62.03	4.99	0.420
Republic of Korea	73.84	79.62	82.96	9.11	0.584
Taiwan	75.76	77.71	79.80	4.04	0.260
South Asia	**21.54**		**30.38**		
Bangladesh	19.81	23.59	28.07	8.26	1.757
Bhutan	16.39	25.42	34.71	18.32	3.823
India	25.55	27.67	30.01	4.46	0.808
Maldives	25.84	27.71	40.10	14.26	2.222
Nepal	8.85	13.43	18.62	9.77	3.787
Pakistan	30.58	33.14	35.90	5.32	0.805
Sri Lanka	18.61	15.83	14.31	−4.30	−1.304
Southeast Asia	**31.62**	**38.16**	**41.84**	**10.22**	**1.410**
Brunei Darussalam	65.83	71.15	75.65	9.82	0.698
Cambodia	12.60	16.91	20.11	7.51	2.365
Indonesia	30.58	42.00	44.28	13.70	1.868
Lao People's Democratic Republic	15.44	21.98	33.18	17.75	3.901
Malaysia	49.79	61.98	72.17	22.38	1.873
Myanmar	24.71	27.80	33.65	8.94	1.556
Philippines	48.59	47.99	48.90	0.31	0.032
Singapore	100.00	100.00	100.00	–	0.000
Thailand	29.42	31.14	33.96	4.54	0.720
Timor-Leste	20.84	24.26	28.12	7.28	1.509
Viet Nam	20.26	24.49	30.38	10.12	2.047
Oceania	**70.70**	**70.39**	**70.22**	**−0.48**	**−0.034**
Australia/New Zealand	**85.29**	**86.91**	**88.62**	**3.33**	**0.192**
Australia	85.40	87.17	89.11	3.71	0.213
New Zealand	84.74	85.68	86.20	1.46	0.085
Melanesia	**19.92**	**18.96**	**18.38**	**−1.53**	**−0.400**
Fiji	41.61	47.91	51.86	10.25	1.107
New Caledonia	59.54	59.20	57.36	−2.18	−0.186
Papua New Guinea	14.99	13.20	12.53	−2.47	−0.895
Solomon Islands	13.68	15.71	18.55	4.87	1.536
Vanuatu	18.72	21.70	25.56	6.85	1.571

(continued)

Table 3.3 (continued)

Major area, region, economy	1990	2000	2010	Absolute Percent Change (1990–2010)	Average Annual Percent Change (1990–2010)
Micronesia	**62.60**	**65.60**	**68.06**	**5.46**	**0.419**
Guam	90.80	93.10	93.17	2.37	0.129
Kiribati	34.99	42.96	43.90	8.91	1.141
Marshall Islands	65.05	68.36	71.76	6.71	0.492
Micronesia (Fed. States of)	25.82	22.33	22.66	–3.16	–0.650
Nauru	100.00	100.00	100.00	–	0.000
Northern Mariana Islands	89.73	90.16	91.34	1.61	0.089
Palau	69.59	69.96	83.39	13.79	0.908
Polynesia	**40.07**	**41.16**	**42.40**	**2.33**	**0.283**
American Samoa	80.95	88.77	92.97	12.03	0.695
Cook Islands	57.72	65.19	75.31	17.59	1.339
French Polynesia	55.86	52.38	51.44	–4.42	–0.411
Niue	30.90	33.07	37.54	6.65	0.979
Pitcairn	–	–	–	–	–
Samoa	21.20	21.98	20.23	–0.97	–0.234
Tokelau	–	–	–	–	–
Tonga	22.70	23.01	23.43	0.72	0.157
Tuvalu	40.66	46.02	50.39	9.73	1.079
Wallis and Futuna Islands	–	–	–	–	–

Source: Data from UN DESA 2009 World Urbanization Prospects: The 2009 Revision, File 2 Percentage of Population Residing in Urban Areas by Major Area, Region and Country, 1950–2050, POP/DB/WUP/Rev.2009/1/F2

3.2.1.3 2010–2050 and Beyond

In the future we expect globalization-driven growth to continue as the region's urbanization and urban population growth rates continue to decrease, but we also expect a doubling of the already large urban population size. The urban share for Asia will grow from 42.2 % in 2010 to 64.6 % in 2050 (Table 3.4). During this period, the Asia-Pacific urban population will grow from 1.54 billion to 2.9 billion people (reference from table). By 2050, urban populations in India, Pakistan and Bangladesh combined (1.2 billion) will become 15 % larger than that of China (1.03 billion).

We will see almost all nations in the region reaching urbanization levels of over 50 % with the average for the region at 64.9 %. The most urbanized regions are expected to be in East Asia (over 74 %), Australia and New Zealand (over 93 %) and Micronesia (over 80 %). South Asia is predicted to reach an urbanized level of more than 55 % and for Southeast Asia more than 65 % will live in dense settlements.

The growth of large cities is expected to continue, although the percentage of populations living in these cities is expected to decrease. From 2010 to 2025, the number

Table 3.4 Asia and the Pacific urbanization levels, 2010–2050 (percent)

Major area, region, economy	2010	2030	2050	Absolute Percent Change (2010–2050)	Average Annual Percent Change (2010–2050)
World	**50.46**	**58.97**	**68.70**	**18.24**	**0.774**
Asia and the Pacific	**40.56**	**52.02**	**64.93**	**24.37**	**1.183**
East Asia	**50.17**	**63.73**	**74.34**	**24.16**	**0.988**
China	46.96	61.91	73.23	26.28	1.117
China, Hong Kong SAR	100.00	100.00	100.00	–	0.000
China, Macao SAR	100.00	100.00	100.00	–	0.000
Dem. People's Republic of Korea	60.22	65.74	74.53	14.32	0.535
Japan	66.83	72.98	80.08	13.24	0.453
Mongolia	62.03	71.56	79.53	17.50	0.623
Republic of Korea	82.96	87.67	90.83	7.88	0.227
Taiwan	–	–	–	–	–
South Asia	**30.38**	**40.55**	**55.14**	**24.75**	**1.501**
Bangladesh	28.07	41.04	56.41	28.34	1.760
Bhutan	34.71	49.97	64.17	29.46	1.548
India	30.01	39.75	54.23	24.22	1.490
Maldives	40.10	60.06	73.12	33.01	1.513
Nepal	18.62	31.74	47.56	28.94	2.372
Pakistan	35.90	45.62	59.37	23.48	1.266
Sri Lanka	14.31	19.55	31.34	17.03	1.979
Southeast Asia	**41.84**	**52.85**	**65.44**	**23.59**	**1.124**
Brunei Darussalam	75.65	82.33	87.21	11.56	0.356
Cambodia	20.11	29.20	43.83	23.72	1.967
Indonesia	44.28	53.70	65.95	21.67	1.001
Lao People's Democratic Republic	33.18	53.07	68.03	34.85	1.811
Malaysia	72.17	82.21	87.85	15.68	0.493
Myanmar	33.65	48.09	62.87	29.22	1.575
Philippines	48.90	58.33	69.36	20.46	0.877
Singapore	100.00	100.00	100.00	–	0.000
Thailand	33.96	45.77	59.96	25.99	1.431
Timor-Leste	28.12	39.89	54.92	26.80	1.687
Viet Nam	30.38	44.18	58.99	28.61	1.673
Oceania	**70.22**	**71.38**	**74.81**	**4.58**	**0.158**
Australia/New Zealand	**88.62**	**91.25**	**93.37**	**4.75**	**0.131**
Australia	89.11	91.86	93.84	4.73	0.129
New Zealand	86.20	88.14	90.88	4.68	0.132
Melanesia	**18.38**	**23.81**	**34.87**	**16.49**	**1.613**
Fiji	51.86	61.66	72.20	20.34	0.831
New Caledonia	57.36	62.68	71.95	14.59	0.568
Papua New Guinea	12.53	18.18	29.75	17.23	2.186
Solomon Islands	18.55	29.20	44.31	25.76	2.200
Vanuatu	25.56	38.01	53.48	27.92	1.863

(continued)

Table 3.4 (continued)

Major area, region, economy	2010	2030	2050	Absolute Percent Change (2010–2050)	Average Annual Percent Change (2010–2050)
Micronesia	**68.06**	**73.27**	**80.03**	**11.97**	**0.406**
Guam	93.17	94.20	95.45	2.28	0.061
Kiribati	43.90	51.66	63.91	20.01	0.943
Marshall Islands	71.76	78.77	84.62	12.86	0.413
Micronesia (Fed. States of)	22.66	30.27	44.43	21.77	1.697
Nauru	100.00	100.00	100.00	–	0.000
Northern Mariana Islands	91.34	93.33	94.89	3.55	0.095
Palau	83.39	92.01	94.49	11.11	0.313
Polynesia	**42.40**	**48.82**	**59.90**	**17.49**	**0.867**
American Samoa	92.97	95.62	96.81	3.83	0.101
Cook Islands	75.31	84.94	89.32	14.02	0.428
French Polynesia	51.44	56.62	67.42	15.98	0.678
Niue	37.54	49.43	63.06	25.52	1.305
Pitcairn	–	–	–	–	–
Samoa	20.23	23.96	36.64	16.41	1.496
Tokelau	–	–	–	–	–
Tonga	23.43	30.40	44.50	21.07	1.617
Tuvalu	50.39	61.48	72.45	22.05	0.912
Wallis and Futuna Islands	–	–	–	–	–

Source: Data from UN DESA 2009 World Urbanization Prospects: The 2009 Revision, File 2 Percentage of Population Residing in Urban Areas by Major Area, Region and Country, 1950–2050, POP/DB/WUP/Rev.2009/1/F2
There are no predictions for urban population increases for these periods

of mega-cities will increase to 15 (Table 3.5), however, the share of the total urban population living in large cities will increase only slightly from 40.6 % to 41.2 %. By 2025, approximately 58.8 % of the region's urban population will live in cities of less than one million. Approximately 12.8 % of the total urban population will live in mega-cities. This trend is important as currently much research and policy attention is devoted to the larger cities, but those that are less than one million will be the locations where the lion's share of the Asia-Pacific urban population is predicted to live.

3.3 Urbanization and Climate in Asia and the Pacific

The growth of cities in Asia and the Pacific has implications for land and energy use, and climate. Generally, the trends include greater urban land usage, more energy and changes in local and region climates. While these are negative trends as measured against a sustainable trajectory, final conclusions must include two

Table 3.5 Asia and the Pacific distribution of urban population and urban agglomeration by size, 2010–2025 (population in thousands)

Urban agglomeration size	2010	2020	2025	Percent Total 2025	Absolute Change (2010–2025)	Percent Total Change (2010–2025)
Ten million and larger						
Number	10	14	15	5.9	5	7.9
Population	174,117	241,151	266,889	12.8	92,772	16.9
Larger than five million and less than ten million						
Number	18	22	22	8.7	4	6.3
Population	131,930	156,950	163,402	7.8	31,472	5.7
Larger than one million and less than five million						
Number	163	210	217	85.4	54	85.7
Population	320,134	432,193	432,193	20.7	112,059	20.4
Less than one million						
Population	916,154	1,072,254	1,229,178	58.8	313,024	57.0
Total number urban agglomeration over one million	191	246	254		63	
Total urban population	1,542,335	1,902,548	2,091,662		549,327	

Source: Data from World Urbanization Prospects: The 2009 Revision, File 12 Population of Urban Agglomerations with 750,000 Inhabitants or More in 2009, by Country, 1950–2025, POP/DB/WUP/Rev.2009/2/F12

qualifications. First, the trends we present would potentially be worse if the populations were not concentrated in dense settlements. That is, land use change would be more dramatic, energy use higher and contributions to global climate change more spectacular if the same populations were spread out evenly over the landscape. Second, the trends observed need not necessarily continue. Given more sensitive urban design and other mitigation measures, evidence suggests that the climate impacts of urban living can be lowered.

3.3.1 Urbanization and Land and Energy Use

According to a recent estimate the combined trends of increasing populations in dense settlements along with decreasing densities indicates that the developing world will triple the urban land taken by cities with more than 100,000 by 2030 (Angel et al. 2005). Of these, urban land use changes in China and India are growing at the fastest rates (Seto et al. 2011).

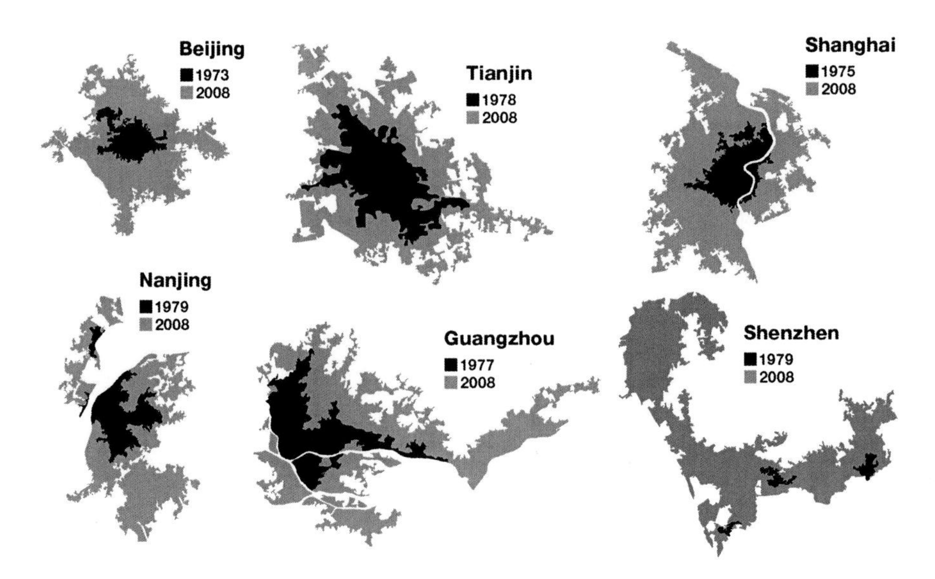

Fig. 3.1 Examples of growth in Chinese cities, 1970s–2008 (Source: Zhang et al. 2012)

Examples of rapidly expanding individual urban areas have been identified in China. Between 1973 and 2008, the average increase in urban land in a sample of 60 cities (4 municipalities, 28 provincial capitals, 2 special administrative regions and 26 other well-known cities) throughout the country was approximately 2.5 times (Wen 2010). Some cities (such as Shenzhen & Houkou) underwent spectacular growth multiplying their land area by more than a factor of 20. From the 1980s to 2005, the urban lands of Beijing-Tianjin-Tangshan Region, Yangtze River Delta City Region and the Pearl River Delta City Region grew by factors of 1.15, 1.04 and 2.63, respectively (Liu 2011) (Fig. 3.1).

The development of land use and building changes has accompanied increases in energy consumption. To meet this need, total energy production in Asia increased by 125 % from 1971 to 1990 and by 90 % from 1990 to 2007. These increases were higher than any other region. Energy production has been largely driven by growth in China, which was approximately 125 % and 105 % during the above periods, respectively. By 2007, China was producing 1.8 billion tonnes oil equivalent, while the rest of Asia was producing 1.2 billion tonnes oil equivalent (OECD 2010). Moreover, energy production in China exceeded that of the Middle East and the Former Soviet Union.

Much of this energy is related to urbanization. As cities grow in size and complexity the energy demands to keep them running smoothly increase. Many large cities appropriate energy in the form of electricity. The growth of urban Asia and the Pacific accompanied growth in electricity consumption. In 2009, Asia included two of the four highest non-OECD consumers of electricity; China and India. Among these four nations, China held the largest share of electricity consumption at 39.4 %. This share reflects rapid increases in electricity provisioning in the country. Between

1973 and 2009, electricity consumption in China increased at an average annual growth rate of 9 % (IEA 2011). Most electricity consumption, particularly in developing countries, occurs in cities. For example, in India, the country's urban residents consume 87 % of the nation's electricity (Sawin and Hughes 2007).

Moreover, urban areas also facilitate the use of motor vehicles. Vehicle usage has increased dramatically in Asia, but not equally amongst nations. East, South and Southeast Asia's share of global automobiles on the road increased from 12.7 % in 1985 to 21.8 % in 2009 (an increase from 62.1 to 210.4 billion vehicles). Car ownership rates in China have been growing at 12 % per annum and in India at 9 % per annum; and include two- and three-wheeled vehicles. Asia produces 95 % of global two- and three-wheeled vehicles, which constitutes 75 % of the world's stocks and China is the fastest growing market for these vehicles (World Energy Council 2011).

Transportation fuel consumption patterns over time suggest that different Asian nations are following distinctive paths (Barter 1999) with Hong Kong, Singapore and Japan on a low consumption path, Republic of Korea and Taiwan at the intermediate level, and Thailand and Malaysia following a more Western, high transport fuel consumption, trajectory (Marcotullio and Marshall 2007).

The largest and most recent industrializing nations to embrace the automobile may follow the high transport fuel consumption trend. From 2005 to 2009, India increased the number of cars on the road by 25 % and the number of motor vehicles in China doubled. In 2009, car sales in China exceeded those in the USA (Ward's 2010). Some estimates suggest that by 2040, future automobile consumption in India and China alone will double the total number of vehicles currently on the road, i.e., adding an additional 800 million automobiles to the global car population (The Economist 2006; Wilson et al. 2004).

3.3.2 Urban Climate

Changes in land to urban uses and increases in energy demand help to change local climate. Atmospheric scientists, such as Landsberg (1981), observed that urban climatic conditions include lower radiation, more cloudiness, higher precipitation, higher temperatures and more particulates, gaseous admixtures and other contaminants than non-urban climates. These characteristics arguably make urban climates unique. Three important characteristics of urban climates include the emergence of urban heat islands, changes in precipitation and changes in ambient air quality. Examples of these specific urban climates are evident in the Asia and the Pacific region.

3.3.2.1 Urban Heat Island

Inhabitants of urban areas are subject to climatic conditions that represent a significant modification of the pre-urban climatic state including the well-known urban heat island (UHI) phenomenon (Oke 1973), arising from the modification of radiation, energy and momentum exchanges resulting from the built form of the city, together with the emission of heat, moisture and pollutants from human activities. Urban temperatures are typically 3–4 °C higher than surrounding areas due to UHI (Oke 1997), but can be as high as 11 °C warmer in urban "hotspots" (Aniello et al. 1995). Dark surfaces such as asphalt roads or rooftops, however, can reach temperatures 30–40 °C higher than surrounding air (Frumkin 2002). The UHI effect is one of the major problems of the twenty-first century (for a review see McKendry 2003; Landsberg 1981; Rizwan et al. 2008).

The UHI phenomenon has been studied in many places in the Asia-Pacific region; for example, Osaka and Bangkok (Taniguchi et al. 2009), Nanjing (Zeng et al. 2009), Shanghai (Tan et al. 2010) and Beijing and Wuhan (Ren et al. 2007). For the purpose of estimating the effect of urban warming over the past 100 years, Seoul, Tokyo, Osaka, Taipei, Manila, Bangkok, and Jakarta were selected as target cities. UHI was calculated by subtracting the temperature data of the four grids around the city from the observational temperature data in the city. In doing so, all urban areas demonstrated an increasing UHI effect with Osaka demonstrating the largest increase from approximately 2.4 °C in 1901 to almost 3 °C after 1981. The increases in Seoul, Tokyo, and Taipei were approximately between 1 °C and 2 °C. Jakarta and Bangkok exhibited smaller increases and Manila and Bangkok experienced rapid increases after 1961 (Kataoka et al. 2009). The point stressed here is that not only does the UHI effect exist but it is increasing in cities across the region.

UHI effects have not been identified as contributing to global warming (Parker 2004; Peterson 2003; Alcoforado and Andrade 2008). These and other studies indicate that effects of urbanization and land use change on land-based temperature records are negligible (0.006 °C per decade) as far as hemispheric- and continental-scale averages are concerned (Trenberth et al. 2007). At the same time, as cities increase in size and number, the UHI effect may play a role in regional climate (Kaufmann et al. 2007).

A further study presented evidence for a significant urbanization effect on the regional climate in southeast China. In this case, the region experienced rapid urbanization and estimates suggest a warming of mean surface temperature of 0.05 °C per decade. The spatial pattern and magnitude of estimates are consistent with those of urbanization characterized by changes in the percentage of urban population and in satellite-measured greenness (Zhou et al. 2004).

Another study examined the trends of urban heat island effects in east China and finds a significant influence of urbanization on surface warming over the region. Overall, UHI effects contribute 24.2 % to regional average warming trends (Yang et al. 2011). These results are consistent with a recent 50 year study that found most temperature time series in China affected by UHI (Li et al. 2004). Evidence, although only recently emerging, suggests that UHI in the region contributes to regional climate change.

3.3.2.2 Changes in Precipitation

Urbanization affects humidity, clouds, storms and precipitation. Studies have described shifts in precipitation patterns in and around cities compared to less densely populated areas (for a review see Souch and Grimmond 2006; Shepherd 2005). The exact mechanisms by which these urban precipitation patterns emerge are poorly understood (Lowry 1998). Unique aspects of urban areas that might affect precipitation levels include high surface roughness that enhances convergence; UHI effects on boundary layers and the resulting downstream generation of convective clouds; generation of high levels of aerosols that act as cloud condensation nuclei sources; and urban canopy creation and maintenance processes that affect precipitation systems.

No matter what the mechanism, urban areas are seen as cloudier and wetter, with heavier and more frequent precipitation within metropolitan areas than those outside, but within the same region (Lei 2011; Changnon 1979). Average increases of 28 % in monthly rainfall rates have been identified within 30–60 km downwind of cities (Shephard et al. 2002).

While most studies on urban precipitation have focused on the USA and Europe, several analyses have been conducted in Asia and the Pacific. Meng et al. (2007) identified increased strength in thunderstorms associated with tropical cyclones as they moved over Guanghzhou City, China. Inamura et al. (2011) simulated the effects of Tokyo on heavy rainfall indicating precipitation increases. These studies have confirmed trends found in other locations. On the other hand, researchers have also found anomalies in regional urban rainfall patterns. For example, Wang et al. (2009a) identified changes in the patterns of rainfall within Beijing associated with rapid urbanization. These changes, however, were restricted to the winter months, when rainfall increased in those areas undergoing the most rapid growth. During other seasons, rainfall patterns did not change significantly. Another study that examined rainfall data in the Pearl River Delta of China from 1988 to 1996 indicated a negative correlation with urbanization, causing an "urban precipitation deficit" during the dry season (Kaufmann et al. 2007). These authors hypothesize that given the Pearl River Delta's extreme urbanization rates the negative effects of built-up areas on precipitation may overwhelm effects, which could boost precipitation. These studies indicate that while urban climates in the Asia-Pacific region may demonstrate patterns similar to those of the now developed world, they may also provide for unique conditions that affect climate in new ways.

3.3.2.3 Air Pollution

The composition of the atmosphere over urban areas differs from undeveloped areas (Pataki et al. 2006). Importantly, urban air contains pollutants. Ambient air pollution refers to gases, aerosols and particles that harm human well-being and the environment. Cities have been seen as sources of air pollution, but upon closer examination air pollution is primarily a function of fuel consumption and land use changes. The impacts of air pollution on human health are discussed further in Sect. 5.3.

Once emitted the dispersion and dilution of air pollutants are strongly influenced by meteorological conditions, especially by wind direction, wind speed, turbulence and atmospheric stability. Topographical conditions and urban structures, like street canyons for example, have a great effect on these parameters. Cities that develop in valleys often undergo atmospheric inversions, which trap pollution and enhance effects. Air pollution has multiple health, infrastructure, ecosystem and climate impacts (Molina and Molina 2004).

While urban air pollution is a ubiquitous problem, trends vary by development status. In industrialized countries the 'classic' air pollutants, such as carbon monoxide, sulphur dioxide and total suspended particulates are decreasing dramatically, while nitrogen oxides and non-methane volatile organic compounds have reached a plateau or demonstrate weakly decreasing trends (Holdren and Smith 2000). In the developing world, air pollutants are increasing (UNEP/WHO 1993). However, the greatest problems with air pollution are often associated with cities in middle income countries (McGranahan and Murray 2003).

Increasingly, motor vehicle traffic is a major air pollution source (Mage et al. 1996). Motor vehicles emit carbon monoxide, hydrocarbons, nitrogen oxides and toxic substances including fine particles and lead. Secondary pollution, such as ozone, is a product of these primary pollutants, which react together in the atmosphere under the sun's energy. Given the trends in automobile usage, even in developing countries, automobiles are a source of air pollutants (Walsh 2003).

Most urban air pollution attention has focused on mega-cities (Mayer 1999; Molina and Molina 2004; Gurjar et al. 2004, 2008; Butler et al. 2007). High levels of air pollution emissions are associated with energy production and fuel consumption in mega-cities in China (He et al. 2002) and India (Kandlikar and Ramachandran 2000). These studies indicate that, in the 1990s, Chinese mega-cities such as Beijing, Shenyang, Xian, Shanghai, and Guangzhou; and Indian mega-cities cities such as Delhi and Mumbai were among the most polluted cities in the world (UNEP/WHO 1993). In a recent study using a multi-pollutant index (for total suspended particles, sulphur dioxide and nitrogen dioxide) researchers ranked the top ten mega-cities with the lowest air quality level. The ranking includes seven Asia-Pacific cities of Dhaka (1), Beijing (2), Karachi (4), Jakarta (5), Delhi (7), Shanghai (8) and Kolkata (9) (Gurjar et al. 2008).

It isn't the largest cities in the world that have the worst pollution levels, however. A recent global study that examined air pollution trends in over 8,000 cities suggests that urban nitrogen oxides, non-methane volatile organic compounds, carbon monoxide and sulphur dioxide emission levels were highest in Asia (Sarzynski 2012). When ranked by the largest total contribution of emissions, the top emitters of various compounds included Tokyo, Taipei, Seoul, Shanghai, Jakarta, Shenzhen, Ulsan and Tianjin. When examining the largest emitters per capita, however, several smaller, less well known, cities such as Anugul and Sidhi, India; Chengguan, Fengzhen Luzhai and Wulumuqi China; Dumai, Indonesia; Port Dickson, Malaysia; Pohang, Republic of Korea; and Rayong, Thailand ranked among the highest for different compounds. This suggests, as some have argued, that some of the smaller cities in the region suffer from some of the worst environmental challenges (Hardoy et al. 2001).

Urban air pollution can have regional effects. Emissions from cities may play a role in regional climate impacts as high levels of fine particulate matter can scatter and/or absorb solar radiation (Molina and Molina 2004). The visible manifestation of this regional air pollution is a brownish layer or haze pervading many areas of Asia (UNEP and C4 2002; Ramanathan and Crutzen 2002). Hotspots for this phenomenon, commonly known as *atmospheric brown clouds*, in Asia include South Asia, East Asia and the Indonesian region. Through the examination of temperature records in urbanized regions of China and India affected by the haze, researchers have demonstrated a significant cooling effect since the 1950s (Kaiser and Qian 2002; Menon et al. 2002). These effects are consistent with the predicted effects of elevated soot levels and fine particulate matter, despite general warming observed for most for the globe. Recent research suggests that the carbonaceous aerosols are from both biomass burning (slash-and-burn land clearance; waste burning in agriculture and forestry; and residential wood combustion for heating and cooking) and from urban fossil fuel combustion, establishing a role for urban activities in the source of these clouds (Gustafsson et al. 2009). The persistence of the haze has significant implications to the regional and global water and energy budget and health (see Chaps. 4 and 5).

3.3.3 Urbanization and Global Climate

Urban areas are major contributors of GHGs (Dhakal 2010). Estimates vary (Dodman 2009; Satterthwaite 2008), but the general consensus is that urban areas are responsible for approximately 72 % of global anthropogenic GHG emissions (IEA 2008). Given the recognition of cities as important contributors to these trends, research has developed that attempts to isolate the urban role in regional and global climate change (Bader and Bleischwitz 2009; Lebel et al. 2007).

An important distinction in understanding GHG emissions from urban areas includes "direct" and "indirect" emissions. More often, local inventories include estimations of GHGs related to activities of government, businesses and residents emitted from within the urban area, known as "direct" emissions. Measurements may also include emissions from activities located outside local jurisdictions but closely related to economic activities that are conducted within jurisdictions, known variously as "indirect" emissions (US EPA 2011). For example, power production and waste disposal may be conducted outside of cities, but relate to the energy and waste disposal needs of urban residents, businesses and governments. Traditional emissions inventories count only emissions that are produced within the study area, regardless of where the related good or service is ultimately consumed, thus placing the full responsibility for emissions reduction within the site of emission production.

More recent work attempts to include a consumption component. As such, urban protocol research has appropriated the term "scope" used for corporate emissions inventories (WRI 2002). Various scopes define the location of embodied energy-related emissions; for example, Scope 1 emissions are directly emitted from within an urban area, while Scope 2 includes emissions that are related to urban activities, particularly energy consumption, but emitted outside urban areas (in thermal power plants).

Several attempts have been made to estimate GHGs and standardize emission protocols across a number of cities (Kennedy et al. 2009a, b; Hillman and Ramaswami 2010; Sovacool and Brown 2010; Hoornweg et al. 2011). A number of GHG emission studies have been performed on cities in the region (Schulz 2010; Phdungsilp 2010; Dhakal 2009; Dhakal and Imura 2004; Marcotullio et al. 2011, 2012).

There have been studies that examined GHG emissions across the region using gridded GHG emission data, urban boundaries and thermal power plant locations to estimate, at the regional scale, urban contributions of GHGs (Marcotullio et al. 2011, 2012). The analysts found that amongst sources identified, energy production is the dominant source of GHG emissions; with evidence that peri-urban areas are significant sources of GHG emissions. Finally, as demonstrated by studies in air pollution, the largest emitting urban areas are not the highest per capita emitters.

3.3.3.1 Urban GHGs by Source

For calculating the GHG emissions from urban areas Kennedy et al. (2009b) following the Intergovernmental Panel on Climate Change (IPCC), identify a set of standardized sources including energy (stationary combustion, mobile combust and fugitive sources); waste; industrial processes and product use; and Agriculture, Forestry and other Land uses (AFOLU). We follow this format and present data on GHG emissions from cities in the region by source.

(a) *Energy production*

The energy sector typically includes stationary combustion, mobile combustion and fugitive sources. The energy source category includes emissions of all GHGs resulting from these activities. We separate mobile combustion and concentrate emissions from emissions for stationary sources including electricity and district heating.

Studies of urban GHGs in the Asia and Pacific region emphasize energy. For example, Mitra et al. (2003) emphasized the importance of the energy sector for Delhi and Calcutta. Ajero (2002) included GHG emissions from energy production in a study of Metro Manila. Sovacool and Brown (2010) examined energy use in buildings and industry across several cities in Asia. Dhakal (2009) separated energy production and transport in an analysis of emissions from four Chinese cities (Beijing, Shanghai, Tianjin and Chongquin).

Evidence supports the notion that energy consumption is an important contribution to GHG emissions from cities in the Asia and Pacific region. For four cities in China, electricity production alone accounted for between 34 % and 41 % of emissions (Dhakal 2009). In Metro Manila, electricity consumption accounted for 40 % of total emissions (Sovacool and Brown 2010). In 2000, urban Asia energy production and GHG emissions released directly from within urban areas accounted for 61.7 % of all emissions from urban areas (Table 3.6). This conservative figure (which doesn't account for all thermal power plant

Table 3.6 Asia total and urban GHG emissions by sector, 2000

Asia total and urban GHG emission by sector, 2000 (million metric tons)					
	Total GHG emissions		Urban GHG emissions		Urban share of total
Sector	(amount)	(percent)	(amount)	(percent)	(percent)
Agriculture	2,460	17.49	145	3.37	5.88
Energy	6,743	47.94	2,648	61.71	39.27
Industry	1,564	11.12	492	11.46	31.46
Transportation	1,238	8.80	432	10.06	34.89
Residential	1,293	9.20	321	7.48	24.84
Waste	766	5.45	253	5.89	33.00
Total	14,065	100.00	4,292	100.00	30.51

Source: Marcotullio et al. (2012). A geography of urban greenhouse gas emission in Asia. *Global Environmental Change*, *22*(4), 944–958

emissions), strongly suggests that energy production is a dominant source of emissions for Asian cities. That is, only 71 % of all thermal power plants in the region were located within urban areas, leaving emission from 29 % outside cities and not accounted for even though, the electricity consumption may be within urban areas.

(b) *Transportation*

GHG emissions from mobile sources are both important contributions to total emission levels and challenging to estimate. Generally, analysts have observed an inverse relationship between urban ground transportation energy use and population density (Newman and Kenworthy 1989, 1999). On the other hand, air transport GHG emissions are sometimes ignored by urban GHG analysts. Aviation emissions, however, are an increasingly important contribution to total GHG emissions, particularly in the Asia-Pacific region.

Transportation-related GHG emissions are a low but growing portion of total GHG emissions for Asia (Table 3.6). In 2000, for the entire region, transportation accounted for 8.8 % of total GHG emissions. In 2000, transportation-related emissions in urban Asia accounted for a higher percentage of total urban GHG emissions than for the region; approximately 10.6 %. This level is similar to that found by analysts for individual cities. For example, transport accounted for 8 % of total energy consumption in Beijing and 10 % in Shanghai (APERC 2007). Others have identified higher levels for individual cities. For example, transportation accounted for 37 % of Tokyo's energy consumption and 25 % of Seoul's energy consumption in 1998 (Dhakal and Imura 2004). According to Phdungsilp (2010), in 2005 Bangkok's transportation sector accounted for 60 % of energy demand. Estimates suggest that for Dhaka, transportation accounts for 25–30 % of total emissions (Alam and Rabbani 2007).

As mentioned earlier, urbanization is bound up with automobile usage and hence energy consumption and GHG emissions. In general, as nations urbanize, their citizens increasingly shift from non-motorized transport such as bicycling and walking to motorized passenger transport. With increasing urban expansion, more urban dwellers move to the city outskirts and employment areas also shift. Within this peri-urbanization, travel distance tends to lengthen, which has been noted in Asia and the Pacific (APERC 2007). Moreover, in rapidly developing countries, demand for private transportation far outstrips infrastructure supply, creating congestion, high levels of accidents and increasing pollution and GHG emissions (Vasconcellos 2001). With increasing infrastructure provision, we expect transportation GHG emissions in the region to rise rapidly (Marcotullio and Marshall 2007).

(c) *Industrial processes and product use*

The industrial processes and product use category includes GHG emissions from industrial products that are not primarily for energy purposes. A wide range of industrial processes and products emits GHGs that are not the result of combustion. Three broad categories for non-energy industrial use include: (i) feedstock, (ii) reducing agents, and (iii) non-energy products such as lubricants, greases, waxes, bitumen and solvents.

In 2000, for all of Asia, industrial processing and product use accounted for approximately 11.2 % of total anthropogenic GHG emissions. Importantly, these emissions varied across the region as identified by locations of manufacturing. Urban Asia industrial processes and product use GHG emissions accounted for 11.5 % of all urban emissions (Table 3.6). Researchers identified selected industrial cities with high-energy use. For example, in Bangkok in 2005, industry accounted for 31 % of total energy consumption (Phdungsilp 2010) and 25 % of Thailand's carbon dioxide emissions came from manufacturing and construction (Corfield 2008a), dominated by industries in the Bangkok Metropolitan Region. Industry dominated carbon dioxide emissions in Beijing, Shanghai, Tianjin and Chongqing, although industrial emissions have been declining in share for Beijing (from 65 % to 43 %) and Shanghai (from 75 % to 64 %) over the past 20 years (Dhakal 2009).

(d) *Waste and wastewater*

Waste GHG emissions are those from waste management activities (Kennedy et al. 2009b). Much of these emissions are in the form of methane from landfills, dumps and wastewater treatment. Global methane emissions from wastewater treatment under anaerobic conditions and from municipal solid waste landfills are estimated to range from 8 % to 11 % and 3–19 % of global anthropogenic methane emissions, respectively (Wunch et al. 2009; IPCC 1996). The waste sector as a whole accounts for just under 4 % of global GHG output. In the future a large proportion of the GHG emission from urban wastewater is expected to be from cities in developing countries, although researchers note that much will depend upon whether a methane recovery system in place or not .

In 2000, for all of Asia, waste accounted for approximately 5.6 % of total GHG emissions. During that year, in urban Asia, waste accounted for approximately 5.9 % of all GHG emissions (Table 3.6). In individual cities the share of

GHG emissions from the waste sector can be higher. Bangkok waste-related GHG emissions, for example, accounted for approximately 11.5 % of total GHG emissions (Kennedy et al. 2009b). In some rapidly growing cities, such as selected cities in India, (Mumbai, Delhi, Kolkata and Chennai), municipal solid waste is growing faster than population, raising concern over GHG emissions (Jha et al. 2008).

(e) *Residential*
Residential emissions include those GHG emission not related to energy production. Importantly, they include biomass burning for heat and cooking. There is some data available for this important aspect of local contributions to GHG emissions. In 2000, Asian residential emissions accounted for 9.2 % of all GHG emissions in the region and in urban Asia, residential emissions accounted for approximately 7.3 % of all emissions (Table 3.6). This suggests that much of the residential GHG emissions occur outside the urban areas.

(f) *Agriculture, forestry and other land use (AFOLU)*
The world's forests have a substantial role in the global carbon cycle (Nabuurs et al. 2007) and urban areas hold significant amounts of carbon in their forests (Nowak and Crane 2002); but, in general, urban GHG emissions from forestry as well as from agriculture are considered low (Kennedy et al. 2009b).

Agriculture emissions include carbon dioxide, methane and nitrous oxide from agricultural activities. In 2000, Asian agriculture (excluding those from forestry and other land uses) accounted for 17.9 % of all GHG emissions in the region. However, in urban areas, GHG emission shares were much lower. In urban Asia in 2000, agriculture (excluding those from forestry and other land uses) accounted for approximately 3.37 % of total urban emissions. Agricultural activities in Delhi and Metro Manila accounted for 2 % and 9 % of total emissions, respectively (Sovacool and Brown 2010). For urban India as a whole, agricultural GHG emissions from within urban areas accounted 7.2 % of all Indian urban GHG emissions (Marcotullio et al. 2011).

3.3.3.2 Urban GHGs Attributed to Aspects of Dense Settlement

While urban areas are major contributors to climate change, there is a large variation in GHG emissions amongst cities. Some of the variation is due to urban factors or those related directly to population size, density and growth rate of cities. Several studies have emphasized various features of cities and their impact on energy consumption and GHG emissions (Sadownik and Jaccard 2001; Lefevre 2009; Permana et al. 2008; Li 2011; Lebel et al. 2007).

Population size is the most important contributing factor to overall urban emissions, but the largest urban GHG emitters per capita are those cities with high levels of emissions and smaller populations, rather than the largest cities. Asia, however, is known for its urban density. The distribution of dense cities in Asia varies from 518 persons/km^2 to over 1,711 persons/km^2. The pattern of emissions per capita highlights the potential effect of density on energy efficiency. As mentioned earlier, there is a common understanding that the lower density urban areas use more transportation-related energy than the higher density cities (Newman and Kenworthy 1999; Weisz and Steinberger 2010) and studies have verified this relationship (Parshall et al. 2010). Australian analysts have identified the differential dependence on the private automobile among cities of varying densities with the highly sub-urbanized urban areas having the highest dependence and therefore highest transportation energy consumption (Lenzen et al. 2008).

Density can also affect the delivery of higher density fuels and electricity and hence affect GHG emissions. For example, in a study of households in China and India, researchers found that urban households have greater access to electricity grids and modern fuels, appliances and equipment and therefore the energy-use patterns differ among rural and urban households with urban households averaging higher percentages of coal, liquid petroleum gas (LPG) and electricity usage (Pachauri and Jiang 2008). Research also suggests that the least dense Asian urban centres have GHG emission per capita levels over twice the regional average and over three times that of the lowest emission group; the medium high density urban areas (Marcotullio et al. 2012).

As mentioned, many studies point out that while cities have high GHG emissions levels, those emitted directly outside cities are even higher. That is, GHG emissions per capita levels in suburban areas are higher than those in core parts of cities. This is not typically found in developing parts of the world however, although there is some evidence in Asia that peri-urban areas are important direct GHG emitters (Marcotullio et al. 2011, 2012). The geographic characteristics are defined by 20, 40 and 80 km distances from urban extents. The data suggest that uniformly a significant share of GHG emissions is released from areas immediately beyond urban extents (up to 20 km outside the boundaries of urban areas) (Table 3.7). Within Asia as a region, approximately 43 % of all GHG emissions are released in these peri-urban areas. In East Asia, the share of GHG emissions released at this distance reaches 47.4 % of total GHG emissions.

Given the higher share of GHG emissions, it is not surprising that the per capita emission levels typically released from the 20 km buffer areas are higher than those from the urban areas. For the region as a whole, the average GHG emissions per capita released from within 20 km of urban extents is 4.59 tonnes per capita. This difference is true for all sub-regions. Moreover, for the region as a whole the per capita levels in the 20 km buffer are only surpassed by the levels of GHG emissions per capita of those furthest from urban areas, 5.29 tonnes per capita. That is, the areas with highest per capita emissions are those in the most rural. In these areas, emissions are largely due to agriculture and the share of population is typically small (ranging from 2.3 % to 18 % of the total by sub-region).

Finally, urban growth rate can influence GHG emissions. The relationship in Asia suggests that slower growing cities are higher GHG emitters than those growing faster. This could be due to the fact that the slower growing urban areas are the largest and hence the biggest emitters. At the same time, the smaller cities are the most rapidly growing or demonstrate the highest relative growth rates but low total emissions (Marcotullio et al. 2012).

Table 3.7 Asian distribution of population and GHG emissions by distance from urban areas, 2000 (& of total and metric tonnes per capita)

Asian distribution of population and GHG emissions by distance from urban areas, 2000 (percent of total and metric tonnes per capita)

Sector	Within Urban extent	Urban extent −20 km	20–40 km	40–80 km	Remainder	Region
Asia						
Population	34.84	36.07	16.47	8.88	3.74	
Total CO2 equivalents	30.51	43.23	12.76	8.32	5.17	
CO2 equivalents/capita	3.35	4.59	2.97	3.59	5.29	3.83
East Asia						
Population	45.34	33.73	12.04	6.61	2.28	
Total CO2 equivalents	34.50	47.35	9.86	5.38	2.91	
CO2 equivalents/capita	4.11	7.58	4.42	4.39	6.89	5.40
South Asia						
Population	23.83	42.56	22.03	9.08	2.50	
Total CO2 equivalents	23.20	43.87	20.12	9.65	3.16	
CO2 equivalents/capita	2.01	2.12	1.88	2.19	2.61	2.06
Southeast Asia						
Population	28.59	33.96	15.12	13.26	9.07	
Total CO2 equivalents	23.71	37.26	14.35	15.19	9.48	
CO2 equivalents/capita	2.23	2.95	2.55	3.08	2.81	2.69
Central Asia						
Population	40.66	20.96	11.53	9.04	17.81	
Total CO2 equivalents	26.84	32.56	9.21	12.98	18.40	
CO2 equivalents/capita	5.60	13.18	6.78	12.19	8.77	8.49
West Asia						
Population	50.97	16.09	14.27	12.94	5.73	
Total CO2 equivalents	31.26	26.34	13.54	14.46	14.41	
CO2 equivalents/capita	4.01	10.72	6.21	7.31	16.47	6.55

Source: Data from Marcotullio et al. (2012). A geography of urban greenhouse gas emission in Asia. *Global Environmental Change*, *22*(4), 944–958

3.3.3.3 GHG Emissions Attributed to Non-Settlement Aspects

There are a number of non urban factors that can influence GHG emission patterns from cities. Perhaps the most important among socio-economic and biophysical factors are wealth and climate. There is a general consensus in the literature that cities with higher income typically have larger global environmental impact than cities of lower income (McGranahan 2005; McGranahan et al. 2005). This relationship is understood to include GHG emissions and the results have been verified by researchers comparing results across a number of cities globally (Kennedy et al. 2009b). It has also been identified within the region (Bai and Imura 2000; Marcotullio 2005). Moreover, at the household level, energy consumption for higher income residents is higher in India and China than lower income households (Pachauri 2004; Pachauri and Jiang 2008). In Asia, the highest GDP per capita group of cities (those with more than 1990 US$5,027 per capita) have the highest GHG emissions per capita, suggesting that as GDP per capita increases, GHG emissions per capita do as well.

This relationship, however, may be more complex than a simple rise in GHG emissions with increasing wealth. Some have found that selected cities in lower income countries, Shanghai in China for example, have higher GHG emissions per capita than cities in high income countries like Tokyo in Japan (Dhakal and Imura 2004). Hence, the relationship between GHG emission levels and wealth may be more complex than a simple positive correlation. Indeed, research has demonstrated that a more complex relationship exists between wealth and GHG emissions in the region. That is, for Asia, a rise in income is an inverted relationship, such that GHG emissions fall after a specific level, but then rise again with higher levels of income; a cubic relationship (Marcotullio et al. 2012).

Another non-urban factor that has been found to be important in influencing urban GHG emissions is climate, including temperature (Kennedy et al. 2009b). In Asia, there is a relationship between the heating degree-days (HDDs) experienced by urban centres and GHG emissions. HDDs are the number of days that a city experiences temperatures below a certain level, requiring heating in buildings. Cities that are located in areas of high HDDs have higher GHG emissions per capita than cities with low HDDs. This is also true of elevation. Cities at higher elevation require more heating than those at lower elevation and therefore use more energy and emit more GHGs.

3.4 Urban Climate Change Hazards and Vulnerabilities in Asia and the Pacific

Globally, temperatures have increased approximately 0.74 °C over the past century and that 11 of the 1995–2006 years rank among the 12 warmest years in the historical record of global surface temperatures (IPCC 2007b). Current trends suggest that for the next two decades a warming of about 0.2 °C per decade is projected. The scientific consensus is that the global climate is changing due to human influence on the system in a variety of ways (Goudie 2006; Houghton 2009). Besides warming, the most fundamental climate-related physical changes to the Earth system include: (i) changes in precipitation levels, (ii) sea-level rise, and (iii) increased variability of weather, including extreme events (such as tropical cyclones).

The impact of climate change on urban areas has been an area of growing interest (IPCC 2007b; UN-Habitat 2011a, b; Prasad et al. 2008). Research demonstrates the potential risks from climate change are significant including flooding, coastal erosion, saltwater intrusion into groundwater supplies, elevated temperatures and heat waves, drought, disease outbreaks, landslides, and increasing damage from tropical storms and cyclones (APN 2010; Nicholls 1995). Furthermore urban risk from climate change is exacerbated by man-made hazards, including subsidence caused by groundwater withdrawal; patterns of urbanization and development. Such as, expansion of impervious surfaces aggravating storm water runoff, man-made structures obstructing drainage; reduction of river channel capacity (APN 2010; Haruyama 1993; World Bank 2010). Moreover, a large portion of the urban population lives in areas that will be significantly affected by climate related changes: coastal, arid and mountain zones.

Specific studies have examined the number of people and value of property at risk, how various climate-related changes harm urban management, quality of life, physical infrastructure and urban markets, and whether or to what extent people and urban systems are vulnerable and resilient to climate change-related risks and hazards. Both disaster and hazard reduction research and climate change adaptation research communities focus on analyses of the underlying causes of exposure and vulnerability and have goals of integrating findings into planning and management (Solecki et al. 2011).

3.4.1 Urban Risks and Hazards Associated with Climate Change

Examinations of urban climate change risks and vulnerabilities focus on the human and economic assets potentially exposed to harm from various climate change-related hazards Some of this research is future scenario-based (i.e., populations at risk from sea-level rise). Alternatively, studies that examine the impacts of climate change examine previous and contemporary events that are or could be related to climate change (the impacts of landslides for example may be enhanced with climate change). Understanding the threats and impacts of climate change on the region's cities is all the more important given the large and increasing urban populations in the region. Moreover, analysts suggest Asia is a climate change 'hotspot' and Asian cities are particularly vulnerable to future climate change harms (Yusuf and Francisco 2009; World Wildlife Fund 2009).

3.4.1.1 Increased Temperatures

Temperatures in the Asia-Pacific region are changing and the likely range of global average surface temperature change is projected to be 2.4–6.4 °C in 2090–2099 relative to 1980–1999 (IPCC 2007b). Generally, there is consensus among climate scientists that future warming will be defined by summer heat waves of longer duration and greater intensity and frequency in East Asia with fewer very cold days in East Asia and South Asia (Christensen et al. 2007).

We understand that climate change will have a wide variety of human health impacts. The World Health Organization (WHO) attributes a mortality rate of more than 150,000 annually since the 1970s to climate-induced diseases, and projects that twice this number will die due to climate change by 2030 (McMichael et al. 2004) (see also Sect. 5.3).

Increasing temperatures and specifically heat waves, however, are particularly important for cities because they act in concert with the UHI phenomenon to increase demand for water and cooling, exacerbate air pollution and heat stress; and increase risk of mortality (Shimoda 2003; Prasad et al. 2008; Beniston and Diaz 2004; Cruz et al. 2007).

Increased temperatures cause heat stress. Exposure to extreme hot weather is associated with increased morbidity and mortality, compared to an intermediate 'comfortable' temperature range (Curriero et al. 2002). The European heat wave of 2003, for example, killed 70,000 people (Robine et al. 2007) demonstrating the impacts of increases in extreme maximum temperature events.

The safest temperature range is closely related to mean temperature for an individual city, so it varies around the world (Patz et al. 2005). An international study of cities found the temperature threshold for heat-related deaths ranged from 16 °C to 31 °C and heat thresholds were generally higher in cities with warmer climates (McMichael et al. 2008). Despite the fact that many cities in the Asia-Pacific region lie in tropical and sub-tropical zones, many have recorded incidences of UHI and higher death rates during heat waves. From 2000 to 2009 there were 60 extreme temperature events, which claimed over 8,700 lives (International Federation of Red Cross and Red Crescent Societies 2010). Heat waves in Shanghai elevated deaths in the city, particularly during the 1998 and 2003 heat waves (Tan et al. 2007, 2010). Heat-related deaths are not uncommon among the sick and elderly in Ho Chi Minh City (Asian Development Bank 2010).

The ability of individuals to combat excessive heat is a function of age (Åström et al. 2011). The elderly and the young are typically more at risk. As Asian ages and populations in cities become older, heat stress-related mortality is projected to increase. One model suggests that with future climate change many countries are projected to experience a four- to five-fold increase in excess mortality due to heat stress with China and India experiencing large losses in absolute terms (Takahashi et al. 2007).

As mentioned earlier, cities in Asia already suffer from high levels of air pollution. Rising temperature can also lead to increased air pollution and associated incidence of disease and higher temperatures can lower air quality through increasing the incidence of smog events (UN-Habitat 2011a). Changes in concentrations of ground-level ozone (a secondary pollutant), for example, have been projected with increasing temperatures (Ebi 2010). Exposure to raised concentrations of ozone, the main component of smog, is associated with higher hospital admissions for a variety of respiratory conditions and also early mortality (Confalonieri et al. 2007). This is particularly important for residents of cities in valleys that are susceptible to temperature inversions. Over 10 % of the Asia-Pacific urban population live in mountainous cities, which are typically located in valleys and therefore susceptible to inversions.

In the future, regions that are heavily urbanized will be more adversely affected by temperature-related climate changes than rural ones (Costello et al. 2009). Urban populations, therefore, will suffer higher exposure to these hazards than rural populations creating higher risks, especially for those with pre-existing respiratory disease (Ayres et al. 2009). Heat stress in combination with UHI and air pollution in developing cities in Asia will likely further enhance respiratory and cardiovascular illnesses (Patz et al. 2000) and the joint impact of these health-related stresses may be greater than the sum of their effects (Satterthwaite et al. 2007).

3.4.1.2 Sea-Level Rise

Scientists suggest that sea level has been rising around the world at an average rate of 1.8 mm per year between 1961 and 2003. This change has largely come about through increases in ocean temperatures. As sea temperatures have increased the volume of ocean water has expanded. This phenomenon is considered to be an important factor currently attributed to sea-level rise. At the same time, melting ice sheets may become more important in the future (Church et al. 2008). Across the twentieth century sea level rose by an estimated 0.17 m, although with significant regional variation (IPCC 2007b).

The IPCC predicts a further rise of 0.26–0.59 m of global sea-level rise by 2100 (Meehl et al. 2007). Independent estimates of future sea level, however, indicate that global seas could rise over 1 m by 2100 (Overpeck and Weiss 2009; Pfeffer et al. 2008; Vermeer and Rahmstorf 2009). Higher sea levels have dramatic implications for coastal cities.

Over one quarter of the urban population lives in coastal ecosystems (McGranahan, et al. 2005) and many of the world's mega-cities are located in coastal areas (Nicholls 1995). The term coastal is considered in a broad sense in light of the local geomorphological, ecological and economic characteristics (Klein et al. 2003). The global urban population living in areas of low elevation coastal zones (LECZs) may be as high as 352 million with approximately two-thirds in Asia (McGranahan et al. 2007). In Asia and the Pacific approximately 29 % of the urban population lives in coastal ecosystems of varying elevation. The potential threat of sea-level rise to these populations depends on a number of factors including a city's geographic location and features, and the infrastructure and socio-economic characteristics of the residents (de Sherbinin et al. 2007; UN-Habitat 2011b).

Recent assessments suggest that coastal mega-cities in the Asia-Pacific region are particularly vulnerable to climate change (Fuchs 2010; Hanson et al. 2011; Nicholls et al. 2008). Amongst 136 port cities, one study found that 38 % (52 ports) are located in Asia and many of these are located in deltaic regions. Cities in deltaic regions tend to have higher risk of coastal flooding (Nicholls et al. 2008). This same study estimated that by 2070, 90 % of the total estimated asset exposure in these large port cities will be concentrated in eight nations, six of which are in Asia (China, USA, India, Japan, Netherlands, Thailand, Viet Nam and Bangladesh). For populations, approximately 90 % of the exposure in the 2070s will be contained in 12 countries, eight of which are in Asia (China, USA, India, Japan, Netherlands, Thailand, Viet Nam, Bangladesh, Myanmar, Egypt, Nigeria and Indonesia). Altogether,

Asian cities houses 65 % of the globally-exposed population living beneath the 100-year water level mark (Hanson et al. 2011).

Sea-level rise acts together with a number of other factors to put both property and people at risk. For example, as sea levels rise, there is inundation from flooding and storm damage (including wind), wetland loss and change, erosion, saltwater intrusion and rising water tables (Nicholls and Tol 2006). An important economic, ecological and human risk factor associated with sea-level rise is coastal erosion. Arthurton (1998) noted that although rarely catastrophic, coastal erosion may be a substantial hazard for some mega-cities. The study by Panya Consultants (2009) analysing climate change impacts for the Bangkok Metropolitan Region (BMR) suggests coastal erosion in the Upper Gulf of Thailand is a critical deterrent to the sustainable development of the BMR, and forced retreats are already occurring in Bangkok's coastal area (World Bank 2010). In Mumbai, the cost of coastal erosion was estimated at US$2.5 million/km for capital works for protecting prime waterfront property (Asian Development Bank 2007). Furthermore, a recent study that examined the impact of several different factors associated with sea-level rise for Jakarta estimated that by 2100, costs could run in excess of 1.2 % of the country's GDP (Ward et al. 2011).

3.4.1.3 Changes in Hydrology

Increases in temperature predicted by climate scientists will have associated changes in precipitation. While some areas will see increases, there will be areas that experience decreases in average precipitation, changes in seasonal distribution and a general increase in the spatial variability of precipitation. Gleick (2010) summarizes the evidence from global climate models as "dry areas will get drier, while wet areas will get wetter."

It is expected that there will be an increase in the frequency of intense precipitation events associated with monsoon rains in parts of East and South Asia. For Australia and New Zealand precipitation will likely decrease in southern Australia in winter and spring. Precipitation is also expected to decrease in south-western Australia in winter, while precipitation is predicted to increase in the west of New Zealand's South Island. Changes in rainfall in northern and central Australia are uncertain (Christensen et al. 2007).

Increases in rainfall in areas of the region that are already experiencing high rainfall patterns will potentially have a large number of significant impacts. These include adverse effects on the quality of surface and groundwater, increased risk of infectious, respiratory and skin diseases, disruption of commerce transportation and daily activities, loss of property and increased landslides, and the resultant loss of life and increased morbidity from all these changes.

The Asia-Pacific region suffers from flooding. Floods are among the most costly and damaging disasters and their frequency and severity has generally increased in the last decades (McCarthy et al. 2001). While flooding is associated with high levels of rainfall, it is also associated with sea-level rise and storm surges and may be exacerbated by other hazards such as land subsidence (World Bank 2010).

Between 2000 and 2009, of the 1,739 reported flood events globally, 655 (38 %) occurred in Asia. These events affected 892 million people, killed 36.8 million and caused US$85 billion in damages in Asia (International Federation of Red Cross and Red Crescent Societies 2010). Cities in Asia, and particularly those of China, India and Thailand, were ranked as being the most vulnerable (in terms of population and asset exposure) to future coastal flooding (Nicholls et al. 2008). The World Bank (2010) identified Bangkok, Ho Chi Minh City and Manila as "hotspots" in Asia. In Bangkok, existing flood protection is considered inadequate for even a 30-year event and exposes 30–35 % of Bangkok's land area and approximately one million people to inundation (Fuchs 2010). This predicted risk was tragically realized during November 2011 when the city experienced its worst floods in over half a century. The impact was so great that the Thai MPs called for a study to examine the relocation of the capital to a less flood-prone province (Bangkok Post 2011).

History has demonstrated that these are not the only cities at risk to flooding in the region, however. In 1988, floods in Bangladesh inundated 52 % of the country and covered 85 % of Dhaka for several weeks. It is estimated that of the six million inhabitants of the city, between 2.2 and 4 million were affected. The total death toll was reported to exceed 2,300 with 150 deaths in Dhaka. Ten years later, in 1998, floods again affected Dhaka, impacting 30 % of housing in the metropolitan area (International Federation of Red Cross and Red Crescent Societies 2010). On 26 July 2005, Mumbai received 944 mm of rain in 14 h (Bhagat et al. 2006), resulting in flooding that caused 500 fatalities and US$2 billion in damage (Hallegatte et al. 2010). Flash flooding in Iloilo, the Philippines, in 2008 affected 152 of its 180 *barangays* and up to 500 people were killed, while 261,335 were affected (International Federation of Red Cross and Red Crescent Societies 2010). Other significant impacts of recent urban flooding have been felt throughout the region (Table 3.8).

In the future, however, impacts may be more significant. By 2050, it is likely that there will be increased flooding events affecting Bangkok, Ho Chi Minh City and Manila. For example, researchers predict that 62 % of Ho Chi Minh City's population will be affected by a 1-in-30 year event. In Manila, Bangkok and Ho Chi Minh City, costs of damage from climate change-related flooding are estimated to range from 2 % to 6 % of the regional GDP. Therefore, a 1-in-30 year flood in Manila could cost between US$900 million and US$1.5 billion, given current flood control infrastructure (World Bank 2010). By the 2080s, the costs of the Mumbai flood event of 2005 will more than double for the city and total losses (both direct and indirect) associated with a 1-in-100 year event could triple compared with the current situation (US$690–US$1890 million) (Hallegatte et al. 2010). Moreover, the IPCC (2007b) predicts increased flooding over the next two or three decades from glacier melt in the Himalayas. Alam and Rabbani (2007) noted how melting glaciers will add to existing damage in Dhaka caused by river floods and excessive rainfall during the monsoon.

For much of Asia, an increase in annual precipitation will also exacerbate landslides. Landslides including those related to dump collapses are currently significant threats in the region. In Ho Chi Minh City, for example, June through

Table 3.8 Climatic changes, possible impacts, and potential urban planning-related consequences

Climatic change	Possible Impacts	Potential urban planning-related consequences
Increased temperatures	Groundwater depletion	Water shortages
	Water shortages	Distress migration to cities/towns due to droughts in rural areas
	Drought	Interruption of food supply networks and higher food prices
	Degraded air quality (smog)	Potential energy price increases (e.g., from reduced hydro-electricity generation in places where it exists)
		Exaggerated urban heat island effect
		Increased energy demands for cooling
		Need for higher and/or additional wastewater treatment
		Population health impacts (e.g., increased mortality during heat waves, decreased access to food/nutrition)
Increased precipitation	Increased flooding	Interruption of food supply networks
	Increased risk of landslides or mudslides on hazard slopes	Property damage (homes and businesses)
		Disruption of livelihoods and city/town economies
		Damage to infrastructure not designed to standards of occurrences being experienced
		Distress migration to cities due to floods in rural areas
		Displacement and population movement from informal settlements built on steep slope hazard lands, etc.
		More favorable breeding grounds for pathogens (e.g., mosquitoes and malaria)
		Population health impacts (increased incidences of water-borne diseases like cholera)
Sea-level rise	Coastal flooding	Displacement and population movement from coastal flood areas
	Salt water intrusion into groundwater supplies in coastal areas	Property damage (homes and businesses)
	Increased storm surge hazard	Damage to infrastructure not designed to standards of occurrences being experienced
		Disruption of livelihoods and city/town economies
		Population health impacts (injuries, increased mortality and illness)

(continued)

Table 3.8 (continued)

Climatic change	Possible Impacts	Potential urban planning-related consequences
Increased extreme weather episodes (storms, cyclones, hurricanes)	More intense flooding Higher risk of landslides/ mudslides on hazard slopes	Property damage (homes and businesses) Damage to infrastructure not designed to standards of occurrences being experienced Population health impacts (injuries, increased mortality, distress) Disruption of livelihoods and city/town economies Interruption of food supply networks

Source: UN-HABITAT (2011a). *Planning for climate change, a strategic values-based approach for urban planners*. Nairobi: UN-Habitat

August is landslide season. The city has identified 42 hotspots in high danger of landslides (Nahn 2010). In 2000, a landslide in Mumbai killed at least 60 people in slum districts in the east of the city (BBC 2000). In 2000, heavy rains in Quezon City from typhoons caused a 15-m hill in the dump to collapse, burying hundreds of homes, killing 288 people and displacing several 100 families. In 2006, a landslide buried the entire Barangay Guinsaugon and affected another 80 barangays, causing 154 deaths, with 968 reported missing, 3,742 displaced and 18,862 affected (World Bank 2011).

While increased precipitation creates significant risks for cities, so does water scarcity. Currently, modelled global results show that 150 million people live in cities with perennial water shortage, defined as having less than 100 L per person per day of sustainable surface and groundwater flow within their urban extent (McDonald et al. 2011). Many of these urban areas are located in arid ecosystems. The Asia-Pacific region has a large proportion of semi-arid drylands (Safriel et al. 2005) and the urban population living in these areas exceeds 35 % of the total urban population in the entire region. Moreover, of this population, approximately 56 % live in semi-arid, arid or hyper-arid regions. Some of these areas have already become fully arid as a result of climate change. For example, a new desert is reportedly forming on the eastern edge of China's Qinghai-Tibet Plateau, which has traditionally been a grassland area used for herding (World Bank 2007b).

Water availability in dry lands is projected to decline from the current average of 1,300 m^3 per person per year (in 2000), which is already below the threshold of 2,000 m^3 required for minimum human well-being (Thirlwell et al. 2007). Water stress is likely to increase in these areas. According to UNFCCC (2007) climate change will bring increasing water stress to over 100 million people due to decreases in freshwater availability in Central, South, East and Southeast Asia, particularly in large river basins such as Changjiang.

Change in freshwater availability is not only due to lack of precipitation, but also due to seasonal differences that accompany warming. For example, as mentioned, global warming is causing glacier melting in the Himalayas. While in the short term this runoff increases the likelihood of flooding, erosion, mudslides and glacier lake outburst flooding (GLOFs) in many South Asian nations; in the longer term, this may also lead to a rise in the snowline and the disappearance of many glaciers thus reducing water storage. Large populations rely on the seven main rivers in South, Southeast and East Asia fed by glacier melt runoff from the Himalayas. Throughout Asia one billion people could face water shortages leading to drought and land degradation by the 2050s (Christensen et al. 2007; Cruz et al. 2007).

It is not surprising then that of the 162 million urbanites that researchers predict to undergo perennial water shortage by 2050, the majority will be in Asia (94 million) (McDonald et al. 2011). Indeed, water shortage is one of the most serious potential threats arising from climate change in the region (Parry et al. 2007). For example, the National Capital Region of Delhi is facing severe water shortfalls and municipal demand is currently competing with irrigated agriculture. Water shortages are exacerbated by high levels of leakage from water supply systems, local pollution of water bodies and groundwater, saltwater intrusion of aquifers and inadequate infrastructure (Marcotullio 2007; Rodolfo and Siringan 2006; de Sherbinin et al. 2007; APN 2010). Delhi's water is transported 300 km and unaccounted for water losses are over 40 % in the city (Revi 2009). A snapshot of recent reports that highlight current city water stress is presented in Table 3.9.

3.4.1.4 Increased Frequency and Intensity of Tropical Storm Events

Tropical cyclones are intense storms that originate over tropical waters and can sustain winds exceeding 64 knots. In North America they are called hurricanes; in the western North Pacific, typhoons; in India they are called cyclones; and in Australia tropical cyclones. Between 2000 and 2009, 416 windstorms were recorded in Asia, which caused over 160,000 deaths and economic losses estimated at US$ 366 billion (International Federation of Red Cross and Red Crescent Societies 2010).

At the urban level, the increase in storms and their increasing intensity has been recorded. The number of tropical storms affecting Ho Chi Minh City is reported to have increased over recent years (Asian Development Bank 2010). The flooding in Mumbai in 2005 was related to a cyclone, which left more than 1,000 dead (de Sherbinin et al. 2007). The tropical storm Ketsana, which hit Manila and the surrounding area, created flood waters reaching nearly 7 m and killed hundreds of residents (World Wildlife Fund 2009).

Increased intensity of tropical storm events in the future will increase coastal flooding with all the related knock-on effects (McCarthy et al. 2001). With an increase of 4 °C one estimate suggests average annual insured losses from tropical cyclones affecting China alone will increase by 32 % to reach approximately US$0.5 billion; 100-year losses will increase by 9 % to reach US$1.340 billion, and 200-year losses will increase by 17 % to reach approximately US$1.8 billion (Dailey et al. 2009).

Table 3.9 Recent reports of water stress in cities in Asia and the Pacific

Recent reports of water stress in Asia and the Pacific cities	
The news article, titled – 'Greater **Jakarta** on alert for drought' – reports on a prolonged dry spell that threatened to disrupt supplies to Greater Jakarta (Jakarta Post 2011)	Unusually low monsoon rainfall resulted in a 30 % cut in the water supply in **Mumbai**, described as the 'worst water shortages in its history'; water shortages in **Delhi** and other cities were also mentioned (BBC 2009)
With reference to **Calcutta** – 'Droughts have been more frequent in the last few decades and are projected to get worse, which will lead to even more salt-water intrusion and thus deteriorate surface and groundwater quality' (WWF 2009)	'China's £1.1bn desalination plant is just the latest megaproject in its increasingly desperate race against water shortages … '**Tianjin** has a chronic shortage. Drought, overuse and pollution have left its population of ten million with just a 10th of the water of the average global citizen' (Guardian 2011)
'Extreme climate events can have serious impacts on **Karachi**'s water supply. While droughts, such as in 1999–2001, cause water shortages in the city, extreme monsoon rainfalls can cause flooding and ensuing outbreaks of waterborne diseases' (WWF 2011)	'Although freshwater is naturally abundant in the metropolitan **Shanghai** area, the city experiences high water stress due to the rising demand of 23 million inhabitants' (WWF 2011)

Sources:
BBC. (2009). Mumbai faces acute water shortage. 7 July 2009. Retrieved from http://news.bbc.co.uk/2/hi/8138273.stm
Guardian. (2011). Can the sea solve China's water crisis? The Guardian newspaper. 24 January 2011. Retrieved from http://www.guardian.co.uk/environment/2011/jan/24/china-water-crisis
Jakarta Post. (2011). Greater Jakarta on alert for drought. 15 September 2011. Retrieved from http://www.thejakartapost.com/news/2011/09/15/greater-jakarta-alert-drought.html
WWF. (2009). Mega-stress for Mega-cities. A Climate Vulnerability Ranking of Major Coastal Cities in Asia. World Wide Fund for Nature
WWF. (2011). Big Cities. Big Water. Big Challenges. Water in an Urbanizing World. World Wide Fund for Nature, Deutschland

3.4.2 Basic Urban Service Challenges Associated with Climate Change

This section provides an overview on the potential impacts of climate change to urban service delivery, given current physical trends. The focus is on selected services including transportation, water supply, sanitation, food, energy and industry. All these services will be impacted by changes in temperature, precipitation, storm events and sea level rise.

3.4.2.1 Transportation

Climate change has a number of impacts on transportation systems that threaten to disrupt movement within cities. Weather conditions, including storms and floods, have immediate impacts on travel and can cause service interruptions to both public and private transportation (for example via subways, flooded roads, etc.). Heavy precipitation events can also cause due damage in the form of erosion of supporting transportation infrastructure such as highways trusses, seaports, bridges and airport runways. In coastal cities this is a particularly significant aspect of climate change as it can combine with seal-level rise and inundate highways and transit systems. Salt water can also corrode infrastructure. In the worst case scenarios, entirely new sub-systems are needed in free-from-flooding zones. High temperatures can damage roads, rails and transit vehicles as well as support systems (deforming rails that result in slower traffic). Prolonged high daily temperatures degrade paved roadways used by heavy vehicles and necessitate frequent repairs. Both steel and concrete can buckle under high temperatures. High winds can reduce the use of upper decks on bridges and impede traffic speeds (Transportation Research Board 2008; Mehrotra et al. 2011).

Transportation costs due to weather-related risks in the region are already significant and will increase with climate change. During the great flood of 1998 both inter- and intra-city bus links from eastern Dhaka were shut down because of inundation and an estimated 384 km of paved roads were inundated (Alam and Rabbani 2007). In 2007, 34 % of the total transport infrastructure was damaged in Bangladesh, including roads, bridges and culverts, road approaches, highways, buildings and railways infrastructures (rail line and bridges), costing about US$363 million (Bangladesh Ministry of Food and Disaster Management 2007). The 2006 flood in Aceh, Indonesia, caused US$35 million to the transportation sector, which was 24 % of total infrastructure costs from the storm (World Bank 2007a).

In Hyderabad, India, climate extremes have adversely impacted transportation infrastructure in several ways. Floods have directly resulted in infrastructure damage, the breakdown of transport networks and the slowdown of services. Heat waves have caused direct damage to electronic devices, and gradual temperature increases working in a more concealed way have slowly damaged railway and road infrastructures (Mehrotra et al. 2011).

3.4.2.2 Water Supply and Sanitation

Climate change can affect the availability, treatment and distribution of urban water supplies and the ability to sanitize wastewater (Major et al. 2011; Case 2008). Warm temperatures can help to degrade water supply infrastructure in similar ways to transportation infrastructure. Warm temperatures can lead to a degradation of water quality through increased biological and chemical activity. For example, heat waves can increase water solubility and, therefore, enhance concentrations of contaminants including algae, microbes and parasites. Warm temperatures also increase evaporation from reservoirs thus potentially requiring changes or additions to infrastructure (Thirlwell et al. 2007). Increases in temperatures also affect water demand as the population uses more water for personal services (drinking, showering, watering lawns, etc.).

Increasing storm events can lead to street, basement and sewer flooding and high levels of storm water run-off. More intense rainstorms will increase runoff into receiving water bodies. If water levels overwhelm treatment plants, the result will increase combined sewer overflows. The resulting cascade of events will include high nutrient loads and eutrophication of water bodies. Those downstream may suffer from taste and odour problems and loading of pathogenic bacteria and parasites in supplies. Intense rainstorms will also increase erosion and stream sediment load, decreasing the quality of water and the life of reservoirs. Higher loading in rivers may disrupt freight traffic and require increases in the frequency of dredging (Major et al. 2011).

As a result of climate change the timing and type of precipitation may change, causing disparities between supply and demand. Reduced snowfall results in less water stored in snowpack and glaciers. On the other hand, if the same amount of precipitation occurs as rain, rather than snow, water will move through the hydrological cycle more quickly. This places demands on reservoir capacity, dams and supply systems. Reduced snowfall can also impact the quality of rivers and the ecosystem services they deliver (Major et al. 2011).

In coastal cities all of these impacts are experienced along with several other important challenges. Coastal cities that rely on groundwater supplies may find increased saltwater intrusion as sea level rises and more coastal dwellers turn to groundwater if surface waters are inadequate to meet demand (Hanson et al. 2011). An increase in sea level may also increase the probability of flooding of wastewater plants. Higher sea levels can inundate fresh and saline wetlands and threaten the stability of canals and embankments, which can impact water via supply and quality as well as cause flooding. Saltwater intrusion is a major threat to Shanghai's water supply; the Qingcaosha reservoir was constructed in an attempt to safeguard drinking water supplies against salinity, rising demand and pollution (Engel et al. 2011). During the dry season, saltwater intrusion of 100 km or more was reported from the Bay of Bengal (Allison et al. 2003).

Asian cities currently have a variety of water supply and sanitation challenges including inadequate provision of water supply and sanitation services, inadequate drainage, ground water overdraw and land subsidence, increasing water consumption and high leakage rates (Marcotullio 2007). Climate change in the region is expected to exacerbate these challenges as well as pose new ones (IPCC 2007b; APN 2010).

3.4.2.3 Food

Climate change can impact urban food systems. The basic urban food system is composed of several elements including the production, distribution, processing, consumption and waste. Food production (agriculture) may be affected by climate change in a variety of ways. Temperature changes will have an impact on the types and yields of crops through, *inter alia*, exceeding temperature thresholds for crops, increasing plant metabolism, changing soil moisture and shifting pest types and abundance. Changes in precipitation levels will affect irrigation and water inputs. On the one hand, reports suggest that climate change will only have a minor effect on global food supplies (Rosenzweig and Parry 1994) and in some regions change will be beneficial to some areas (Reilly et al. 2003).

In Central and South Asia, however, crop yields are predicted to fall by up to 30 % and reduced soil moisture and evapo-transpiration may increase land degradation and desertification, increasing the risk of hunger in several countries. At the same time, agriculture may expand in productivity in northern areas of the region (IPCC 2007b).

Food distribution systems may be affected by various climate-related changes. In a study of 18 low- and middle-income countries urban food insecurity equalled or exceeded rural levels in 12 (66 %) of those (Ahmed et al. 2007). Lack of urban connections to agricultural production sites will lower food availability and increase food prices. Rising prices impact urban populations, particularly the poor, reducing their calorific intake and the diversity and nutritional value of foods consumed (Cohen and Garrett 2010). Maintaining two-way flows of food between cities and rural areas is crucial to India's development and climate change can have dramatic impacts on this important link (Revi 2009).

Moreover, a nutritional transition has accompanied urbanization and economic development (Popkin 1994, 1999). Diet changes have also been observed and, interestingly, within Japanese urban populations meat has become an important component of Tokyo residents' diet overtaking fish consumption for the first time in the early 2000s. Such diet changes may have important implications for the environmental footprint of cities triggering land use change and environmental pollution (Gadda and Gasparatos 2009).

Nutritional transition is defined by changes in diet with rising income. Higher added-value processed foods, meats and dairy products are consumed in larger amounts with increasing incomes. Higher income households' demand for ready-to-eat and easily prepared foods is higher than poorer households. This is important for urban areas, as they concentrate wealth. As urban areas demand more food and more diverse food products, the complexity of urban food systems increases and with it the potential risk of being disrupted. For example FAO (2000) calculated the truckloads of food necessary to feed the burdening populations of cities in Asia by 2010. They estimated that Bangkok would need 104,000 additional 10-tonne truckloads each year; Beijing would have over 302,000; Mumbai, 313,000; Dhaka, 205,000; Jakarta, 205,000; Karachi 217,000; and Shanghai almost 360,000. Moreover, they estimated that food losses between the production and retail stages ranged from 10 % to 30 % and are caused by a combination of on-farm, transport, distribution and spoilage problems, which are greater in urban than rural areas. Given the large amount of food necessary for these areas and the increasing distance needed to get food to the urban market, climate-related disruptions in the transportation system create risk for urbanites.

3.4.2.4 Energy Production, Transmission and Distribution

Climate change is likely to impact urban energy systems by affecting infrastructure and demand for energy. Energy infrastructure is susceptible to various aspects of climate change for transportation, water supply and sanitation. Hammer et al. (2011) argued that climate effects on energy systems include those on resource production and delivery, power generation and transmission and distribution systems.

Resource production includes fuel stocks, which are typically located outside urban areas, but which are affected by climate change. Drilling platforms and refineries are vulnerable to weather events. For example, during 2008 heavy snowfall in central and southern China blocked rail networks and highways used for delivering coal to power plants in these regions. Of China's 31 provinces, 17 were affected including hundreds of millions of people in these provinces who were forced to ration power (Hammer et al. 2011).

Urban poor households use a mix of fuels, including biomass. In developing parts of the Asia-Pacific region, the use of biomass fuels for cooking and heating remains significant. For example, about 25 % of the urban poor in the Philippines use fuel wood primarily for cooking and about 45 % of household procure their fuel sources from their own rural plots (Approtech Asia 2005). In India, Sri Lanka and Thailand, wood harvesting has produced a halo of deforestation around cities (Sawin and Hughes 2007). The extent and renewability of biomass will depend upon changes in temperature and precipitation as well as population growth.

Power plants located in coastal areas are at risk from the impacts of storm surges, floods, etc. The recent nuclear power problems experienced by Japan due to the 2010 earthquake-induced tsunami have resonated with those examining the possibilities of like effects, but due to climate change. For example, Urban and Mitchell (2011) note that disaster risks caused by storm events such as typhoons, etc. and the resulting flooding could pose increasing threats by damaging nuclear power plants and subsequent risks to health and safety.

While hydroelectric plants are very important in the region, a recent report has suggested that over half of existing and planned capacity for major power companies in South and Southeast Asia is located in areas that are considered to be water scarce or stressed (Sauer et al. 2010). This analysis suggests that delays in project execution and loss of output due to water scarcity could be significant. With changes in precipitation occurring, the lifetime use and effectiveness of these plants may come under question.

The combination of urban population growth and changing local weather can place increased demands on energy systems for energy generation and distribution. Increases in demand are the most important economic impacts for energy at the global scale, particularly in tropical areas. Climate warming may increase the demand for space cooling, for example. Air conditioning in the commercial sector already accounts for a greater proportion of final energy demand than in the residen tial sector in developed countries. In Hong Kong, air conditioning accounts for as much as 60 % of total electricity use in the commercial sector (Hunt and Watkiss 2011). In Chinese cities, over the past 15 years, air conditioner ownership rates have increased to exceed an average of one unit per household and in Guangzhou ownership rates have increased from approximately one per every other household to more than two per household during this period (Hammer et al. 2011). The increases in such appliances reflect the increased demand on systems that may also occur in more northern areas with further warming.

3.4.2.5 Industry and Services

Manufacturing is an important component of economic development in the Asia-Pacific region, which is often called the "factory of the world" (UN-Habitat 2011c). Export-led growth, originally based in the industrial sector, but increasing in services, continues to be a significant source of economic growth and employment. Foreign direct investment is typically attracted to the region's cities that are linked to the global economy through transportation and communications networks. Scholars have correlated urban development of the region with the growth of industries and manufacturing (Lo and Yeung 1996). These industrial firms typically locate around the region's cities in doughnut fashion (Marcotullio 2003).

Manufacturing industries that are not directly dependent on natural resources and tourism generally would not be affected by climate unless key infrastructure is destroyed by events such as floods or landslides, or unless shipments are affected. However, manufacturers are influenced by climate change in two other ways. First, they would be affected through the impact of government policies pertaining to climate change such as carbon taxes (thereby increasing the cost of inputs). Second, they could be affected through consumer behaviour that in turn is affected by climatic variations. For example, less cold-weather clothing and more warm-weather clothing might be ordered (Wilbanks et al. 2007). Not enough is known, however, of the potential impact to industries.

3.4.3 Greater Risk and Vulnerability for the Urban Poor

The vulnerability or security of individuals and urban areas is determined by the condition of natural resources and infrastructure as they change with climate, availability of financial, social and natural resources, and by the entitlement of individuals and groups to call on these resources (Adger et al. 2003; IPCC 2007a). That is, vulnerability is not simply defined by exposure to physical risks and hazards and the damage that is done, but also includes the capacity of communities to adapt. This capacity allows for individuals, households and cities to adjust to climatic changes, to mediate related damage, to cope with changes and take advantage of opportunities. Capacity to adapt is embodied in both the physical infrastructure within urban systems as well as urban socio-economic structures.

One of the main constraints on adaptive capacity for cities in the region is the number of those poor and marginalized, the elderly, and the young. These groups typically have lower access to formal services and depend in greater part on ecosystems services. Therefore, they are highly exposed to climate change effects and do not have high adaptive capacity. For example, the urban poor typically live in vulnerable ecological zones (high slopes, flood prone areas, waste dumps, etc.) and sometimes without land tenure. Even small perturbations in the supply of water sources affect these populations given that they largely depend upon informal arrangements (local private markets). Hence, not only do these groups have lower capacity to adapt, they are more vulnerable to the effects of climate change than other groups. Cities that have large population shares of the poor, elderly and young often have insufficient infrastructure to support the basic needs of these people and, therefore, suffer from an "adaptation deficit" (Satterthwaite et al. 2007).

As such, some have considered that not only are the effects of climate change predicted to be especially strongly in the town and cities of the least developed countries (Dodman 2011), it is those that face the highest risk to climate change events that have the most constraints on their capacity to adapt and, ironically, contribute the least to climate change (Bicknell et al. 2009; Costello et al. 2009).

While the Asia-Pacific region has experienced a decrease in the number of those in poverty; the total numbers that remain in a state of poverty continues to be of grave concern. Notwithstanding the significant drop of those in poverty over the past decades, there are still significant concerns for the poor. Unfortunately, due to the financial crisis, poverty reduction has slowed since 2008. Asia and the Pacific remains home to the largest number of those in poverty (Wan and Sebastian 2011). A recent report that examined climate variability and the adaptive capacity of 11 cities in Asia found the most vulnerable were those of lower income. The rating from most vulnerable to least included Dhaka, Jakarta, Manila, Calcutta, Phnom Penh, Ho Chi Minh City, Shanghai, Bangkok, Hong Kong, Kuala Lumpur and Singapore (World Wildlife Fund 2009). Dhaka was cited as having a large poor population located a few metres above sea level. Jakarta, Manila, Calcutta and Phnom Penh all have relatively low adaptive capacity.

3.5 Urban Mitigation and Adaptation Strategies in Asia and the Pacific

Urbanization of Asia and the Pacific includes billions of people and has unfolded in ways that both contribute to climate change and place much of this population at risk of climate change-related hazards. Given the low adaptive capacity of many cities much of the urban population is left vulnerable.

Continued growth without responding to climate change is unwise. A recent UN-Habitat (2011c) report calls for future urban development in the region to include "green and low carbon development," which was also the major focus of the recent 2012 Rio+20 conference. Otherwise, not only will the prosperity of the region suffer, but if certain "tipping points" or physical system thresholds are exceeded, evidence indicates that the Earth system may uncontrollably flip into an irreversible state. There is a real possibility that we will see a range of major large-scale events that will be beyond human management (Pearce 2007). Given the scale of urbanization, economic activity, GHG emissions and adaptation deficits; local, regional and global consequences could be significant.

Mitigation and adaptation are two responses to increasing climate change. The goals of these actions are to prevent "dangerous climate change" (Hansen et al. 2006) and avoid the costs of changing atmospheric dynamics and the cascading effects that will detrimentally affect high quality human development into the future. This section overviews mitigation and adaptation issues and action in Asia and the Pacific and although they are presented separately, these actions can be complementary, substitutable or independent of each other.

3.5.1 Urban Mitigation Strategies

Mitigation strategies and measures contribute to the stabilization of GHG concentrations in the atmosphere. This includes any action taken by individuals, corporations or governments to reduce GHG emissions. The goal of mitigation is to minimize the effects on global climate change and prevent "dangerous" anthropogenic interference with the climate system.

Mitigation efforts over the next two to three decades will have a large impact on opportunities to achieve lower stabilization levels. Studies indicate that there is substantial economic potential for the mitigation of global GHG emissions that could offset the projected growth of global emissions or reduce emissions below current levels (IPCC 2007c).

Mitigation measures have both sector and non-sector applications. A sector application might include methane emission reductions through waste management or sewage treatment, while a non-sector mitigation measure includes activities such as education, training and public awareness related to climate change.

3.5.1.1 Urban Design

Attention to evolution of urban form and layout has substantial potential to influence GHG budgets (Swart and Raes 2007; Hamin and Gurran 2009). Planning new urban areas can, from the start, include commuter rail, thus reducing congestion and transport-related emissions. Consideration of where people work, study and play can reduce problems of access that lead to sprawl, low density, dispersed and decentralized urban development (Alberti 2005; Ewing 1997) and long commutes (Hayashi et al. 2004).

Climate-appropriate architecture and placement of a building on a site can greatly reduce energy consumption. Insulation, trees and placement of windows have impacts on needs for air-conditioning, heating and lighting (Whitelegg and Williams 2000). Solar hot water heaters, heat pumps and fluorescent lighting are examples of key component technologies. For the city of Kyoto, Japan improvements in just household and commercial energy efficiency could cut 1990 emissions in half by 2030 (Jarvis 2003; Mokhtarian 2002). Makati City, Philippines has developed elevated walkways that connect tall buildings to reduce traffic and congestion (Prasad et al. 2008).

Vegetation in and around buildings and transport roots may contribute to climate control, reducing UHI effects (Barter 2000). Green spaces and parks can make a modest contribution to carbon sequestration, but more studies are needed to identify the effectiveness of this measure (Chin 2000; Pataki et al. 2006). Examples of the greening of cities have increased in the region over the last few decades. Singapore, known as the region's "Garden City" through a programme launched by Lee Kuan Yew, encouraged the planting of trees and the incorporation of grass verges, green corridors along roadways, trees, and parks into city developments. Emphasis has been placed on local flora, and encouraging ecological sensitive afforestation (Corfield 2008b; Prasad et al. 2008). Seoul has re-claimed the Cheonggyecheon River in the middle of the city, including the removal of a 5.6 km elevated highway. The new park provides tourism, recreation and UHI reduction benefits (Cho 2010). In Shanghai, the city's plan stresses the importance of incorporating greenbelts into downtown Shanghai within the layout of urban development. By 2008, the city's parks and urban green spaces reached 22,000 ha. In Shanghai, urban greenery rose from 1 m^2 per capita in 1990 to 12.5 m^2 per capita in 2008. Temperatures have dropped by 5 % since the increase in urban greenery and the widespread use of green roofs within the city. Moreover, Shanghai envisions that 30 % of the city will be covered by greenery by 2020 (Solecki 2011).

Urban spatial planning will often need to consider both adaptation and mitigation issues simultaneously as there may be complementarities as well as trade-offs (Sari and Salim 2005). For example, street trees and green spaces cool buildings but take space. Many adaptation actions appear to require more space and a less densely-built environment; modularity may be a solution. Coordination of urban land use and transport infrastructure is a key planning and governance issue.

3.5.1.2 Transportation

As transportation is a large and growing contributor to total GHG emissions, national governments are using a variety of mitigation measures in this sector including, *inter alia*, mandatory fuel economy and biofuel blending of fuels, taxes on vehicle purchases, registration, use and motor fuels and investments in public transportation infrastructure (IPCC 2007c).

At the local level, meeting the green mobility needs of people and for transport of goods within and among cities is a key issue for reducing GHG emissions. Different cities have taken different approaches reflecting differences in initial densities, roles in national economies and histories of investment in public mass-transit systems (Santos et al. 2010). Some cities in the region, like Singapore, have largely escaped the vehicle-trap, whereas others like Bangkok and Kuala Lumpur remain dominated by these modes of transport.

Well-designed systems are likely to be multi-modal (Lebel et al. 2007). In densely built-up areas non-motorized transport remains plausible if pedestrian areas are safe and convenient (Liu and Deng 2011). Mass-transit systems are particularly important for commuting to and from the workplace. Addressing mobility needs while reducing emissions is more challenging when it comes to how people shop or spend leisure time (Dhakal 2010).

Restraints or disincentives on private vehicle ownership along with promotion and investments in public transport are key (Savage 2006; Lebel 2004; McGranahan and Satterthwaite 2003). Singapore, for example, has used a mixture of incentives and regulatory instruments to control traffic congestion and emissions from cars, electronic road pricing, efficient public transport, and taxing car purchase and use (Solecki and Leichenko 2006; Barter 2005). Other cities in the region, such as Jakarta, Kunming and Seoul, have introduced "Bus Rapid Transit" and dedicated lanes (Hook 2005; Pucher et al. 2005). In Beijing and Shanghai, two- and three-wheeled motor vehicles were banned, stimulating the rapid rise in electronic bikes (E-bikes) (World Energy Council 2011). Making transport systems more sustainable will require a mix of policy instruments (Lebel et al. 2007).

3.5.1.3 Energy Production and Demand

The energy sector is the largest contributor of anthropogenic GHG emissions globally. National governments have attempted a number of different mitigation measures including the reduction of fossil fuel subsidies, taxes or carbon charges on fossil fuels, feed-in tariffs for renewable energy technologies, tax incentives and obligations for renewable energy technologies and energy producer subsidies and waste management regulations (IPCC 2007c). In Southeast Asia, while Singapore has the largest number of diverse activities for climate change mitigation, Thailand appears to have one of the most advanced national energy efficiency policies (including development of solar, wind, biomass, biogas, hydro, bio-fuels, geothermal and fuel cells) (Yuen and Kong 2009). Thailand's policies are driven by the government's intent on becoming a regional energy hub (Australian Business Council for Sustainable Energy 2005).

At the urban level, researchers have demonstrated that residential and commercial electricity consumption typically grows with wealth (Pachauri and Jiang 2008). People purchase more appliances to make homes and work places comfortable and convenient (Dhakal 2010). The levels of associated emissions depend on how and where that electricity is produced. Typically, and as noted previously, energy is produced outside but consumed inside urban areas; and inventories and reduction measures need to take these relationships into account (McGranahan 2005; Bai et al. 2010).

Electricity production is an important source of GHGs and the types and quality of fuels used to make electricity have major impact on emissions. In China, many cities are highly dependent on low-grade coal. Alternative energy sources such as wind, solar or biomass are options for some cities. The China low carbon city programme recognizes that exact mix of policies and targets should reflect local conditions. Singapore has adopted more efficient technologies such as combined-cycle gas turbines in gas-fired power plants, improving overall power generation from 37 % in 2000 to 44 % in 2004 (Prasad et al. 2008).

Policies to implement green energy sourcing are in effect in some cities in the region. Albay Province, Philippines, is developing geothermal energy and already the renewable source generates 25 % of its energy (Prasad et al. 2008). A solar city programme in Dezhou, China, has stimulated over 100 solar enterprises to install over 3 million m^2 of solar water heaters, which now account for 16 % of the national market. With this success, the city is now in the process of developing the China Solar Valley project, which is one of the biggest solar power projects in the country (ICLEI et al. 2009). In Rizhao, China, 99 % of households in the central districts use solar water heaters. Most of the traffic signals, street lights and parks are illuminated through photovoltaic cells. In addition, 6,000 households have solar cooking facilities and more than 60,000 greenhouses are heated by solar panels, thus reducing overhead costs for farmers (Bai 2007).

Another option for municipalities is to recycle and capture waste energy, for example, from the incineration of solid waste or through the capture of emissions from landfills. In cities with major industrial activities there may be other opportunities for waste heat capture. Waste management and methane recovery programmes, for example, have been initiated in Naga City, Philippines; Thungsong, Thailand; and Ratnagiri, India (ICLEI et al. 2009).

To lower demand actions at the local level, cities in the region have been addressing public lighting and building efficiency. In Makati City, Philippines; and Guntur and Jabalpur, India, programmes to replace street lights with more energy-efficient systems with programmable controls has reduced electricity consumption by up to 35 % (Prasad et al. 2008; ICLEI et al. 2009).

Energy efficient commercial and residential buildings are an important mitigation contribution (Levine et al. 2007). In Beijing, the China Agenda 21 Demonstration Energy-Efficient Office Building, which hosts the Chinese Ministry of Science and Technology has been built with a number of energy conservation features, including both building envelope and mechanical system measures, a cross-shaped building design was used to maximize day-lighting potential, windows were located on the north and south facades to better control solar heat gain, passive solar, photovoltaics and geothermal power systems. In India, the CII-Godrej Green Business Center (CII-Godrej GBC) building in Andhra Pradesh is the first Platinum-rated LEED-certified building outside the US, and was the most energy-efficient building in the world at the time it was rated. These are but a few of the energy efficient buildings springing up around the region (Hong et al. 2007).

Well-designed urban areas provide opportunities to install high quality energy distribution systems that are managed 'intelligently' with respect to peak and lower patterns of energy demand. Cities typically emphasize multiple local benefits when pursuing emission reductions.

3.5.1.4 Urban Development and Climate Change Mitigation Opportunities

While cities are sources of GHG emissions, urban areas are more carbon efficient because they offer greater access to public transport, shorter commuting distances to regularly used services, and smaller, more compact settlements and thus closer social connections than rural areas. Cities are nodes of exchange for goods, information and people that don't reside in them. These services are of value well beyond a city's boundary. The emissions embedded in goods and services flow to cities can be large. Carbon management needs to become more integrated, considering emissions and sequestration, and must also be multi-levelled to include emissions associated with energy production and diets that arise beyond city borders but because of consumption patterns within them (Lebel et al. 2007).

3.5.2 Urban Adaptation Strategies

Adaptation strategies are those that facilitate an adjustment in natural or human systems to a new or changing environment. For cities, climate adaptation strategies facilitate mechanisms necessary to address modifications brought upon by climate change. Researchers now believe that adaptation strategies are necessary to address the unavoidable impacts of climate change (IPCC 2007a).

Various types of adaptation are available to individuals and communities including anticipatory and reactive, private and public and autonomous and planned strategies. Moreover, responses include technological, behavioural, managerial and policy measures. In developing nations, some researchers are promoting community-based adaptation plans, which focus on addressing daily hazards, and as an extension also address climate risks (Satterthwaite et al. 2007). Adaptation plans enhance the adaptive capacity of communities and, as a result, reduce climate risk vulnerability.

The UNFCCC provides support for the formulation and implementation of national adaptation programs of action (NAPAs) in least developed countries. To date, 39 NAPAs have been submitted and many of them from the Asia-Pacific region (Bangladesh, Bhutan, Cambodia, Kiribati, Lao PDR, Maldives, Nepal, Samoa, Solomon Islands, Timor-Leste, Tuvalu and Vanuatu). At the same time, however, many nations in the region have prepared adaptation strategies and implemented adaptation policies at different governmental levels.

Some of these policies focus on cities. In India, for example, the Jawaharlal Nehru National Urban Renewal Mission (JNNURM) was initiated in 2005. This mission targets 60 of the largest and most important cities in the country. The government provides US$10 billion for infrastructure development, urban poverty and improvements in urban governance. The Republic of Korea has developed a National Climate Change Adaptation Master Plan through the Framework Act on Low Carbon, Green Growth (LCGG). In this plan, local governments develop 5-year adaptation plans based on the national plan. These and other efforts demonstrate that there are adaptation options available for cities.

While there has been action at multiple levels to enhance adaptive capacity, more extensive adaptation measures are required to reduce vulnerability to future climate change (IPCC 2007a). Moreover, while there is increasing academic literature on adaptation options to climate change, few studies have considered urban areas, which require specific consideration and analytical approaches (for exceptions see Hallegatte et al. 2011; Hunt and Watkiss 2011; Rosenzweig et al. 2011). This subsection overviews adaptation action in urban water resources and infrastructure in coastal areas and urban food, transportation and energy systems.

3.5.2.1 Urban Water Resources and Infrastructure in Coastal Areas

Water resources adaptation could be the lead sector in reducing vulnerability to climate change, particularly in developing countries (Muller 2007). Adaptation strategies that address sea-level rise and increasing storm surges include, *inter alia*, setting up early warning systems; relocation of infrastructure and people away from flood zones; improving drainage and pumping systems in areas that flood; protection of infrastructure through building sea walls and barriers; creation and/or protection of existing natural barriers (marshland, wetlands and beaches); and reduction of wave energy through off-shore barriers. These have been organized into strategies that protect, accommodate or retreat (Leonard et al. 2008).

Adaptation strategies that address water supply include expanded rainwater harvesting, water storage and conservation techniques, water reuse, desalination and water-use and irrigation efficiency and integrated water resource management (World Bank 2011; Prasad et al. 2008).

Some sea-level rise and storm surge responses have been seen in the region. For example, climate change is considered in the design of infrastructure projects such as coastal defence in the Maldives. Local government in Navotas City, Metro Manila has implemented a programme to construct sea walls and pumping stations along the most vulnerable inundation zones (Prasad et al. 2008).

One of the strongest urban responses has been to flooding. Shanghai has developed a flood control project that is designed to both regulate water flow and facilitate water-quality monitoring. The project will provide water-level data in real time, enabling the water authority to see conditions across the entire region as they develop and make decisions to protect areas downstream from flooding or overflow. Hanoi, Viet Nam now has a comprehensive water adaptation programme, which includes actively improving flood preparedness and prevention standards, strengthening the dyke system to protect the right bank of the Red River and monitoring and responding to dyke emergencies (Prasad et al. 2008).

The extreme rainfall event of 2005 has been a lesson for Mumbai. The city is setting up a response mechanism based on real-time monitoring of rainfall at 27 locations in the city to handle recurrences of similar events in the future. The Central Water Power Research Station in Pune is currently in the process of preparing a detailed scale model for carrying out the hydraulic model studies for the Mithi River. This model is intended to provide a basis for long-term planning of Mumbai taking into account the impacts of climate change and sea level rise (Gupta 2006). Flood protection in Dhaka includes major changes in land use in Dhaka West. Improvements of the drainage system was highlighted by the 2008 Bangladesh Climate Change Strategy and Action plan (Alam and Rabbani 2007).

3.5.2.2 Food

One adaptation response to urban food system vulnerability has been to increase urban food production. Increasing urban agriculture reduces potential food price shocks and other disruptions that might arise with climate-related changes (Larsen and Barker-Reid 2009; UNDP 1996). Food may be grown in city open spaces (backyards, school yard, wastelands, rooftops, parks) and more high-tech agricultural spaces (hydroponic operations, aquaponics, greenhouses, vertical farms). Cities in Asia and the Pacific with formal policies and programmes to support urban agriculture include Bangkok, Beijing, Brisbane, Melbourne, Mumbai and Shanghai (Barata et al. 2011). In 2002, Beijing residents grew 55 % of the vegetables consumed in the city (Wang et al. 2009b). The city includes tens of thousands of household farms and the municipal government plans to cultivate gardens on 3 million m^2 of roof space over a 10-year period (Halweil and Nierenberg 2007). Many of those involved in this type of activity are migrants. One estimate suggests that over 600,000 migrants (17 % of the total that year) were engaged in activities directly related to urban agriculture (Wang et al. 2009b).

Urban agriculture occurs in cities without formal programmes, particularly in peri-urban areas. Peri-urban agriculture in Hyderabad, India plays an important role in supporting livelihoods of a diverse group of people from different castes, religions and social classes (Buechler and Devi 2002). Although there are only a few 100 vegetable growers along the Musi River, in a city of seven million, these farmers provide an important diversity of fresh vegetables to the city's markets (Jacobi et al. 2009). Urban agriculture may also indirectly contribute to sustainable management through the demand for urban open spaces, including flood zones, buffer zones, steep slopes, roadsides, river bank and water harvesting (Dubbeling et al. 2009). Many suggest that urban sustainability is tightly linked with the city's ability to feed part of its population (Halweil and Nierenberg 2007).

At the same time, urban agriculture still provides only a small percentage of food for most cities. The same climate challenges that face rural farmers also face urban farmers, and the challenges are greater in urban areas. There are risks from urban agriculture particularly when wastewater is used to irrigate crops or food is grown in contaminated soils (Cole et al. 2008).

3.5.2.3 Transportation

Adaptive transportation strategies include realignment and/or relocation of infrastructure and implementing design standards and planning for roads, rail and other infrastructure to cope with warming and drainage (World Bank 2011). For example, both Taipei's subway entrances were raised to avoid flash flooding and, like Japan, the inter-city trains are on elevated tracks to avoid flooding during the typhoon season (Mehrotra et al. 2011). Generally, it is important for all transportation networks to avoid flood-prone areas and incorporate climate change into all relevant decisions concerning transportation infrastructure (Coffee et al. 2010). For cities exposed to flooding, zoning rules can be used to relocate existing storage yards for buses and train cars out of flood-prone areas (Mehrotra et al. 2011).

During the expansion of transportation infrastructure, inclusion of informal settlements in regular service may require serving marginal land, but can also have co-benefits for economic development and poverty reduction by providing residents with better access to jobs and business opportunities (World Bank 2011).

Green infrastructure can help make transportation more resilient, with direct advantages such as decreasing run-off during rainstorms and indirect advantages such as improving water and air quality (Bloomberg 2010). Research has shown that pervious pavements can lead to a reduced need for road salt application on streets in the winter by as much as 75 % (Foster et al. 2011).

Low-cost adaptation options in the transportation sector include improving transportation customer communications, retrofitting existing bus fleets with white roofs to reduce solar heat gain; ventilation to ensure adequate air circulation; and working with ports and maritime businesses to synchronize shipping schedules around high tides to avoid problems with bridge clearance (World Bank 2011).

Recent research argues that Asia lags in formulating and implementing adaptation strategies for the transportation sector. One study surveyed 21 countries around the region and the results highlight low awareness of climate change impacts on transportation (75 % recorded low awareness by policy-makers). Moreover, the majority of the respondents indicated that there are no existing laws, rules or guidelines to assess environmental and climate change impacts (Regmi and Hanaoka 2011).

3.5.2.4 Energy Transmission and Efficiency

Adaptation strategies for energy systems include underground cabling for utilities, strengthening of overhead transmission and distribution infrastructure, using renewable sources and reducing dependence on single sources of energy (Prasad et al. 2008). The World Bank (2011) argues that energy efficiency, conservation and renewable energy investments serve as important adaptation and mitigation strategies. Conservation and efficiency programmes can reduce peak electricity demand and limit the risk of blackouts, while developing distributed energy systems involving co-generation and local renewable energy can buffer the effects of interruptions in transmission. Crucial components of any urban-related adaptation policies are vulnerability assessments, emergency warning systems and adjustments in design standards to reflect climate impacts on the energy sector.

3.5.3 Urban Climate Change Policy and Governance

Governance is a crucial part of any mitigation and adaptation strategy. A growing number of municipalities, NGOs and civil society organizations have been involved in actions to reduce GHG emissions and formulate and implement adaptation measures. In the bigger picture, responses to climate may best be served at all governance levels including the local, with the urban level an important site for addressing climate change (Bulkeley 2010).

As such there has been much attention given to the planning and development of various strategies and the underlying reasoning as to why these measures are appropriate and why they should be effective for cities in the region (Prasad et al. 2008; UN-Habitat 2011b; World Bank 2011). These normative studies have been important because they not only provide guidelines, but point out best practices and, therefore, highlight on-going activities and showcase successes.

From these types of studies several policy recommendations have been offered for cities in Asia and the Pacific. There are, at least, five areas of policy concern including: development and disaster risk management; capacity building; local participation; financing; and sector specific strategies. This chapter reviews several sector specific strategies. We restrict our review in this section to the cross-cutting issues and provide thoughts on opportunities for urban climate governance.

3.5.3.1 Development and Disaster Risk Management Strategies

Many policy and governance studies emphasize the importance of integrating climate adaptation and mitigation strategies with development and disaster risk management strategies. Although particularly important in developing world cities, focusing on the local development needs when developing and implementing climate policies is considered to be a good idea for all cities. For example, sustainable development policies include lowering the share of the urban population with poor housing. This is not only a development policy, but also a risk reduction policy, as those in poor housing also typically live in high risk areas. Adaptation policies include finding secure tenure for all residents in areas away from potential flood zones or protecting these areas from future exposure to climate related hazards. Pro-poor strategies are a necessary component of adaptation as most of those at risk are in low income communities with low adaptive capacity (Satterthwaite et al. 2007). At the same time, all communities would benefit from a review of land use planning and regulations with an eye on incentives and restrictions related to building in storm surge areas or on building and materials standards that address reducing GHG emissions.

Furthermore, development strategies include planning for the long term economic, social and environmental health of the city. City planning and infrastructure decisions can "lock in" urban form and particular sector strategies for long periods of time, especially when physical investments have extended life spans. City decision makers have been advised to strategically plan for the future and include mitigation and adaptation strategies within these plans (UN-Habitat 2009).

For developed and developing world cities, disaster risk management policies focus on the need for risk assessments for exposed cities and populations within cities, early warning systems and evacuation plans including emergency preparedness and neighbourhood response systems and improved education about climate risks (Kovats and Akhtar 2008). In any context, these policies can double as those for adaptation, if the focus of the disaster is on climate related risks.

3.5.3.2 Capacity Building

Improving capacity within government and civil society is considered a key ingredient for policy and governance. Given global trends in decentralization, this is a major challenge, as national governments are increasingly less able or willing to support local government (Satterthwaite et al. 2007).

As climate change impacts are local, local decision makers and stakeholders need to improve their understanding of climate change, its potential impacts in their context, appropriate options for responses and their ability to effectively collaborate. City governments are responsible for decisions and actions related to the delivery of a wide range of services that can be improved with mitigation and adaptation strategies including: land use planning and zoning; water supply, sanitation, and drainage; housing construction, renovation, and regulation; economic development; public health and emergency management; and transportation and environmental protection. Stakeholders and bureaucrats working within any of these sectors can benefit from capacity building to reduce vulnerability and GHG emissions.

Moreover, rarely does a single city agency have the authority for all planning decisions and investments that shape urban form and responses to climate change, nor do city agencies act alone. Indeed, the literature calls for agencies to engage in "joined-up thinking" (Revi 2009). Climate change mitigation and adaptation in cities requires collaborative problem-solving and coordination across sectors. This is also true in the governance context. As municipalities are often called upon to act as conveners of a wide range of partners, collaborative skills that help to bring together broad partnerships that include other governments, local communities, non-profit organizations, academic institutions, and the private sector are considered critical to success (World Bank 2011).

3.5.3.3 Participation

Participation is a key ingredient in good governance and, therefore, effective development of mitigation and adaptation policies. For example, a city land-use planning process can be an arena in which stakeholders make and institutionalize key decisions relating to the energy sector, such as where to site energy infrastructure and how to address traffic congestion. With local knowledge through participating civil society organizations, energy and other vital infrastructure can be planned to effectively reduce populations living in vulnerable locations.

Participation from all local groups that are exposed to climate change hazards is particularly important for adaptation measures. While climate change is a global phenomenon, adaptation hinges on the quality of local knowledge, local capacity and the willingness to act (Satterthwaite et al. 2007). It is, therefore, necessary to have local citizens participate in adaptation strategy development and implementation.

Those that have been and are currently exposed to climate-related risks are already coping. While these actions are not formal they are in many cases effective, although only as immediate responses. To develop long term effective strategies the knowledge and experience as well as the needs of these groups must be included in all urban climate policies. Moreover, to develop adaptation strategies and maintain effective mitigation strategies the affected, including the poor, need to be brought into the formal process.

3.5.3.4 Insurance and Financing

Insurance and finance help to reduce economic risk and are key to infrastructure and housing development. Interestingly, the global insurance industry has demonstrated increasing interest in climate change research and in climate change-related urban vulnerabilities (Munich Re Foundation 2011; Dailey et al. 2009; Swiss Re 2006; Lloyd's 2008). The consensus goal of the industry is to take a more leading role in understanding and managing the impacts of climate change. Insurers realize that understanding climate-related risk is necessary to inform underwriting strategy (from the pricing of risk to the wording of policies), for business counselling purposes (including business development and planning) and to help to create meaningful, tangible partnerships to mitigate risk (Lloyd's 2008).

While this may be effective in developed cities, where much of the insurance risk exposure is located, there is a desperate need for insurance in cities of developing nations (Wilbanks et al. 2007). Approximately 99 % of households and businesses in low income nations do not have disaster insurance (Satterthwaite et al. 2007). In this case, and particularly for the urban poor, macro- and micro- climate insurance may provide some relief (Pierro and Desai 2008).

3.5.3.5 Opportunities for Urban Policy and Governance Research

Despite the large and growing literature on climate change adaptation policies, the reality is that urban climate governance has largely focused on mitigation (Satterthwaite et al. 2007). Even with mitigation strategies, cities have failed to pursue systematic and structural approaches to GHG reductions and instead prefer to implement no-regret measures on a case by case basis (Alber and Kern 2008). That is, while research and transnational municipal networks have advocated a systematic approach to urban mitigation strategies including GHG emission assessment, target setting and performance monitoring, many cities have failed to implement this approach.

Most measures are undertaken in the energy sector (with notable exceptions such as London and New York City) and GHG emissions reductions are typically restricted to municipal sources. A review of the literature suggests that two key issues lie at the heart of the gaps between the rhetoric and reality of urban responses; institutional capacity of municipalities and the political economy within which such approaches are framed and implemented (Bulkeley 2010). The first issue includes constraints such as jurisdictional boundary limitations or municipal resources and the second issue includes limitations of political leadership or the lack of complementarity of climate change policies and other social and economic goals. The author of the review argues, therefore, that opportunities for understanding the lack of progress in both systematic mitigation policy and adaptation measures exist. These include a focus on the relationship between public and private authority, the ways in which the policy problem of climate change is constituted and the basis of policy interventions in the infrastructures that mediate human-environment relations. Engaging issues of power underpinning urban outcomes, for example, may provide new insights into the influences of climate governance and environmental justice within and out with the city.

3.6 Conclusions: Resilient Cities in Asia and the Pacific

3.6.1 Urbanization and Resilience

Climate resilient cities are those that can withstand climate effects and not change dramatically. That is, they include biophysical and socio-economic sub-systems that can withstand floods, intense storms, heat waves, droughts, etc., and continue to develop in a fairly predictable manner. Cities that are not resilient change dramatically to new states with new relationships emerging both within the socio-economic subsystem and between the socio-economic and biophysical sub-systems. Resilient cities are sustainable cities, in that they can function with shocks. Resilience can only be achieved when urban areas move along a more sustainable pathway. The goal of policy makers and stakeholders for their individual urban centres, urban regions as well as nations in the face of climate change is to enhance resilience. In the present review of cities in the Asia-Pacific region, we identify some important aspects that impinge on this goal. Addressing uncertainties, research gaps and policy needs related to climate change and urbanization will help make cities in the region more resilient.

3.6.2 Uncertainties

Uncertainty includes imperfect knowledge of both the current trends and conditions the region is experiencing and future predictions. In this case, there is considerable uncertainty as to the exact types and intensities of changes in climate that the region will experience, the cascading effects of these changes as well as the current trends. Those estimating regional change (from population to GHG emissions to climate) know all too well the limitations of the currently available data and the models. For example, in this study we use UN population statistics to describe urbanization. These statistics, however, are not standardized across countries, meaning that the definition of urban varies by nation. As discussed in the present Chapter, estimates of GHG emissions from cities vary, in large part, because researchers use different information and different methods to calculate emissions.

Some response from the scientific community has been to provide ranges for future estimates and to qualify statements about specific current trends. These are important qualifications that allow readers to assess the validity of the data. Another way is to provide the scientific consensus on issues, stating whether the community of researchers agrees and to what level on specific issues.

Unfortunately, due to the newness of many of the aspects under study that we present here, it is difficult to identify a community consensus in all issues. We have, however, attempted to identify the major trends that the region is undergoing as key findings. These findings, we believe, most researchers and scholars working in this area would agree upon.

Notwithstanding the uncertainties in the data and methods, however, there are good reasons to fault on the side of caution. Termed the precautionary principle, decision makers and stakeholders would benefit in the face of uncertainty to weigh the desirable actions on the side of the lowest risk of harm. In this way, decisions are prudent and protections relaxed when scientific findings provide sound evidence that no, or acceptable, harm will result. Moreover, and related to this concept, despite uncertainties, the no-regrets policy approach argues that actions must be taken now to build capacity and adapt to long term climate changes.

3.6.3 Research Gaps

As mentioned throughout the present Chapter there are areas in research on climate and urbanization that need further attention. We have identified five important research gaps.

The first is the lack of information and study on the urbanization dynamics and socio-ecological relationships in medium and smaller settlements. Much of the findings that we identified are through studies of the largest and perhaps best managed cities in the region. Less is known of the smaller urban centres.

Second, there is more need for data collection and monitoring, particularly spatial information in terms of urban socio-economic changes and linkages to climate. Importantly, researchers need to continue to work on urban GHG protocols so that cross urban differences can be better evaluated.

Third, there is much work to be performed in understanding the options for and costs and impacts of sector-wide mitigation strategies at the urban level. In this regard, not only do we need to further evaluate strategies for different sectors, but better understand any potential trade-offs between different sectors and between mitigation and adaptation measures.

Fourth, there is a need to develop a better understanding of the costs of adaptation, particularly for small- and medium-sized communities and also for vulnerable populations. Given that these communities are already burdened by weather-related hazards, the task for researchers has become more urgent.

Finally, as mentioned above, there needs to be a better understanding of the factors behind urban climate governance. While transitions have occurred, the exact mechanisms as to the triggers and contexts are not well understood.

3.6.4 Policy Needs

We have already presented several areas where the research community would like to see policy movement. In this section we highlight some issues that emerge from the findings of the present Chapter. These include four main policy arenas.

First, given the large urban population and the predicted changes to come, urban climate policy must be developed at all levels of governance. Cities cannot alone address the challenges they face. Policy support is needed at the national and international levels. The urbanization of the Asia-Pacific region has global implications for economic growth and climate.

Second, and related to the first, is that more attention is needed to develop effective mitigation measures across different sectors. These measures must be tailored for each individual urban context and be part of an appropriate strategic/flexible systems planning effort. Short-, medium- and long-term strategies are needed to develop responses that will not "lock-in" carbon intensive trajectories.

Third, the cities, nations and global institutions need to place more effort in developing adaptation strategies under a no-regrets policy framework. The large vulnerable urban population must be included in the development of these policies and plans. Moreover, policies focused on reducing everyday risk in these communities will go a long way to reducing climate risk and increasing adaptive capacity.

Fourth, there should be more emphasis on retaining functioning ecosystems within and around urban areas as part of mitigation and adaptation strategies. As mentioned at several points in this Chapter, using ecosystem services offers options for GHG mitigation and given that the poor often depend upon these services, they should be part of adaptation strategies as well.

References

Adger, W. N., Huq, S., Brown, K., Conway, D., & Hulme, M. (2003). Adaptation to climate change in the developing world. *Progress in Development Studies, 3*(3), 179–195.

Ahmed, A. U., Vargas Hill, R., Smith, L. C., Wiesmann, D. M., Frankenberger, T., Gulati, K., … & Yohannes, Y. (2007). *The world's most deprived: Characteristics and causes of extreme poverty and hunger*. Washington, DC: International Food Policy Research Institute (IFPRI).

Ajero, M. Y. (2002). *Emissions patterns of Metro Manila, Proceedings of the IGES/APN Mega-City Project Conference*. Kitakyushu: Proceedings of the IGES/APN Mega-City Project Conference, 23–25 January.

Akamatsu, K. (1962). Historical pattern of economic growth in developing countries. *The Developing Economies, 1*, 3–25.

Alam, M., & Rabbani, M. D. G. (2007). Vulnerabilities and responses to climate change for Dhaka. *Environment and Urbanisation, 19*, 81.

Alber, G., & Kern, K. (2008). Governing climate change in cities: Modes of urban climate governance in multi-level systems. *Proceedings of the organization for economic co-operation and development conference, competitive cities and climate change*. Paris: OECD.

Alberti, M. (2005). The effects of urban patterns on ecosystem function. *International Regional Science Review, 28*(2), 168–192.

Alcoforado, M. J., & Andrade, H. (2008). Global warming and the urban heat island. In J. M. Marzluff, E. Shulenberger, W. Endlicher, M. Alberti, G. Bradley, C. Ryan, U. Simon, & C. ZumBrunnen (Eds.), *Urban ecology, an international perspective on the interactions between humans and nature* (pp. 249–262). Dordrecht: Springer.

Allison, M. A., Kahm, S. R., Goodbred, S. L., & Kuehl, S. A. (2003). Stratigraphic evolution of the late Holocene Ganges-Grahmaputra lower delta plain. *Sedimentation Geology, 155*, 317–342.

Amsden, A. H. (1989). *Asia's next giant: South Korea and late industrialization*. New York: Oxford University Press.

Angel, S., Sheppard, S. C., & Civco, D. L. (2005). *The dynamics of global urban expansion*. Washington, DC: Transport and Urban Development Department, The World Bank.

Aniello, C., Morgan, K., Busbey, A., & Newland, L. (1995). Mapping micro-urban heat islands using Landsat Tm and a GIS. *Computers & Geosciences, 21*(8), 965–969.

APERC. (2007). *Urban transport energy use in the APEC region, trends and options*. Tokyo: Asia Pacific Energy Research Center (APERC).

APN. (2010). Climate change vulnerability assessment and urban development planning for Asian coastal cities. Scientific capacity building for climate impact and vulnerability assessments (SCBCIA). Asia-Pacific Network for global change research. http://www.apn-gcr.org/resources/items/show/1698

Approtech Asia. (2005). *Enabling urban poor livelihoods policy making: Understanding the role of energy services, Philippine country report*. Manila: Approtech.

Arthurton, R. S. (1998). Marine-related physical natural hazards affecting coastal megacities of the Asia-Pacific region – Awareness and mitigation. *Ocean and Coastal Management, 40*, 65–85.

Asian Development Bank. (1997). *Emerging Asia: Challenges and changes*. Hong Kong: Asian Development Bank/Oxford University Press.

Asian Development Bank. (1999). *Asian Development Outlook*. Hong Kong: Asian Development Bank/Oxford University Press.

Asian Development Bank. (2007). India: Preparing the sustainable coastal protection and management project. Technical assistance report, Project number: 40156. Retrieved from http://www.adb.org/Documents/TARs/IND/40156-IND-TAR.pdf

Asian Development Bank. (2010). *Ho Chi Minh City adaptation to climate change. Summary report*. Manila: Asian Development Bank.

Åström, D. O., Forsberga, B., & Rocklöv, J. (2011). Heat wave impact on morbidity and mortality in the elderly population: A review of recent studies. *Maturitas, 69*(2), 99–105.

Australian Business Council for Sustainable Energy. (2005). *Renewable energy in Asia: The Thailand report*. Carlton: Australian Business Council for Sustainable Energy.

Ayres, J. G., Forsberg, B., Annesi-Maesano, I., Dey, R., Ebi, K. L., Helms, P. J., … & Environment and Health Committee of the European Respiratory Society. (2009). Climate change and respiratory disease: European Respiratory Society position statement. *European Respiratory Journal, 34*(2), 295–302.

Bader, N., & Bleischwitz, R. (2009). Measuring urban greenhouse gas emissions: The challenge of comparability. *Surveys and Perspectives Integrating Environment & Society, 2*(3), 7–21.

Bai, X. (2007). Cityscape: Rizhao, solar-powered city. In Worldwatch Institute & L. Starke (Eds.), *State of the World 2007, our urban future* (pp. 108–109). Washington, DC: W. W. Norton and Company.

Bai, X., & Imura, H. (2000). A comparative study of urban environment in East Asia: Stage model of urban environmental evolution. *International Review for Environmental Strategies, 1*(1), 135–158.

Bai, X., McAllister, R. R. J., Beaty, R. M., & Taylor, B. (2010). Urban policy and governance in a global environment: Complex systems, scale mismatches and public participation. *Current Opinion in Environmental Sustainability, 2*(3), 129–135.

Bangkok Post. (2011, 15 November). PT MPs to propose new capital city. Retrieved from http://www.bangkokpost.com/breakingnews/266403/pt-mps-to-propose-new-capital-city

Bangladesh Ministry of Food and Disaster Management. (2007). *Consolidated damage and loss assessment, lessons learnt from the flood 2007 and future action plan*. Bangladesh: Disaster Management Bureau.

Barata, M., Ligeti, E., De Simone, G., Dickinson, T., Jack, D., Penney, J., … & Zimmerman, R. (2011). Climate change and human health in cities. In C. Rosenzweig, W. D. Solecki, S. A. Hammer, & S. Mehrotra (Eds.), *Climate change and cities, first assessment report of the urban climate change research network* (pp. 179–213). Cambridge: Cambridge University Press.

Barter, P. (1999). An international comparative perspective on urban transport and urban form in Pacific Asia: The challenge of rapid motorization in dense cities. Dissertation, Institute for Sustainability and Technology Policy, Division of Social Sciences, Humanities and Education, Murdoch University, Perth.

Barter, P. (2000). Urban transport in Asia: Problems and prospects for high-density cities. *Asia-Pacific Development Monitor, 2*(1), 33–66.

Barter, P. A. (2005). A vehicle quota integrated with road usage pricing: A mechanism to complete the phase-out of high fixed vehicle taxes in Singapore. *Transport Policy, 12*(6), 525–536.

BBC. (2000, July 14). Hopes fade for landslide victims. BBC News. Retrieved from http://news.bbc.co.uk/2/hi/south_asia/831267.stm

Beniston, M., & Diaz, H. (2004). The 2003 heat wave as an example of summers in a greenhouse climate? Observations and climate model simulations for Basel, Switzerland. *Global and Planetary Change, 44*(1–4), 73–81.

Bernard, M., & Ravenhill, J. (1995). Beyond product cycles and flying geese: Regionalization, hierarchy and the industrialization of East Asia. *World Politics, 47*(2), 171–209.

Bhagat, R. B., Guha, M., & Chattopadhyay, A. (2006). Mumbai after 26/7 deluge: Issues and concerns in urban planning. *Population and Environment, 27*(4), 337–349.

Bicknell, J., Dodman, D., & Satterthwaite, D. (Eds.). (2009). *Adapting cities to climate change, understanding and addressing the development challenges*. London: Earthscan.

Bloomberg, M. R. (2010). *NYC green infrastructure plan, a sustainable strategy for clean waterways*. New York: Office of the Mayor.

Brandon, C. (1994). Reversing pollution trends in Asia. *Finance & Development, 31*(2), 21–23.

Buechler, S., & Devi, G. (2002). Livelihoods and wastewater irrigated agriculture – Musi River in Hyderabad City, Andhra Pradesh, India. *Urban Agriculture Magazine, 8*, 14–17.

Bulkeley, H. (2010). Cities and the governing of climate change. *Annual Review of Environment and Resources, 35*, 229–253.

Butler, T. M., Lawrence, M. G., Gurjar, B. R., van Aardenne, J., Schultz, M., & Lelievald, J. (2007). The representation of emission from megacities in global emissions inventories. *Atmospheric Environment, 42*, 703–719.

Case, T. (2008). Climate change and infrastructure issues. *Drinking Water Research, Special Issue: Climate Change, 18*(2), 11–14.

Changnon, S. A. (1979). Rainfall changes in summer caused by St. Louis. *Science, 205*(4404), 402–404.

Chin, A. (2000). Sustainable urban transportation: Abatement and control of traffic congestion and vehicular emissions from land transportation in Singapore. *Environmental Economics and Policy Studies, 3*, 355–380.

Cho, M.-R. (2010). The politics of urban nature restoration: The case of Cheonggyecheon restoration in Seoul, Korea. *International Development Planning Review, 32*(2), 145–165.
Christensen, J. H., Hewitson, B., Busuioc, A., Chen, A., Gao, X., Held, I., … & Whetton, P. (2007). Regional climate projections. In S. Solomon, D. Qin, M. Manning, Z. Chen, M. Marquis, K. B. Averyt, M. Tignor, & H. L. Miller (Eds.), *Climate change 2007: The physical science basis. Contribution of working group I to the Fourth Assessment Report of the Intergovernmental Panel on Climate Change* (pp. 847–940). Cambridge/New York: Cambridge University Press.
Church, J. A., White, N. J., Aarup, T., Wilson, W. S., Woodworth, P. L., Domingues, C. M., … & Lambeck, K. (2008). Understanding global sea levels: Past present and future. *Sustainability Science, 3*(1), 9–22.
Coffee, J. E., Parzen, J., Wagstaff, M., & Lewis, R. S. (2010). Preparing for a changing climate: The Chicago climate action plan's adaptation strategy. *Journal of Great Lakes Research, 36*(2), 115–117.
Cohen, M. A. (1996). The hypothesis of urban convergence: Are cities in the North and South becoming more alike in an age of globalization? In M. A. Cohen, B. A. Ruble, J. S. Tulchin, & A. M. Garland (Eds.), *Urban future: Global pressures and local forces* (pp. 25–38). Washington, DC: Woodrow Wilson Center Press.
Cohen, M. J., & Garrett, J. L. (2010). The food price crisis and urban food (in)security. *Environment and Urbanisation, 22*(2), 467–482.
Cole, D., Lee-Smith, D., & Nasinyama, G. (2008). *Healthy city harvests: Generating evidence to guide policy on urban agriculture*. Lima: International Potato Center (CIP)/Makerere University Press.
Confalonieri, U., Menne, B., Akhtar, R., Ebi, K. L., Hauengue, M., Kovats, R. S., … & Woodward, A. (2007). Human health. In M. L. Parry, O. F. Canziani, J. P. Palutikof, P. J. van der Linden, & C. E. Hanson (Eds.), *Climate change 2007: Impacts, adaptation and vulnerability. Contribution of working group II to the Fourth Assessment Report of the Intergovernmental Panel on Climate Change* (pp. 391–431). Cambridge: Cambridge University Press.
Corfield, J. (2008a). Thailand. In *Encyclopedia of global warming and climate change* (pp. 961–963). Thousand Oaks: Sage.
Corfield, R. S. (2008b). Singapore. In *Encyclopedia of global warming and climate change* (pp. 898–900). Thousand Oaks: Sage.
Costello, A., Abbas, M., Allen, A., Ball, S., Bell, S., Bellamy, R., … & Patterson, C. (2009). Managing the health effects of climate change. *The Lancet, 373*(9676), 1693–1733.
Cruz, R. V., Harasawa, H., Lal, M., Wu, S., Anokhin, Y., Punsalmaa, B., … & C. E. Hanson (Eds.), *Climate change 2007: Impacts, adaptation and vulnerability. Contribution of working group II to the Fourth Assessment Report of the Intergovernmental Panel on Climate Change* (pp. 469–506). Cambridge: Cambridge University Press.
Curriero, F. C., Heiner, K., Zeger, S., Samet, J. M., & Patz, J. A. (2002). Temperature and mortality in 11 cities the eastern United States. *American Journal of Epidemiology, 155*, 80–87.
Dailey, P., Huddleston, M., Brown, S., & Fasking, D. (2009). *The financial risks of climate change; examining the financial implications of climate change using climate models and insurance catastrophe risk models*. London: Association of British Insurers and AIR Worldwide Corp. and the Met Office.
Daniere, A. (1996). Growth, inequality and poverty in South-east Asia, the case of Thailand. *Third World Planning Review, 18*(4), 373–395.
de Sherbinin, A., Schiller, A., & Pulsipher, A. (2007). The vulnerability of global cities to climate change. *Environment and Urbanisation, 19*(1), 39–64.
Deolalikar, A. B., Brillantes, A. B., Jr., Gaiha, R., Pernia, E. M., & Racelis, M. (2002). *Poverty reduction and the role of institutions in developing Asia* (Economics and research development working paper series, no. 10). Manila: Asian Development Bank.
Dhakal, S. (2009). Urban energy use and carbon emissions from cities in China and policy implications. *Energy Policy, 37*(11), 4208–4219.

Dhakal, S. (2010). GHG emission from urbanisation and opportunities for urban carbon mitigation. *Current Opinion in Environmental Sustainability, 2*(4), 277–283.

Dhakal, S., & Imura, H. (2004). *Urban energy use and greenhouse gas emissions in Asian mega-cities, policies for a sustainable future*. Tokyo: Institute for Global Environmental Strategies.

Dick, H. W., & Rimmer, P. J. (1998). Beyond the Third World City: The new urban geography of South-east Asia. *Urban Studies, 35*(12), 2303–2321.

Dicken, P. (1992). International production in a volatile regulatory environment: The influence of national regulatory policies on the spatial strategies of transnational corporations. *Geoforum, 23*(3), 303–316.

Dodman, D. (2009). Blaming cities for climate change? An analysis of urban greenhouse gas emissions inventories. *Environment and Urbanisation, 21*(1), 185–201.

Dodman, D. (2011). Box: Urban vulnerabilities in the least developed countries. In C. Rosenzweig, W. D. Solecki, S. A. Hammer, & S. Mehrotra (Eds.), *Climate change and cities, First Assessment Report of the urban climate change research network* (pp. 121–124). Cambridge/New York: Cambridge University Press.

Douglass, M. (2000). Mega-urban regions and world city formation: Globalization, the economic crisis and urban policy issues in Pacific Asia. *Urban Studies, 37*(12), 2315–2335.

Drakakis-Smith, D. (1992). *Pacific Asia*. London: Routledge.

Dua, A., & Esty, D. C. (1997). *Sustaining the Asia Pacific miracle, environmental protection and economic integration*. Washington, DC: Institute for International Economics.

Dubbeling, M., Caton Campbell, M., Hoekstra, F., & van Veenhuizen, R. (2009). Building resilient cities. *Urban Agriculture Magazine, 22*, 3–11.

Ebi, K. L. (2010). Human health. In S. H. Schneider, A. Rosencranz, M. D. Mastrandrea, & K. Kuntz-Duriseti (Eds.), *Climate change science and policy* (pp. 124–130). Washington, DC: Island Press.

Economist, T. (2006). More of everything. *The Economist, 380*(8495), 18–20.

Engel, K., Jokiel, D., Kraljevic, A., Geiger, M., & Smith, K. (2011). *Big cities. Big water. Big challenges. Water in an urbanizing world*. Gland: WWF.

Ewing, R. (1997). Is Los Angeles-style sprawl desirable? *Journal of the American Planning Association, 63*(1), 107–126.

FAO. (2000). Feed Asian cities. In *Proceedings of the food supply and distribution to cities programme regional seminar, November 27–30*. Bangkok: FAO.

Foster, J., Lowe, A., & Winkelman, S. (2011). *The value of green infrastructure for urban climate change adaptation*. Washington, DC: The Center for Clean Air Policy.

Frumkin, H. (2002). Urban sprawl and public health. *Public Health Reports, 117*(3), 201–217.

Fuchs, R. J. (2010). *Cities at risk: Asia's coastal cities in an age of climate change*. Honolulu: East–West Center.

Gadda, T., & Gasparatos, A. (2009). Land use and cover change in Japan and Tokyo's appetite for meat. *Sustainability Science, 4*(2), 165–177.

Gleick, P. H. (2010). Water. In S. H. Schneider, A. Rosencranz, M. D. Mastrandrea, & K. Kuntz-Duriseti (Eds.), *Climate change science and policy* (pp. 74–81). Washington, DC: Island Press.

Goudie, A. (2006). *The human impact on the natural environment*. Oxford: Blackwell.

Gupta, K. (2006). Urban flood resilience planning and management and lessons for the future: A case study of Mumbai, India. *Urban Water Journal, 4*(3), 183–194.

Gurjar, B. R., van Aardenne, J., Lelievald, J., & Mohan, M. (2004). Emission estimates and trends (1990–2000) for megacity Delhi and implications. *Atmospheric Environment, 38*, 5663–5681.

Gurjar, B. R., Butler, T. M., Lawrence, M. G., & Lelievald, J. (2008). Evaluation of emissions and air quality in megacities. *Atmospheric Environment, 42*, 1593–1606.

Gustafsson, O., Krusa, M., Zencak, Z., Sheesley, R. J., Granat, L., Engstrom, E., … & Rodhe, H. (2009). Brown clouds over South Asia: Biomass or fossil fuel combustion? *Science, 323*(5913), 495–498.

Hack, G. (2000). Infrastructure and regional form. In R. Simmonds & G. Hack (Eds.), *Global city regions, their emerging forms* (pp. 183–192). London: Spon Press.

Hallegatte, S. P., Henriet, F., Patwardhan, A., Narayanan, K., Ghosh, S., Karmakar, S., … & Naville, N. (2010). *Flood risks, climate change impacts and adaptation benefits in Mumbai: An initial assessment of socio-economic consequences of present and climate change induced flood risks and of possible adaptation options* (OECD environment working papers, Vol. 27). Paris: OECD. http://dx.doi.org/10.1787/5km4hv6wb434-en.

Hallegatte, S., Henriet, F., & Corfee-Morlot, J. (2011). The economics of climate change impacts and policy benefits at city scale: A conceptual framework. *Climatic Change, 104*(1), 51–87.

Halweil, B., & Nierenberg, D. (2007). Farming the cities. In M. O'Meara Sheehan (Ed.), *The State of the World 2007, our urban future* (pp. 48–65). New York: W. W. Norton and Company.

Hamin, E., & Gurran, N. (2009). Urban form and climate change: Balancing adaptation and mitigation in the U.S. and Australia. *Habitat International, 33*, 238–245.

Hammer, S. A., Keirstead, J., Dhakal, S., Mitchell, J., Colley, M., Connell, R., … & Hyams, M. (2011). Climate change and urban energy systems. In C. Rosenzweig, W. D. Solecki, S. A. Hammer & S. Mehrotra (Eds.), *Climate change and cities, First Assessment Report of the urban climate change research network*. Cambridge/New York: Cambridge University Press.

Hansen, J., Sato, M., Ruedy, R., Lo, K., Lea, D. W., & Elizade, M. M. (2006). Global climate change. *Proceedings of the National Academy of Sciences of the United States of America, 103*(39), 14288–14293.

Hanson, S., Nicholls, R., Ranger, N., Hallegatte, S., Corfee-Morlot, J., Herweijer, C., & Chateau, J. (2011). A global ranking of port cities with high exposure to climate extremes. *Climatic Change, 104*(1), 89–111.

Hardoy, J. E., Mitlin, D., & Satterthwaite, D. (2001). *Environmental problems in an urbanisation world*. London: Earthscan.

Haruyama, S. (1993). Geomorphology of the central plain of Thailand and its relationship with recent flood conditions. *GeoJournal, 31*(4), 327–334.

Hatch, W., & Yamamura, K. (1996). *Asia in Japan's embrace, building a regional production alliance*. Hong Kong: Cambridge University Press.

Hayashi, Y., Doi, K., Yagishita, M., & Kuwata, M. (2004). Urban transport sustainability: Asian trends, problems and policy practices. *European Journal of Transport and Infrastructure Research, 4*(1), 27–45.

He, K., Huo, H., & Zhang, Q. (2002). Urban air pollution in China: Current status, characteristics, and progress. *Annual Review of Energy and the Environment, 27*, 397–431.

Hillman, T., & Ramaswami, A. (2010). Greenhouse gas emission footprints and energy use benchmarks for eight US cities. *Environmental Science & Technology, 44*, 1902–1910.

Holdren, J. P., & Smith, K. R. (2000). Energy, the environment, and health. In UNEP (Ed.), *World energy assessment, the challenge of sustainability* (pp. 61–110). New York: United Nations Development Programme.

Hong, W., Chiang, M. S., Shapiro, R. A., & Clifford, M. L. (Eds.). (2007). *Building energy efficiency, why green buildings are key to Asia's future*. Hong Kong: Asia Business Council.

Hook, W. (2005). *Bus rapid transit planning, institutional reform, and air quality: Lessons from Asia 18*. New York: Institute for Transportation and Development Policy.

Hoornweg, D., Sugar, L., Lorena, C., & Gomez, T. (2011). Cities and greenhouse gas emissions: Moving forward. *Environment and Urbanisation, 23*(1), 207–227.

Houghton, J. (2009). *Global warming, the complete briefing*. Cambridge: Cambridge University Press.

Hunt, A., & Watkiss, P. (2011). Climate change impacts and adaptation in cities: A review of the literature. *Climatic Change, 104*(1), 13–49.

ICLEI, UNEP, & UN-Habitat. (2009). *Sustainable urban energy planning, a handbook for cities and towns in the developing countries*. Nairobi: ICLEI – Local Governments for Sustainability, UNEP and UN-Habitat.

IEA. (2008). *World energy outlook 2008* (p. 578). New Milford: International Energy Agency.

IEA. (2011). *Electricity information*. Paris: International Energy Administration.

Inamura, T., Izumi, T., & Matsuyama, H. (2011). Diagnostic study of the effects of a large city on heavy rainfall as revealed by an ensemble simulation: A case study of central Tokyo, Japan. *Journal of Applied Meteorology and Climatology, 50*(3), 713–728.

International Federation of Red Cross and Red Crescent Societies. (2010). *World disasters report 2010, focus on urban risk*. Geneva: International Federation of Red Cross and Red Crescent Societies.

IPCC. (1996). *IPCC guidelines for national greenhouse gas inventories: Reference manual*. New Delhi: National Physical Laboratory.

IPCC. (2007a). Summary for policymakers. In M. L. Parry, O. F. Canziani, J. P. Palutikof, P. J. van der Linden, & C. E. Hanson (Eds.), *Climate change 2007: Impacts, adaptation and vulnerability. Contribution of working group II to the Fourth Assessment Report of the Intergovernmental Panel on Climate Change* (pp. 7–22). Cambridge: Cambridge University Press.

IPCC. (2007b). Summary for policymakers. In S. Solomon, D. Qin, M. Manning, Z. Chen, M. Marquis, K. B. Averyt, M. Tignor, & H. L. Miller (Eds.), *Climate change 2007: The physical science basis. Contribution of working group I to the Fourth Assessment Report of the Intergovernmental Panel on Climate Change*. Cambridge/New York: Cambridge University Press.

IPCC. (2007c). Summary for policymakers. In B. Metz, O. R. Davidson, P. R. Bosch, R. Dave, & L. A. Meyer (Eds.), *Climate change 2007: Mitigation. Contribution of working group III to the Fourth Assessment Report of the Intergovernmental Panel on Climate Change*. Cambridge/New York: Cambridge University Press.

Jacobi, J., Drescher, A. W., Amerasinghe, P. H., & Weckenbrock, P. (2009). Agricultural biodiversity strengthening livelihoods in peri-urban Hyderabad, India. *Urban Agriculture Magazine, 22*, 45–47.

Jarvis, H. (2003). Dispelling the myth that preference makes practice in residential location and transport behaviour. *Housing Studies, 18*(4), 587–606.

Jha, A. K., Sharma, C., Singh, N., Ramesh, R., Purvaja, R., & Gupta, P. K. (2008). Greenhouse gas emissions from municipal solid waste management in Indian mega-cities: A case study of Chennai landfill sites. *Chemosphere, 71*(4), 750–758.

Jomo, K. S. (1998). Globalization and human development in East Asia. In UNDP (Ed.), *Globalization with a human face, background papers II, human development report 1999* (pp. 81–123). New York: United Nations Publications.

Kaiser, D. P., & Qian, Y. (2002). Decreasing trends in sunshine duration over China for 1954–1998: Indication of increased haze pollution? *Geophysical Research Letters, 29*(21), 2042–2045.

Kandlikar, M., & Ramachandran, G. (2000). The causes and consequences of particulate air pollution in urban India: A synthesis of the science. *Annual Review of Energy and the Environment, 25*, 629–684.

Kataoka, K., Matsumoto, F., Ichinose, T., & Taniguchi, M. (2009). Urban warming trends in several large Asian cities over the last 100 years. *Science of the Total Environment, 407*(9), 3112–3119.

Kaufmann, R. K., Seto, K. C., Schneider, A., Liu, Z., Zhou, L., & Wang, W. W. (2007). Climate response to rapid urban growth: Evidence of a human-induced precipitation deficit. *Journal of Climate, 20*(10), 2299–2306.

Kennedy, C., Steinberger, J., Gason, B., Hansen, Y., Hillman, T., Havranck, M., … & Mendez, G. V. (2009a). Greenhouse gas emissions from global cities. *Environmental Science & Technology, 43*(19), 7297–7302.

Kennedy, C. A., Ramaswami, A., Carney, S., & Dhakal, S. (2009b). Greenhouse gas emissions baselines for global cities and metropolitan areas, *Proceedings from cities and climate change: Responding to an urgent Agenda, 28–30 June*, Marseille: World Bank, Government of France.

Klein, R. J. T., Nicholls, R. J., & Thomalla, F. (2003). The resilience of coastal megacities to weather-related hazards. In A. Kreimer, M. Arnold, & A. Carlin (Eds.), *Building safer cities: The future of disaster risk* (pp. 101–121). Washington, DC: The World Bank.

Kojima, K. (2000). The "flying geese" model of Asian economic development: Origin, theoretical extensions, and regional policy implications. *Journal of Asian Economies, 11*, 375–401.

Kovats, R. S., & Akhtar, R. (2008). Climate, climate change and human health in Asian cities. *Environment and Urbanisation, 20*(1), 165–175.

Landsberg, H. E. (1981). *The urban climate*. New York: Academic Press.

Laquian, A. A. (2005). *Beyond metropolis: The planning and governance of Asia's mega-urban regions*. Baltimore: The Johns Hopkins Press.

Larsen, K., & Barker-Reid, F. (2009). Adapting to climate change and building urban resilience in Australia. *Urban Agriculture Magazine, 22*, 22–24.

Lebel, L. (2004). Transitions to sustainability in production-consumption systems. *Journal of Industrial Ecology, 9*(1), 1–3.

Lebel, L., Garden, P., Banaticla, M. R. N., Lasco, R. D., Contreras, A., Mitra, A., … & Sari, A. (2007). Integrating carbon management into the development strategies of urbanizing regions in Asia: Implications of urban function, form, and role. *Journal of Industrial Ecology, 11*(2), 61–81.

Lefevre, B. (2009). Long-term energy consumptions of urban transportation: A prospective simulation of "transport–land uses" policies in Bangalore. *Energy Policy, 37*(3), 940–953.

Lei, M. (2011). Urbanisation impacts on severe weather dynamical processes and climatology, Department of Earth and Atmospheric Sciences, Purdue University, West Lafayette.

Lenzen, M., Wood, R., & Foran, B. (2008). Direct versus embodied energy – The need for urban lifestyle transitions. In P. Droege (Ed.), *Urban energy transition, from fossil fuels to renewable power* (pp. 91–120). Sydney: Elsevier.

Leonard, K. J., Suhrbier, J. H., Lindquist, E., Savonis, M. J., Potter, J. R., & Dean, W. R. (2008). How can transportation professionals incorporate climate change in transportation decisions? In *Impacts of climate change and variability on transportation systems and infrastructure: Gulf coast study, phase I* (pp. 5-1–5-28). Washington, DC: US Climate Change Program and Subcommittee on Global Change Research, Department of Transportation.

Levine, M., Ürge-Vorsatz, D., Blok, K., Geng, L., Harvey, D., Lang, S., … & Yoshino, H. (2007). Residential and commercial buildings. In B. Metz, O. R. Davidson, P. R. Bosch, R. Dave & L. A. Meyer (Eds.). *Climate change 2007: Mitigation. Contribution of working group III to the Fourth Assessment Report of the Intergovernmental Panel on Climate Change*. Cambridge/New York: Cambridge University Press.

Li, J. (2011). Decoupling urban transport from GHG emissions in Indian cities – A critical review and perspectives. *Energy Policy, 39*(6), 3503–3514.

Li, Q., Zhang, H., Liu, X., & Huang, J. (2004). Urban heat island effect on annual mean temperature during the last 50 years in China. *Theoretical and Applied Climatology, 79*(3), 165–174.

Liu, F. (2011). *Research on urban expansion and its forces of typical regions, China*. Beijing: Chinese Academy of Sciences.

Liu, J., & Deng, X. (2011). Impacts and mitigation of climate change on Chinese cities. *Current Opinion in Environmental Sustainability, 3*(3), 188–192.

Lloyd's. (2008). *360 risk project*. London: Lloyd's.

Lo, F.-C., & Yeung, Y.-M. (Eds.). (1996). *Emerging world cities in Pacific Asia*. Tokyo: UNU Press.

Lowry, W. P. (1998). Urban effects on precipitation amount. *Progress in Physical Geography, 22*(4), 477–520.

Mage, D., Ozolins, G., Peterson, P., Webster, A., Orthofer, R., Vandeweered, V., & Gwynne, M. (1996). Urban air pollution in megacities of the world. *Atmospheric Environment, 30*, 681–686.

Major, D. C., Omojola, A., Dettinger, M., Hanson, R. T., & Sanchez-Rodriquez, R. (2011). Climate change, water and wastewater in cities. In C. Rosenzweig, W. D. Solecki, S. A. Hammer, & S. Mehrotra (Eds.), *Climate change and cities, First Assessment Report of the urban climate change research network* (pp. 113–143). Cambridge/New York: Cambridge University Press.
Marcotullio, P. J. (2003). Globalization, urban form and environmental conditions in Asia Pacific cities. *Urban Studies, 40*(2), 219–248.
Marcotullio, P. J. (2005). *Time-space telescoping and urban environmental transitions in the Asia Pacific*. Yokohama: United Nations University Institute of Advanced Studies.
Marcotullio, P. J. (2007). Urban water-related environmental transitions in Southeast Asia. *Sustainability Science, 2*(1), 27–54.
Marcotullio, P. J., & Marshall, J. D. (2007). Potential futures for road transportation CO2 emissions in the Asia Pacific. *Asia Pacific Viewpoint, 48*(3), 355–377.
Marcotullio, P. J., Albrecht, J., & Sarzynski, A. (2011). The geography of greenhouse gas emissions from within urban areas of India: A preliminary assessment. *Resources, Energy and Development, 8*(1), 11–35.
Marcotullio, P. J., Sarzynski, A., Albrecht, J., & Schulz, N. (2012). The geography of urban greenhouse gas emissions in Asia: A regional analysis. *Global Environmental Change, 22*(4), 944–958.
Mayer, H. (1999). Air pollution in cities. *Atmospheric Environment, 33*, 4029–4037.
McCarthy, J. J., Canziani, O. F., Leary, N. A., Dokken, D. J., & White, K. S. (Eds.). (2001). *Climate change 2001: Impacts, adaption and vulnerability*. Cambridge: Cambridge University Press.
McDonald, R. I., Green, P., Balk, D., Fekete, B. M., Revenga, C., Todd, M., & Montgomery, M. (2011). Urban growth, climate change, and freshwater availability. *Proceedings of the National Academy of Sciences of the United States of America, 108*(15), 6312–6317.
McGee, T. G., & Robinson, I. M. (Eds.). (1995). *The mega-urban regions of Southeast Asia*. Vancouver: University of British Columbia Press.
McGranahan, G. (2005). An overview of urban environmental burdens at three scales: Intra-urban, urban-regional and global. *International Review for Environmental Strategies, 5*(2), 335–356.
McGranahan, G., Marcotullio, P. J., Bai, X., Balk, D., Braga, T., Douglas, I., … & Zlotnik, H. (2005). Urban systems. In R. M. Hassan, R. Scholes & N. Ash (Eds.). *Millennium ecosystem assessment* (pp. 795–825). Washington, DC: Island Press.
McGranahan, G., & Murray, F. (Eds.). (2003). *Air pollution & health in rapidly developing countries*. London: Earthscan Publications Ltd.
McGranahan, G., & Satterthwaite, D. (2003). Urban centers: An assessment of sustainability. *Annual Review of Environment and Resources, 28*, 243–274.
McGranahan, G., Balk, D., & Anderson, B. (2007). The rising tide: Assessing the risks of climate change and human settlements in low elevation coastal zones. *Environment and Urbanisation, 19*(1), 17–37.
McKendry, I. G. (2003). Applied climatology. *Progress in Physical Geography, 27*(4), 597–606.
McMichael, A., Campbell-Lendrum, D., Kovats, S., Edwards, S., Wilkinson, P., Wilson, T., … & Andronova, N. (2004). Global climate change. In M. Ezzati, A. D. Lopez, A. Rodgers & C. J. L. Murray (Eds.) *Comparative quantification of health risks global and regional burden of disease attributable to selected major risk factors, volume 1* (pp. 1543–1650). Geneva: World Health Organization.
McMichael, A. J., Wilkinson, P., Kovats, R. S., Pattenden, S., Hajat, S., Armstrong, B., … & Nikiforov, B. (2008). International study of temperature, heat and urban mortality: The 'ISOTHURM' project. *International Journal of Epidemiology, 37*(5), 1121–1131.
Meehl, G. A., Stocker, T. F., Collins, W. D., Friedlingstein, P., Gaye, A. T., Gregory, J. M., … & Zhao, Z.-C. (2007). Global climate projections. In S. Solomon, D. Qin, M. Manning, Z. Chen, M. Marquis, K. B. Averyt, M. Tignor, & H. L. Miller (Eds.), *Climate change 2007: The physical science basis. Contribution of working group I to the Fourth Assessment Report of the*

Intergovernmental Panel on Climate Change (pp. 747–846). Cambridge/New York: Cambridge University Press.

Mehrotra, S., Lefevre, B., Zimmerman, R., Cercek, H., Jacob, K., & Srinivasan, S. (2011). Climate change and urban transportation systems. In C. Rosenzweig, W. D. Solecki, S. A. Hammer, & S. Mehrotra (Eds.), *Climate change and cities, First Assessment Report of the urban climate change research network* (pp. 145–177). Cambridge/New York: Cambridge University Press.

Meng, W., Yan, J., & Hu, H. (2007). Urban effects and summer thunderstorms in a tropical cyclone affected situation over Guangzhou city. *Science in China Series D Earth Sciences, 50*(12), 1867–1876.

Menon, S., Hansen, J., Nazarenko, L., & Luo, Y. (2002). Climate effects of black carbon aerosol in China and India. *Science, 297*(5590), 2250–2253.

Mitra, A. P., Sharma, C., & Ajero, M. A. Y. (2003). Energy and emissions in South Asian megacities: Study on Kolkata, Delhi and Manila. Paper read at international workshop on policy integration towards sustainable urban energy use for cities in Asia, at East–West Center, Honolulu.

Mokhtarian, P. L. (2002). Telecommunications and travel: The case for complementarity. *Journal of Industrial Ecology, 6*(2), 43–57.

Molina, M. J., & Molina, L. T. (2004). Megacities and atmospheric pollution. *Journal of the Air & Waste Management Association, 54*, 644–680.

Muller, M. (2007). Adapting to climate change: Water management for urban resilience. In J. Bicknell, D. Dodman, & D. Satterthwaite (Eds.), *Adapting cities to climate change, understanding and addressing the development challenges* (pp. 291–307). London: Earthscan.

Munich Re Foundation. (2011). *Urbanisation and megacities*. Retrieved from http://www.munichre-foundation.org/home/Topics/Topics_old/UrbanisationAndMegacities.html

Nabuurs, G. J., Masera, O., Andrasko, K., Benitez-Ponce, P., Boer, R., Dutschke, M., … & Zhang, X. (2007). Forestry. In B. Metz, O. R. Davidson, P. R. Bosch, R. Dave, & L. A. Meyer (Eds.), *Climate change 2007: Mitigation. Contribution of working group III to the Fourth Assessment Report of the Intergovernmental Panel on Climate Change* (pp. 541–584). Cambridge/New York: Cambridge University Press.

Nahn, T. (2010). Translated by Mien, H. Landslides to threat unprepared Ho Chi Minh City. Article published in the English edition of the Saigon Daily. Retrieved from http://www.saigongpdaily.com.vn/Hochiminhcity/2010/5/82184/

Newman, P., & Kenworthy, J. (1989). Gasoline consumption and cities: A comparison of U.S. cities with a global survey. *Journal of the American Planning Association, 55*(1), 24–37.

Newman, P., & Kenworthy, J. (1999). *Sustainability and cities*. Washington, DC: Island Press.

Nicholls, R. J. (1995). Coastal megacities and climate change. *GeoJournal, 37*(3), 369–379.

Nicholls, R. J., & Tol, R. S. J. (2006). Impacts and responses to sea-level rise: A global analysis of the SRES scenarios over the twenty-first century. *Philosophical Transactions: Mathematical, Physical and Engineering Sciences, 364*(1841), 1073–1095.

Nicholls, R. J., Hanson, S., Herweijer, C., Patmore, N., Hallegatte, S., Corfee-Morlot, J., … & Muir-Wood, R. (2008). *Ranking port cities with high exposure and vulnerability to climate extremes: Exposure estimates* (OECD environment working papers, no. 1). Paris: OECD.

Nowak, D. J., & Crane, D. E. (2002). Carbon storage and sequestration by urban trees in the USA. *Environmental Pollution, 116*(3), 381–389.

OECD. (2010). *OECD factbook 2010: Economic, environmental and social statistics*. Paris: Organization of Economic Cooperation and Development.

Oke, T. R. (1973). City size and the urban heat island. *Atmospheric Environment, 7*(8), 769–799.

Oke, T. R. (1997). Urban climate and global change. In A. Perry & R. Thompson (Eds.), *Theoretical and applied climatology* (pp. 273–287). London: Routledge.

Overpeck, J. T., & Weiss, J. L. (2009). Projections of future sea level becoming more dire. *Proceedings of the National Academy of Sciences of the United States of America, 106*(51), 21461–21462.

Pachauri, S. (2004). An analysis of cross-sectional variations in total household energy requirements in India using micro survey data. *Energy Policy, 32*(15), 1723–1735.
Pachauri, S., & Jiang, L. (2008). The household energy transition in India and China. *Energy Policy, 36*(11), 4022–4035.
Panya Consultants. (2009). *Climate change impact and adaptation study for Bangkok metropolitan region*. Bangkok: Panya Consultants Co. Limited.
Parker, D. E. (2004). Large scale warming in not urban. *Nature, 432*(7015), 290–291.
Parry, M. L., Canziani, O. F., Palutikof, J. P., et al. (2007). Technical summary. In M. L. Parry, O. F. Canziani, J. P. Palutikof, P. J. van der Linden, & C. E. Hanson (Eds.), *Climate change 2007: Impacts, adaptation and vulnerability. Contribution of working group II to the Fourth Assessment Report of the Intergovernmental Panel on Climate Change* (pp. 23–78). Cambridge: Cambridge University Press.
Parshall, L., Gurney, K., Hammer, S. A., Mendoza, D., Zhou, Y., & Geethakumar, S. (2010). Modeling energy consumption and CO2 emissions at the urban scale: Methodological challenges and insights from the United States. *Energy Policy, 38*(9), 4765–4782.
Pataki, D. E., Alig, R. J., Fung, A. S., Golubiewski, N. E., Kennedy, C. A., McPherson, E. G., … & Lankao, P. R. (2006). Urban ecosystems and the North American carbon cycle. *Global Change Biology, 12*(11), 2092–2102.
Patz, J. A., McGeehin, M. A., Bernard, S. M., Ebi, K. L., Epstein P. R., Grambsch, A., … & Trtanj, J. (2000). The potential health impacts of climate variability and change for the United States: Executive summary of the report of the health sector of the U.S. National Assessment. *Environmental Health Perspectives, 108*(4), 367–376.
Patz, J. A., Campbell-Lendrum, D., Holloway, T., & Foley, J. A. (2005). Impact of regional climate change on human health. *Nature, 438*(7066), 310–317. doi:10.1038/nature04188.
Pearce, F. (2007). *With speed and violence, why scientists fear tipping points in climate change*. Boston: Beacon.
Permana, A. S., Perera, R., & Kumar, S. (2008). Understanding energy consumption pattern of households in different urban development forms: A comparative study in Bandung City, Indonesia. *Energy Policy, 36*(11), 4287–4297.
Peterson, T. C. (2003). Assessment of urban versus rural in situ surface temperatures in the contiguous United States: No difference found. *Journal of Climate, 16*(18), 2941–2959.
Pfeffer, W. T., Harper, J. T., & O'Neel, S. (2008). Kinematic constraints on glacier contributions to 21st century sea-level rise. *Science, 321*(5894), 1340–1343.
Phdungsilp, A. (2010). Integrated energy and carbon modeling with a decision support system: Policy scenarios for low-carbon city development in Bangkok. *Energy Policy, 38*, 4808–4817.
Pierro, R., & Desai, B. (2008). Climate insurance for the poor: Challenges for targeting and participation. *IDS Bulletin, 39*(4), 123–129.
Popkin, B. M. (1994). The nutrition transition in low-income countries: An emerging crisis. *Nutrition Reviews, 52*(9), 285–298.
Popkin, B. M. (1999). Urbanisation, lifestyle changes and the nutrition transition. *World Development, 27*(11), 1905–1916.
Prasad, N., Ranghieri, F., Shah, F., Trohanis, Z., Kessler, E., & Sinha, R. (2008). *Climate resilient cities, a primer on reducing vulnerabilities to climate change impacts and strengthening disaster risk, management in East Asian cities*. Washington, DC: The World Bank.
Pucher, J., Park, H., Kim, M. H., & Song, J. (2005). Public transport reforms in Seoul: Innovations motivated by funding crisis. *Journal of Public Transportation, 8*(5), 41–62.
Ramanathan, V., & Crutzen, P. J. (2002). New directions: Atmospheric brown "clouds". *Atmospheric Environment, 37*(28), 4033–4035.
Regmi, M. B., & Hanaoka, S. (2011). A survey on impacts of climate change on road transportation infrastructure and adaptation strategies in Asia. *Environmental Economics and Policy Studies, 13*(1), 21–41.
Reilly, J., Tubiello, F., McCarl, B., Abler, D., Darwin, R., Fuglie, K., … & Rosenzweig, C. (2003). U.S. agriculture and climate change: New results. *Climatic Change, 57*(1), 43–67.

Ren, G. Y., Chu, Z. Y., Chen, Z. H., & Ren, Y. Y. (2007). Implications of temporal change in urban heat island intensity observed at Beijing and Wuhan stations. *Geophysical Research Letters, 34*(5), 1–5.

Revi, A. (2009). Climate change risk: An adaptation and mitigation agenda for Indian cities. In J. Bicknell, D. Dodman, & D. Satterthwaite (Eds.), *Adapting cities to climate change: Understanding and addressing the development challenges* (pp. 311–338). London: Earthscan.

Rizwan, A. M., Leung, D. Y. C., & Liu, C. (2008). A review on the generation, determination and mitigation of Urban Heat Island. *Journal of Environmental Sciences, 20*(1), 120–128.

Robine, J., Cheung, S., Le Roy, S., Van Oyen, H., & Herrmann, F. R. (2007). *Report on excess mortality in Europe during summer 2003*. Brussels: EU Community Action Programme for Public Health.

Rodolfo, K. S., & Siringan, F. P. (2006). Global sea-level rise is recognised, but flooding from anthropogenic land subsidence is ignored around northern Manila Bay, Philippines. *Disasters, 30*(1), 118–139.

Rosenzweig, C., & Parry, M. L. (1994). Potential impact of climate change on world food supply. *Nature, 367*(6459), 133–138.

Rosenzweig, C., Solecki, W. D., Hammer, S. A., & Mehrotra, S. (Eds.). (2011). *Climate change and cities, First Assessment Report of the urban climate change research network*. Cambridge: Cambridge University Press.

Rowan, H. (Ed.). (1998). *Behind East Asian growth, the political and social foundations of prosperity*. London: Routledge.

Sadownik, B., & Jaccard, M. (2001). Sustainable energy and urban form in China: The relevance of community energy management. *Energy Policy, 29*(1), 55–65.

Safriel, U., Adeel, Z., Neimeijer, D., Puigdefabregas, J., White, R., Lal, R., … & King, C. (2005). Dryland systems. In R. Hassan, R. Scholes, & N. Ash (Eds.), *Ecosystems and human well-being; current state and trends* (Vol. 1, pp. 623–662). Washington, DC: Island Press.

Santos, G., Behrendt, H., & Teytelboym, A. (2010). Part II: Policy instruments for sustainable road transport. *Research in Transportation Economics, 28*(1), 46–91.

Sari, A., & Salim, N. (2005). *Carbon and the city: Carbon pathways and decarbonization opportunities in greater Jakarta, Indonesia* (USER working paper WP-2005-04). Chiang Mai: Unit for Social and Environmental Research, Chiang Mai University.

Sarzynski, A. (2012). Bigger is not always better: A comparative analysis of cities and their air pollution impact. *Urban Studies, 49*(14), 3121–3138.

Satterthwaite, D. (2008). Cities' contribution to global warming: Notes on the allocation of greenhouse gas emissions. *Environment and Urbanisation, 20*(2), 539–549.

Satterthwaite, D., Huq, S., Pelling, M., Rei, H., & Lankao, P. R. (2007). Adapting to climate change in urban areas, the possibilities and constraints in low- and middle-income nations. In *Human settlements discussion paper series, theme: Climate change and cities – 1*. London: International Institute for Environment and Development.

Sauer, A., Klop, P., & Agrawal, S. (2010). *Over heating: Financial risks from water constraints on power generation in Asia*. Washington, DC: World Resources Institute and HSBC.

Savage, V. (2006). Ecology matters: Sustainable development in Southeast Asia. *Sustainability Science, 1*, 37–63.

Sawin, J. L., & Hughes, K. (2007). Energizing cities. In M. O'Meara Sheehan (Ed.), *State of the World 2007, our urban future* (pp. 90–111). New York: W. W. Norton & Company.

Schmidt, J. D. (1998). Globalization and inequality in urban South-east Asia. *Third World Planning Review, 20*(2), 127–145.

Schulz, N. (2010). Delving into the carbon footprint of Singapore – Comparing direct and indirect greenhouse gas emissions of a small and open economic system. *Energy Policy, 38*(9), 4848–4855.

Setchell, C. A. (1995). The growing environmental crisis in the world's mega-cities, the case of Bangkok. *Third World Planning Review, 17*(1), 1–17.

Seto, K. C., Fragkias, M., Guneralp, B., & Reilly, M. K. (2011). A meta-analysis of global urban land expansion. *PLoS One, 6*(8), 1–9.

Shephard, J. M., Pierce, H., & Negri, A. J. (2002). Rainfall modification by major urban areas: Observations from spaceborne rain radar on the TRMM satellite. *Journal of Applied Meteorology, 41*(4), 689–701.

Shepherd, J. M. (2005). A review of current investigations of urban-induced rainfall and recommendations for the future. *Earth Interactions, 9*(12), 1–27.

Shimoda, Y. (2003). Adaptation measures for climate change and the urban heat island in Japan's built environment. *Building Research & Information, 31*(3–4), 222–230.

Solecki, S. (2011). Box 8.8: Urban land use and urban heat island phenomenon in Shanghai, China. In C. Rosenzweig, W. D. Solecki, S. A. Hammer, & S. Mehrotra (Eds.), *Climate change and cities, First Assessment Report of the urban climate change research network* (p. 239). Cambridge: Cambridge University Press.

Solecki, W. D., & Leichenko, R. M. (2006). Urbanisation and the metropolitan environment: Lessons from New York and Shanghai. *Environment, 48*(4), 8–23.

Solecki, W., Leichenko, R., & O'Brien, K. (2011). Climate change adaptation strategies and disaster risk reduction in cities: connections, contentions, and synergies. *Current Opinion in Environmental Sustainability, 3*(3), 135–141.

Souch, C., & Grimmond, S. (2006). Applied climatology: Urban climate. *Progress in Physical Geography, 30*(2), 270–279.

Sovacool, B. K., & Brown, M. A. (2010). Twelve metropolitan carbon footprints: A preliminary comparative global assessment. *Energy Policy, 38*(9), 4856–4869.

Stubbs, J., & Clarke, G. (Eds.). (1996). *Megacity management in the Asian and Pacific region, volume I and II*. Manila: Asian Development Bank.

Swart, R., & Raes, F. (2007). Making integration of adaptation and mitigation work: Mainstreaming into sustainable development policies? *Climate Policy, 7*(4), 288–303.

Swiss Re. (2006). *The effects of climate change: Storm damage in Europe on the rise*. Zurich: Swiss Reinsurance Company.

Takahashi, K., Honda, Y., & Emori, S. (2007). Estimation of changes in mortality due to heat stress under changed climate. *Risk Research, 10*(3), 339–354.

Tan, J., Zheng, Y., Song, G., Kalkstein, L. S., Kalkstein, A. J., & Tang, X. (2007). Heat wave impacts on mortality in Shanghai, 1998 and 2003. *International Journal of Biometeorology, 51*(3), 193–200.

Tan, J., Zheng, Y., Tang, X., Guo, C., Li, L., Song, G., … & Chen, H. (2010). The urban heat island and its impact on heat waves and human health in Shanghai. *International Journal of Biometeorology, 54*(1), 75–84.

Taniguchi, M., Shimada, J., Fukuda, Y., Yamano, M., Onodera, S.-I., Kaneko, S., & Yoshikoshi, A. (2009). Anthropogenic effects on the subsurface thermal and groundwater environments in Osaka, Japan and Bangkok, Thailand. *Science of the Total Environment, 407*(9), 3153–3164.

Thirlwell, G. M., Madramootoo, C. A., Heathcote, I. W., & Osann, E. R. (2007). Coping with climate change: Short-term efficiency technologies. In *Canada-US water conference*. Ottawa/Washington, DC: Policy Research Initiative of Canada and the Woodrow Wilson Institute.

Transportation Research Board. (2008). *Potential impacts of climate change on U.S. transportation*. Washington, DC: National Resource Council.

Trenberth, K. E., Jones, P. D., Ambenje, P., Bojariu, R., Easterling, D., Tank, A. K., Zhai, P., et al. (2007). Observations: Surface and atmospheric climate change. In S. Solomon, D. Qin, M. Manning, Z. Chen, M. Marquis, K. B. Averyt, M. Tignor & H. L. Miller (Eds.), *Climate change 2007: The physical science basis. Contribution of working group I to the Fourth Assessment Report of the Intergovernmental Panel on Climate Change*. Cambridge/New York: Cambridge University Press.

United Nations (UN)-Habitat. (2009). *Global report on human settlements 2009, planning sustainable cities*. London: Earthscan.

United Nations (UN)-Habitat. (2011a). *Global report on human settlements 2011, cities and climate change: Policy directions*. London/Washington, DC: UN-Habitat and Earthscan.
United Nations (UN)-Habitat. (2011b). *Planning for climate change, a strategic values-based approach for urban planners*. Nairobi: UN-Habitat.
United Nations (UN)-Habitat. (2011c). *State of Asian cities 2010/2011*. Bangkok: UN-Habitat.
United Nations Development Programme (UNDP). (1996). *Urban agriculture, food, jobs and sustainable cities*. New York: United Nations Development Programme.
United Nations Environment Programme (UNEP), and C4. (2002). The Asian brown cloud (ABC). *Environmental Science and Pollution Research, 9*(5), 289–295.
United Nations Environment Programme (UNEP)/World Health Organization (WHO). (1993). *City air quality trends – Volume 2 (GEMS air data)*. Nairobi/Geneva: UNEP and WHO.
United Nations Framework Convention on Climate Change (UNFCCC). (2007). *Climate change: Impacts, vulnerabilities and adaptation in developing countries*. Bonn: United Nations Framework Convention on Climate Change Secretariat.
Urban, F., & Mitchell, T. (2011). *Climate change, disasters and electricity generation*. Brighton: Institute of Development Studies, University of Sussex.
US EPA. (2011). *Inventory of U.S. greenhouse gas emissions and sinks: 1990–2009*. Washington, DC: EPA. 430-R-11-005.
Vasconcellos, E. A. (2001). *Urban transportation, environment and equity; the case for developing countries*. London: Earthscan Publications, Ltd.
Vermeer, M., & Rahmstorf, S. (2009). Global sea level linked to global temperature. *Proceedings of the National Academy of Sciences of the United States of America, 106*(51), 21527–21532.
Walsh, M. P. (2003). Vehicle emission and health in developing countries. In G. McGranahan & F. Murray (Eds.), *Air pollution & health in rapidly developing countries* (pp. 146–175). London: Earthscan Publications, Ltd.
Wan, G., & Sebastian, I. (2011). *Poverty in Asia and the Pacific: An update* (ADB economic working paper series 267). Manila: Asian Development Bank.
Wang, X., Wang, Z., Qi, Y., & Guo, H. (2009a). Effect of urbanisation on the winter precipitation distribution in Beijing area. *Science in China Series D Earth Sciences, 52*(2), 250–256.
Wang, Y., Cai, J., Xie, L., & Liu, J. (2009b). Resilient Chinese cities: Examples from Beijing and Shanghai. *Urban Agriculture Magazine, 22*, 20–21.
Ward, P. J., Marfai, M. A., Yulianto, F., Hizbaron, D. R., & Aerts, J. C. J. H. (2011). Coastal inundation and damage exposure estimation: A case study for Jakarta. *Natural Hazards, 56*(3), 899–916.
Ward's. (2010). *World motor vehicle data 2010*. Southfield: Ward's Automotive Group.
Weisz, H., & Steinberger, J. K. (2010). Reducing energy and material flows in cities. *Current Opinion in Environmental Sustainability, 2*(3), 185–192.
Wen, Q. (2010). *Urban expansion in China – Spatiotemporal analysis using remote sensing data*. Beijing: Chinese Academy of Sciences.
Whitelegg, J., & Williams, N. (2000). Non-motorised transport and sustainable development: Evidence from Calcutta. *Local Environment, 5*(1), 7–18.
Wilbanks, T. J., Romero Lankao, P., Bao, M., Berkhout, F., Cairncross, S., Ceron, J.-P., … & Zapata-Marti, R. (2007). Industry, settlement and society. In M. L. Parry, O. F. Canziani, J. P. Palutikof, P. J. van der Linden & C. E. Hanson (Eds.), *Climate change 2007: Impacts, adaptation and vulnerability. contribution of working group II to the Fourth Assessment Report of the Intergovernmental Panel on Climate Change* (pp. 357–390). Cambridge: Cambridge University Press.
Wilson, D., Purushothaman, R., & Fiotakis, T. (2004). *The BRICS and global markets: Crude, cars, and capital*. New York: Goldman Sachs.
World Bank. (1993). *The East Asian miracle, economic growth and public policy*. New York: World Bank/Oxford University Press.
World Bank. (2007a). *Aceh flood, damage and loss assessment*. Jakarta: World Bank Country Office.
World Bank. (2007b). *East Asia environment monitor: Adapting to climate change*. Washington, DC: The World Bank.

World Bank. (2010). *Climate risks and adaptation in Asian coastal megacities, a synthesis report.* Washington, DC: The World Bank.

World Bank. (2011). *Guide to climate change adaptation in cities*. Washington, DC: The World Bank.

World Energy Council. (2011). *Global transportation scenarios 2050.* London: WEC.

World Wildlife Fund. (2009). *Mega-stress for mega-cities. A climate vulnerability ranking of major coastal cities in Asia*. Gland: WWF.

WRI. (2002). *Designing a customized greenhouse gas calculation tool.* Washington, DC: World Resources Institute.

Wunch, D., Wennberg, P. O., Toon, G. C., Keppel-Aleks, G., & Yavin, Y. G. (2009). Emissions of greenhouse gases from a North American megacity. *Geophysical Research Letters, 36*(L15810), 1–5.

Yang, Y., Hou, Y., & Chen, B. (2011). Observed surface warming induced by urbanisation in east China. *Journal of Geophysical Research, 116*(D14), 1–12.

Yuen, B., & Kong, L. (2009). Climate change and urban planning in Southeast Asia. *Survey and Perspectives Integrating Environment & Society, 2*(3), 1–11.

Yusuf, A. A., & Francisco, H. A. (2009). *Climate change vulnerability mapping for Southeast Asia.* Singapore: Economy and Environment Program for Southeast Asia (EEPSEA).

Zeng, Y., Qui, X. F., Gu, L. H., He, Y. J., & Wang, K. F. (2009). The urban heat island in Nanjing. *Quaternary International, 208*(1–2), 38–43.

Zhang, Z. X., Zhao, X. L., & Wang, X. (2012). *Atlas of land use in China by remote sensing monitoring*. Beijing: Star Map Press.

Zhou, L., Dickinson, R. E., Tian, Y., Fang, J., Li, Q., Kaufmann, R. K., … & Mynen, R. B. (2004). Evidence for a significant urbanisation effect on climate in China. *Proceedings of the National Academy of Sciences of the United States of America, 101*(26), 9540–9544.

Chapter 4
Climate and Security in Asia and the Pacific (Food, Water and Energy)

Lance Heath, Michael James Salinger, Tony Falkland, James Hansen, Kejun Jiang, Yasuko Kameyama, Michio Kishi, Louis Lebel, Holger Meinke, Katherine Morton, Elena Nikitina, P.R. Shukla, and Ian White

Abstract The impacts of increasing natural climate disasters are threatening food security in the Asia-Pacific region. Rice is Asia's most important staple food. Climate variability and change directly impact rice production, through changes in rainfall, temperature and CO_2 concentrations. The key for sustainable rice crop is water management. Adaptation can occur through shifts of cropping to higher latitudes and can profit from river systems (via irrigation) so far not considered. New opportunities arise to produce more than one crop per year in cooler areas. Asian wheat production in 2005 represents about 43 % of the global total. Changes in agronomic practices, such as earlier plant dates and cultivar substitution will be required. Fisheries play a crucial role in providing food security with the contribution of fish to dietary animal protein being very high in the region – up to 90 % in small island

L. Heath (✉)
Climate Change Institute (CCI), The Australian National University, Canberra, ACT 0200, Australia
e-mail: lance.heath@anu.edu.au

M.J. Salinger (✉)
University of Auckland, 3/7 Mattson Road, Pakuranga, Auckland 2010, New Zealand
e-mail: salinger@orcon.net.nz

T. Falkland
Island Hydrology Services, 9 Tivey Place, Hughes ACT 2605, Australia
e-mail: tony.falkland@netspeed.com.au

J. Hansen
The International Research Institute for Climate and Society (IRI), Columbia University Lamont Campus, P.O. Box 1000, 61 Route 9 W, Monell Building, Palisades, NY 10964-1000, USA
e-mail: jhansen@iri.columbia.edu

K. Jiang
Energy Research Institute, National Development and Reform Commission, B1403, GuoHong Mansion, Muxidi Beili jia No.11, Xicheng District, Beijing 100038, China
e-mail: kjiang@eri.org.cn

M.J. Manton and L.A. Stevenson (eds.), *Climate in Asia and the Pacific: Security, Society and Sustainability*, Advances in Global Change Research 56, DOI 10.1007/978-94-007-7338-7_4,

developing states (SIDS). With the warming of the Pacific and Indian Oceans and increased acidification, marine ecosystems are presently under stress. Despite these trends, maintaining or enhancing food production from the sea is critical. However, future sustainability must be maintained whilst also securing biodiversity conservation. Improved fisheries management to address the existing non-climate threats remains paramount in the Indian and Pacific Oceans with sustainable management regimes being established. Climate-related impacts are expected to increase in magnitude over the coming decades, thus preliminary adaptation to climate change is valuable.

Water security has become a defining issue of the twenty-first century for Asia and the Pacific. In the case of the Himalaya-Tibetan Plateau (HTP) region, cross-border conflicts over international water rights have also led to increased geopolitical tensions. For the Pacific, the main sources of freshwater for island communities is very limited being constrained to rainwater, surface water and groundwater. There is a need for a range of effective water management strategies for dealing with water

Y. Kameyama
Centre for Global Environmental Research, National Institute for Environmental Studies, 16-2 Onogawa, Tsukuba-City, Ibaraki 305-8506, Japan
e-mail: ykame@nies.go.jp

M. Kishi
Graduate School of Fisheries Sciences, School of Fisheries Sciences, Hokkaido University, 3-1-1, Minato-cho, Hakodate, Hokkaido 041-8611, Japan
e-mail: mjkishi@nifty.com; kishi@salmon.fish.hokudai.ac.jp

L. Lebel
Unit for Social and Environmental Research (USER), Chiang Mai University, Chiang Mai 50200, Thailand
e-mail: louis@sea-user.org

H. Meinke
Tasmanian Institute of Agriculture, University of Tasmania, Level 2, Life Sciences Building (building no.16), College Road, Sandy Bay Campus, Hobart, TAS 7005, Australia
e-mail: holger.meinke@utas.edu.au

K. Morton
The Australian National University, International Relations Research School of Pacific & Asian Studies, Building 130, Hedley Bull Centre, Canberra, ACT 0200, Australia
e-mail: Katherine.morton@anu.edu.au

E. Nikitina
EcoPolicy Research and Consulting (EcoPolicy), Piluigina st., 4/1/182, Moscow 117393, Russia
e-mail: elenanikitina@bk.ru

P.R. Shukla
Public Systems Group, Indian Institute of Management, Vastrapur, Ahmedabad 380015, Gujarat, India
e-mail: shukla@iimahd.ernet.in

I. White
The Fenner School of Environment & Society, The Australian National University, Building 141, Frank Fenner Building, Canberra, ACT 0200, Australia
e-mail: ian.white@anu.edu.au

security issues ranging from more effective water governance through to enhanced community participation. Flood disasters are the most frequent and devastating and their impacts have grown in the region. For longer term disaster risk reduction planning procedures are required as integral elements for 'good governance' of floods.

Energy security in three major energy-consuming economies in Asia; namely China, India and Japan is crucial, and requires climate change mitigation policies. Both energy efficiency and renewable energy are important factors in solutions to the energy conundrum. Technological innovation and diffusion is an important component for improving energy efficiency, with the promotion of renewable energy requiring financial investment and innovation. However, costs of new technologies are likely to decrease as they become more widely adopted. Demand side management is also need to provide key solutions.

Keywords Climate and energy security • Climate and food security • Climate and water security

4.1 Climate and Food Security: Agriculture and Fisheries

4.1.1 Introduction

The world population is projected to grow from 7 billion today to 8.3 billion in 2030, with about 9 billion in 2050 and 10 billion by 2100. The growth will be concentrated in developing countries where the economic well-being of communities is also increasing. Global food production will, therefore, need to increase by more than 50 % by 2030, and nearly double by 2050 (Beddington et al. 2011). Figures from the Food and Agriculture Organization of the United Nations (FAO) indicate that the cereal stocks-to-utilization ratio in 2008 fell to 20 %, its lowest level in 30 years. The developing countries only recorded an increase of 1.1 % in cereal production in 2008 (FAO 2009b). In fact, if China, India and Brazil are excluded from the group, production in the rest of the developing world actually fell by 0.8 %. The international and regional financial institutions see a drastic reduction in resources allocated to food production, which constitutes the principal livelihood of 70 % of the world's poor (FAO 2009a).

There are 450 million smallholder farms in the world largely in Asia and several issues in recent years are threatening their livelihoods (Sivakumar and Motha 2007). The frequency and intensity of natural disasters including floods, droughts, tropical cyclones (typhoons), heat waves and wild fires have been rising in recent years (see Chap. 2). In 2008, Cyclone Nargis and Typhoon Fengshen caused significant damage to lives and property, and the first decade of the present century was the warmest since the beginning of routine temperature recording globally. Climate variability and change is probably contributing to increasingly frequent weather extremes and ensuing natural catastrophes (IPCC 2012).

Added to the impacts of increasing natural disasters on agriculture, prices for fertilizer, seeds and animal feed have risen by 98 %, 72 % and 60 %, respectively since 2006. The FAO input price index doubled in the first 4 months of 2008, compared to the same period in 2007 while US dollar prices of some fertilizers more than tripled. Small subsistence farmers in developing countries are always particularly hard-hit by soaring input prices as they have to pay more for their seeds, fertilizer and diesel without being able to benefit from higher output prices. After months of rising prices, governments in developing countries had to draw on their budget reserves and households on their savings.

Fisheries play a crucial role in providing food security and opportunities to earn income, particularly in Asia and the Pacific. According to the FAO, fish comprises about 20 % of the animal protein in the diets of over 2.8 billion people. The contribution of fish to dietary animal protein can reach 50 % in the world's poorest regions, and up to 90 % in small island developing states (SIDS) in the Indian and Pacific Oceans. The FAO concluded that, in 2008, the important role of fisheries is threatened by changes to the environment associated with increased emissions of greenhouse gases (GHGs), including higher water temperatures, increases in ocean acidification and changes in storm surges due to altered tropical cyclone (typhoon) patterns.

Fish populations are projected to respond to such variations in different ways. As has been seen from changes in the distribution patterns of migratory fishes associated with the interannual or El Niño-scale variation in the ocean environment, climate change can be expected to affect the reproduction, recruitment and growth of oceanic fish species. Climate change may also have other impacts, including cyclic changes in the production level of marine ecosystems in ways that may favor one species or group over another.

Agriculture and fisheries play a very important role in feeding the large human populations in Asia and the Pacific; with cereal crops of rice and wheat, together with capture fisheries, providing much of the food. Impacts of climate variability on livestock mainly include two aspects: impacts on animals from, for example, increasing heat and disease-related deaths, and impacts on pasture. While it is recognized that livestock is also a main source of animal protein and vulnerable to the additional stress factors of climate change in the Asia-Pacific region, it is out of the scope of the present manuscript.

4.1.2 Agriculture

4.1.2.1 Introduction

In this section, we consider the vulnerability of agriculture in Asia to climate variability and change. These stresses are additional to the difficulties arising from both internal issues, such as the availability of suitable land, and global economic issues that are increasingly affecting the viability of agriculture in Asia. Recognizing the dominance of rice as the staple food in the region, we focus on this crop and note the relevance of the discussion to other crops.

As elaborated in Chap. 2, several countries in tropical Asia have reported increasing surface temperature trends in recent decades. Although there is no definite trend discernible in the long-term mean for precipitation, some countries have shown a decreasing trend in rainfall in the past three decades. These effects add to stresses on agriculture in the region that include natural climate variability and socio-economic stresses.

In developing countries of the humid and sub-humid tropics of Asia, the current state of agriculture is characterized by stagnant yields, recurrent natural disasters, deforestation and land degradation, and poverty. Although there are different results from different studies, most assessments indicate that climate variability will have negative effects on agriculture in the Asian humid and sub-humid tropics. The impacts of climate variability and change are expected to be greatest in developing countries with current degradation of resources, poor access to technologies, and low investments in production. Cereal crop yields will decrease generally with even minimal increases in temperature. For commercial crops, extreme events such as tropical cyclones (typhoons), droughts and floods will lead to larger losses as a result of changes of mean climate.

The adverse effects of climate change on agricultural production are likely to be felt more in lower latitude countries even though the extent of temperature change there is projected to be less than at higher latitudes. These adverse effects are mainly because of a lower adaptive capacity in socio-economic systems. In the face of great uncertainties that agriculture is presented with from changes in technology, economic and social forces, we have little or no understanding of what the added stress of climate change will do. What can and will be done to adapt to change is less in the hands of the farmers themselves, and more dependent upon agro-business and the global political economy.

4.1.2.2 Rice: Asia's Real Treasure

Rice[1] is the world's most important staple food. Although mainly produced in Asia (91 %), it is consumed on all continents and its global importance and consumption is increasing. The limited scope to expand production areas coupled with increasing resource constraints (mainly the lack of, or competing demands for land and water) make it difficult to meet necessary production increases. Climate change further compounds these problems. This constitutes a huge challenge for science, policy and farmers. The provision of effective solutions is complex due to the range of issues that must be integrated when setting research, policy and management priorities. Here, we discuss some of the Research for Development (R4D) activities that are targeted towards solving the challenges of feeding an increasing population from a diminishing resource base (Foresight 2011).

Rice is the staple food crop for about three billion people and feeds roughly half the planet's population. Approximately 750 million of the world's poorest people

[1] While we use rice as a case study, most of the principles we discuss are equally valid for other crops.

Table 4.1 Global rice production (Source: FAOSTATS)

Production	World	Asia	Central and South America	USA	Africa	Europe	Australia
Mt	650	590	23.9	9.0	23.5	3.5	0.2
%	100	90.8	3.7	1.4	3.6	0.5	0.0

depend on rice to survive, according to the International Rice Research Institute (IRRI) (http://beta.irri.org/news/images/stories/ricetoday/5-4/). In 2007, global rice production reached 650 Mt, with 90.8 % of this production coming from Asia (Table 4.1).

Global demand for rice is strong and will rise further as populations increase while the global food crisis continues (Von Grebmer et al. 2008). Meeting this demand will require considerable production increases. Since the 1960s there has been a strong, near-linear increase in production, leading to a more than threefold increase in annual global production volume, made possible by the green revolution (mainly in Asia) and expansion of production area in nearly all regions (Fig. 4.1).

During the last decade some concerning trends have emerged. Not only did the El Niño event of 2002 lead to the largest annual production decline ever (29 Mt or 5 % less than in 2001), but there is also mounting concern that the potential for further production increases is limited, especially in Asia. A report by Mukherji et al. (2009) from the International Water Management Institute (IWMI) provides some sobering facts for Asia. The report concludes that "Investments to raise yields and productivity from irrigated land must be key elements of a strategy to produce the extra food needed, while safeguarding the environment from additional stresses. Alternative options, such as upgrading rain-fed farming and increasing international trade in food grains must also contribute, but they will need to be supplemented by a significant increase in production from irrigated agriculture." With a looming food crisis, demand for rice will continue to be strong, putting increasing pressure on the already stretched production resources.

4.1.2.3 The Supply Challenges

Meeting the ever-increasing demand for rice is already putting substantial pressure on resources, particularly land and water. Concerning land, further substantial expansion of paddy rice systems seems unlikely, given the strongly competing demands for the same areas from other food and energy crops, or non-agricultural land use such as urban expansion (Pielke et al. 2007). In regard to water, the situation is equally precarious, given the limited and dwindling water resources globally (Barnaby 2009; Rodell et al. 2009). A recent International Food Policy Research Institute (IFPRI) report estimates that by 2025, water scarcity could cause annual global losses of 350 million metric tons of food production - slightly more than the entire current U.S. grain crop – if urgent

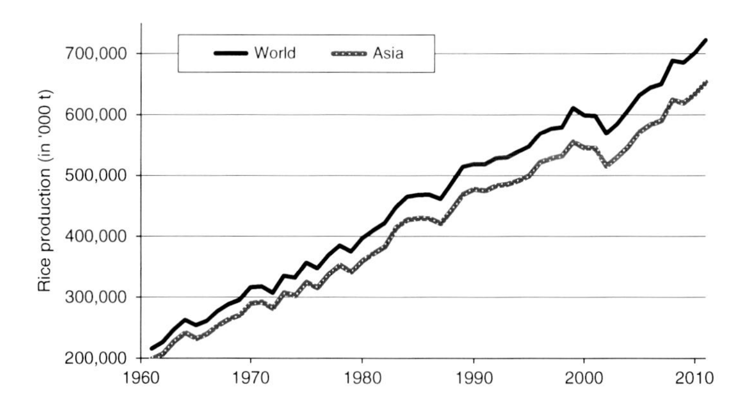

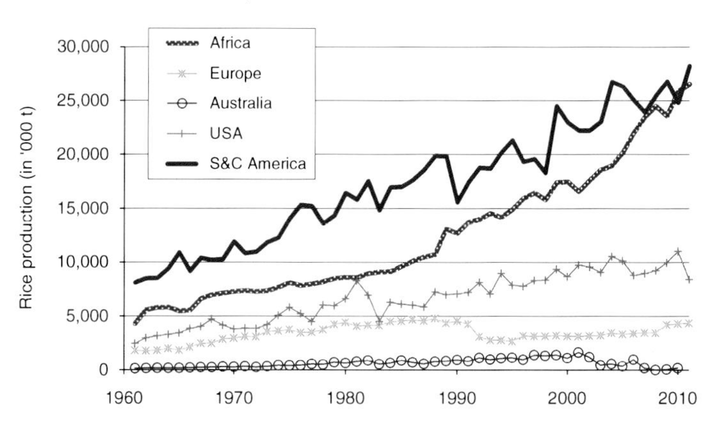

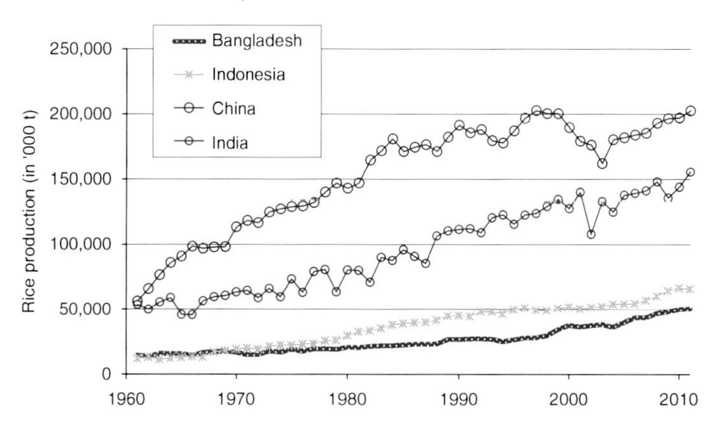

Fig. 4.1 Global and regional rice production since 1961 (Created with data from FAO; FAOSTAT Database)

measures are not taken now (http://climate-l.iisd.org/news/new-issue-of-ifpri-forum-outlines-ways-to-prevent-global-water-crisis/). This implies that future production increases must be achieved via productivity gains in terms of production per unit area and production per unit of water; producing more with less must therefore become the mantra of the rice industry.

4.1.2.4 The Environmental Challenges and Opportunities

Such productivity increases have to be achieved sustainably. For rice, this means that the resource base, i.e. the land, must be maintained; access to labor secured; and the use of other resources such as water, fertilizers and pesticides must not impact negatively on the environment and the people living in the region. Global changes are leading to increasingly limited and variable water supplies for most regions, while the frequencies of temperature extremes have already increased and continue to negatively impact production, particularly in warmer regions.

Wassmann et al. (2009) highlighted heat, drought, flooding and salinity as the key risk factors that need to be managed in order to increase rice production. However, climate change can also have positive effects, and related opportunities need to be identified and converted into productivity or efficiency gains. Specifically for rice, Shimono et al. (2007) found in their free-air CO_2 enrichment (FACE) experiments that elevated CO_2 can considerably reduce the incident of lodging (i.e. the disruption of shoots or roots) under high nitrogen supply due to shortened and thickened lower internodes. Such carbon dioxide-induced physiological changes could substantially reduce lodging-related yield losses in regions susceptible to damaging winds. At the lower end of the temperature spectrum some regions so far regarded as marginal for rice production could become increasingly important. In this respect, suitable cultivars need to be identified and best management practices need to be designed.

4.1.2.5 The Scale Challenges

Sustainable productivity increases can be achieved via a wide range of intervention actions ranging from breeding (either selective breeding or genetic modifications), improved matching of physiological traits to the environment, better management practices, improvements in irrigation technology, and to the development of local, national and global policies that encourage productivity gains. In contrast to the green revolution, this time there are no obvious technological ‘winners’. Productivity increases have to come from a combination of efforts and technologies that can be tailored to specific regional, bio-physical, economic and societal circumstances. Systems approaches based on simulation modeling offer a way to bridge this gap by providing tractable, quantifiable solutions that can be evaluated in terms of their desirability by multiple stakeholders and across scales (Gaydon et al. 2012a, b).

4.1.2.6 Responding to the Impacts of Climate Variability and Change

Climate variability and change directly impact rice production, mainly via changes in rainfall, temperature and CO_2 concentrations, indirectly through inundation associated with salinity intrusion in coastal regions, and added pressure on land resources (Wassmann et al. 2009). Although we briefly discuss each of these parameters sequentially, we stress the importance of managing the combined impacts of climate variability and change.

In terms of rainfall and water supply, water is the most critical resource for rice production: 1 kg of paddy rice requires approximately 2,500 l of water, with about 55 % accounting for evapotranspiration and the rest for run-off, drainage or leakage. Hence, projected changes in rainfall patterns could be very significant. The first decade of the twenty-first century has experienced three El Niño events (2002, 2004 and 2009). All-Indian rice yields in 2002 were reduced by 23 % compared to 2001 production levels (5 % reduction across all of Asia; Fig. 4.1), while in Australia this drought marked the end of an era for the rice industry, providing some insights into possible shifts in rice-growing regions due to climate change (Schneider 2009; Wassmann et al. 2009; Dunn and Gaydon 2011; Gaydon et al. 2011a, b). This highlights that coping strategies to better manage rice under water scarcity are urgently needed.

For non-productive water losses, many engineering or management options are available. These options range from better irrigation systems (improved irrigation channels, separate drainage systems) to fundamental changes in the way rice is grown (Rice Today, Vol. 9, 2009 with comments by Colin Chartres, Director General of IWMI at http://www.iwmi.cgiar.org/news_room/pdf/Taipei_Times.pdf.)

Among the promising new management practices are Alternative Wetting and Drying (AWD) whereby the paddy is allowed to dry, but irrigation water is reapplied before water limitations start to impact on yields; and aerobic rice systems, where especially developed rice varieties are grown in well-drained soil, just like dryland crops. This can save up to half of the normal water requirement while, with good management, yields between 4 and 6 tonnes/ha can be routinely achieved. Although this practice results in lower yields per unit area, it substantially increases the yield per unit of water.

Departing from the established paradigm of growing rice under continuously flooded conditions (a peculiar system of conserving soil fertility and facilitating management of weeds, pests and diseases) involves much more than just fine-tuning hydrology management. It requires adapting the plant to an ecosystem in which the commonly grown high yielding varieties (HYV) and hybrids are ecologically less competitive. Existing germplasm must be replaced by material deriving resilience from improved general adaptation, probably involving a host of traits that need to be identified and obtained from within or beyond the species. As was the case for the green revolution, radically new water-saving rice systems require profound innovation at both genetic and agronomic levels.

To sustain high transpiration and conversion efficiencies, transpiration use efficiency can principally be improved by two different means: by altering the crop's physiology or its genetic basis through breeding; or through breeding and crop management measures that ensure maximum yield for a given amount of transpiration (for example, by avoiding yield-reducing stress factors such as high temperatures during anthesis or pest or disease pressure). Both pathways will require knowledge of local micro-meteorological conditions and therefore model-aided geographical zoning from both a thermal and a water perspective. Reducing transpiration, per se, is not useful because it reduces growth almost proportionally and increases canopy temperature.

Modeling rice-based systems is considered a pivotal technology for innovation in R4D and, in order to design profitable and sustainable, climate-robust rice-based systems, research and monitoring activities must be supplemented by systems modeling (Hammer et al. 2002). Although several institutions such as the International Rice Research Institute (IRRI) and Wageningen University and Research Centre (WUR) have a long tradition in rice modeling, their current model, Oryza2000 (Bouman and van Laar 2006), requires substantial developmental work to handle the simulation of rice crops in complex farming systems, in rotation with other dryland and irrigated crops and pastures. A merging of two systems led to a prototype of APSIM-Oryza (Zhang et al. 2004; Bouman et al. 2004), a model configuration capable of modeling several crops sequentially. Recently, Gaydon et al. (2009) and Gaydon et al. (2012a, b) included new functionality that handles the transition from flooded to aerobic conditions.

In addressing modeling techniques for temperature impacts and stresses, traditionally low temperatures were regarded as the key yield-limiting factor for rice production (Dingkuhn and Miézan 1995; Luquet et al. 2008). However, the combination of climate change, expansion into new production areas, the increasing importance of dry season irrigated rice, the application of AWD technology and the emerging aerobic rice systems have increased the significance of high temperature constraints on production. Although the species rice is highly adaptable to diverse environments, even short spells of excessive heat during sensitive growth stages can lead to substantial yield losses. However, none of the available rice crop models are currently able to simulate this.

In the study of genetic analysis of tolerance traits for drought and thermal stresses, drought avoidance and drought tolerance mechanisms include early heading (Fujita et al. 2007), the development of short duration varieties (Kumar and Abbo 2001) and root characteristics such as thickness, rooting depth, root density, root pulling force and root penetration ability (Nguyen et al. 1997). Osmotic adjustment (Hsiao et al. 1984) and membrane stability (Tripathy et al. 2000) are also important physiological criteria for selection under water limitations. Leaf characteristics such as glaucousness (Ludlow and Muchow 1990; Richards et al. 1986), leaf size (Henson 1985) and leaf pubescence (Sandquist and Ehleringer 2003) are potentially important traits for avoiding excessive heat and regulating leaf temperature. Recently, a new class of medium statured, moderately drought-tolerant ‘aerobic rice varieties’ has been developed with high harvest indices (HI), improved lodging resistance and input responsiveness (Atlin et al. 2006). The selection for such traits should help plant survival under drought stress conditions.

Effects of water scarcity on crop growth and yield can be highly variable depending on the severity, timing and the duration of the dry spell and genetics and phenotyping of metabolic and photosynthetic pathways are being studied to address this. Breeding for drought tolerance needs model-based strategic planning that can account for local environmental conditions. For example, the impacts of certain types of drought could be mitigated via improved morphological traits such as rooting depth and root distributions (Manschadi et al. 2008). 'Supercharging' photosynthesis is another option for potential, substantial increases in yield (Mitchell and Sheehy 2006). At least in principle, it is possible to insert the C4 biochemical pathway into rice while simultaneously modifying leaf anatomy. Yin and Struik (2008) incorporated equations for C3 and C4 photosynthesis into a diffusional conductance model running within the crop model GECROS in order to evaluate the impact of the successful introduction of the full C4 system into rice.

The implications for breeding incorporating CO_2 and photosynthesis modeling are being considered. Under elevated CO_2 conditions, photosynthesis increases at single leaf and canopy level (mainly before anthesis) while respiration does not seem to be affected. As a result, more sinks can be committed, new organs are formed and biomass partitioning is modified. Despite the potential increase in yield and better competitiveness against certain weeds under elevated CO_2, rice productivity is predicted to decline under future climate due to the down-regulation of photosynthesis, the stagnation of nitrogen uptake under elevated CO_2, and the detrimental effect of increased temperature. Future crops will be exposed to further environmental challenges, resulting in strong interactions influencing factors such as nitrogen uptake.

For managing rice-based systems, a better understanding of management options is needed. Rice can shift to higher latitudes and can profit from river systems (via irrigation) so far not considered. To a lesser extent this is also true for altitude (for example, in mountainous regions of Himalaya). New opportunities might arise to produce more than one crop per year in areas where the off-season used to be too cold (for example, in sub-Himalayan river basins). Mapping this production potential and matching it with regional constraints will be an important first step in the targeted improvement of transpiration use efficiency.

4.1.2.7 Wheat in Asia

Similar principles to those in the preceding section are applicable to wheat production. Asian wheat production in 2005 represents about 43 % of the global total with the largest production in China (15 %), India (11 %) and the Russian Federation (11 %) (Table 4.2). For monsoon Asia the production is concentrated in northeast China and the Indo-Gangetic Plain in India and Pakistan.

From 1961 to 2000, wheat production increased fivefold (Fig. 4.2) in Asia but it is only about half the weight in production compared with rice. During the first decade of this century Asian wheat production slowed, with the 2002 El Niño event also impacting on production.

Table 4.2 Wheat production: world, regional and national time series (million tonne per year)

Region	2000	2001	2002	2003	2004	2005
Africa	14	18	17	22	22	21
Asia	255	247	254	245	256	268
Europe	184	202	212	154	220	208
World	586	590	575	561	632	631
Country	**2000**	**2001**	**2002**	**2003**	**2004**	**2005**
China	100	94	90	86	92	97
India	76	70	73	66	72	72
Russian Federation	34	47	51	34	45	48
Australia	22.1	24.3	10.1	26.1	21.9	25.1
Pakistan	21.1	19.0	18.2	19.2	19.5	21.6

Source: FAOSTATS

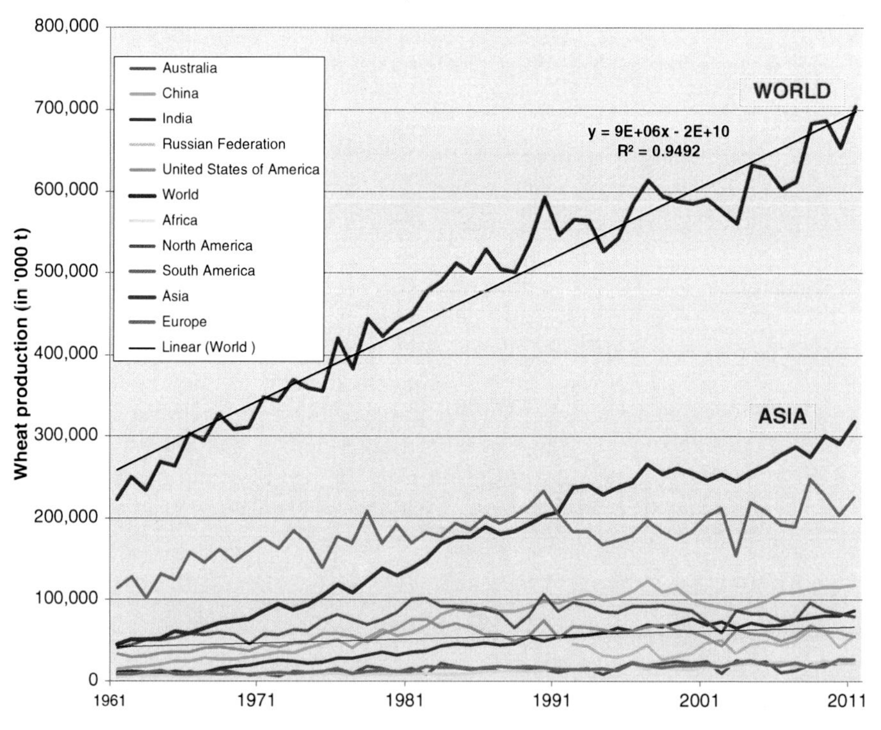

Fig. 4.2 Global and regional wheat production since 1961 (Created with data from FAO; FAOSTAT Database)

4.1.2.8 Adaptation and Mitigation

There are many general adaptation and mitigation strategies in Asia. Changes in agronomic practices; such as earlier plant dates, cultivar substitution or microclimate modification required to cool the environment, are being implemented. Further, convergence of existing cropping system models allow more studies on 'what-if' scenarios to be conducted. Particularly, we need more detailed analyses on rainy season commencement and rainfall distribution and their impacts on production.

The development of physiological-based animal models with well-developed climate components is needed urgently to cover gaps in knowledge and for future projections. To improve carbon sequestration from agriculture and forestry, adopting permanent land cover, utilizing conservation tillage, reducing fallow land in summer, incorporating rotations of forage and improving nutrient management with fertilizers are some strategies that can be considered.

Thus, potentially effective adaptive strategies recommended for the tropics include improving monitoring of crops and climate for management purposes; implementing sustainable agriculture and forestry practices utilizing efficient water conservation strategies; developing innovative new technologies alongside traditional methods; and seeking the active participation of local communities.

4.1.3 Fisheries

4.1.3.1 Introduction

Human activities affect the ocean, not only through global warming but also through pollution from industrial activities and increased fishing pressure resulting from population growth. Some marine species may be on their way to extinction with a consequent loss of diversity. Notably, after World War II, the diversity and/or biomass of marine ecosystems has shown a rapid change, e.g., Japanese herring collapsed, stocks of pacific sardine dramatically changed, short-lived species like squid or jellyfish increased, while long-lived species like whales decreased. Figure 4.3 shows recent trends of world fisheries.

4.1.3.2 Physical Characteristics of Climate Change in the Oceanic Environment

The ocean plays a very significant role in regulating the climate because its heat capacity is about 1,000 times greater than that of the atmosphere. The world's oceans have warmed substantially over the past decades with oceans absorbing approximately 80 % of the total increase in the Earth's heat (Bindoff et al. 2007). Additionally, changes in ocean biogeochemistry can feed back into the climate system, for example, through changes in the uptake or release of gases such as carbon dioxide. While the global trend is one of warming, significant decadal variations have been observed in the global time series (Trenberth et al. 2007).

Observations of changes in water temperature and salinity indicate that warming has occurred measurably over approximately 700 m of the upper global ocean, penetrating deeper into the Atlantic Ocean than in the Pacific, Indian and Southern Oceans, because of the North Atlantic deep overturning circulation (Levitus et al.

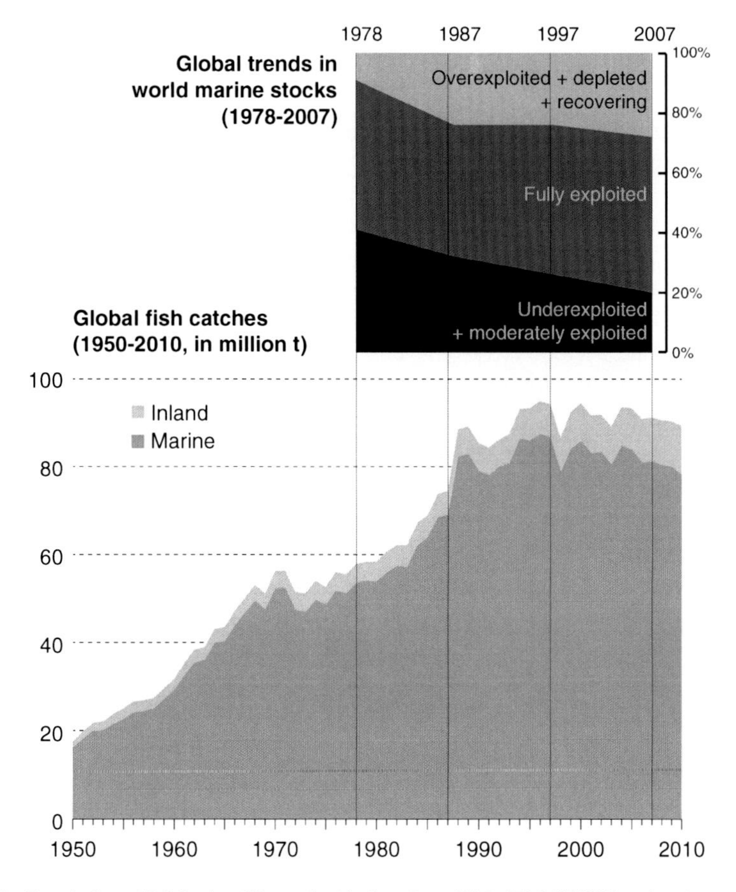

Fig. 4.3 Trend of world fisheries (Created with data from FAO; FAOSTAT Database)

2005). Subtropical seas, for example the Japan/East China Sea, are also warming. Projected patterns of atmospheric warming show greatest temperature increases over land and at high northern latitudes, and less warming over the southern oceans and North Atlantic (Meehl et al. 2007).

In the past 50 years, near-surface waters in the oceans' evaporative regions show increases in salinity in almost all ocean basins. Exceptions are high latitudes where a decreasing trend is observed due to greater precipitation, higher run-off, ice melting and advection. While salinity has increased in surface waters of the subtropical North Atlantic Ocean, freshening has occurred in the Pacific, except in the subtropical gyre, where salinity is increasing (Antonov et al. 2002).

The combined effect of temperature and salinity changes reduces the density of surface waters, resulting in an increase in vertical stratification and a reduction in surface mixing. In most of the Pacific Ocean, surface warming and freshening act in the same direction and contribute to reduced mixing (Lough et al. 2011; Sarmiento et al. 2004; Watanabe et al. 2005; Ganachaud et al. 2011).

For changes in upwelling systems, Bakun & Weeks (2008) predicted that differential warming between oceans and landmasses would, by intensifying the alongshore wind stress on the ocean surface, lead to acceleration of coastal upwelling. Contrastingly, Vecchi et al. (2006) suggest that because the poles will warm more dramatically than the tropics, the upwelling-favorable Trade Winds should weaken. With a weakening of open-ocean upwelling and an absence of enhanced coastal upwelling, the overall effect of global warming could be to decrease global biological productivity.

About half the CO_2 released by human activities between 1800 and 1994 is stored in the ocean and oceans take up emissions of about one third of modern CO_2 (Sabine et al. 2004; Feely et al. 2004). Continued uptake of atmospheric CO_2 has decreased the pH of surface seawater by 0.1 units in the last 200 years. The impacts of these changes will vary regionally and within ecosystems. It is expected to be most severe for shell-borne organisms, tropical coral reefs and cold-water corals in the Southern Ocean (Orr et al. 2005), and the entire water column in some regions of the subarctic North Pacific may become under-saturated with respect to aragonite (Feely et al. 1984).

Impacts on other marine organisms and ecosystems are less certain than the physical changes because the mechanisms shaping sensitivity to long-term moderate CO_2 exposures is not yet fully understood. Changes in pH may affect marine species in ways other than through calcification, e.g., ocean acidification has been found to reduce sperm motility and fertilization success of the sea urchin, and to affect oxygen transport and respiration systems of oceanic squid (Pörtner et al. 2005). It should be noted that the adaptability of most species to the change of seawater pH is unknown. Observations have shown that corals from the southwestern Pacific have adapted to 50-year cycles of large variations in pH, suggesting that adaptation to long-term pH change might be possible in some coral reef ecosystems.

The most obvious driver of interannual variability is the El Niño Southern Oscillation (ENSO). It is an irregular oscillation of 3–7 years involving warm and cold states that arise from the interaction between the atmosphere and ocean. Although ENSO is a global event, its major signal occurs in the equatorial Pacific with varying intensities between events. El Niño events are associated with abnormal patterns of rainfall, changes in easterly winds across the tropical Pacific, and consequent changes in marine ecosystems (Lehodey et al. 2006).

The most noticeable teleconnections over the Northern Hemisphere are the North Atlantic Oscillation (NAO) and the Pacific-North American (PNA) patterns (Minobe 1999). Like the NAO, the PNA is an internal mode of atmospheric variability. The PNA is related to an index of North Pacific sea surface temperatures variability, known as the Pacific Decadal Oscillation (PDO). The NAO and PNA accounts for approximately 35 % of the climate variability during the twentieth century (Quadrelli and Wallace 2004).

Changes in climate variability patterns in the North Pacific are often referred to as regime shifts. The index generally used to identify the shifts is based on the Pacific

Decadal Oscillation (PDO; Mantua et al. 1997; Mantua and Hare 2002). The 1977 regime shift led to changes in surface wind stress, cooling of the central Pacific, warming along the west coast of North America, a deepening of the mixed layer and decreases in Bering Sea sea-ice cover. Ecosystem signals included a doubling of chlorophyll-*a* in the central North Pacific (Miller et al. 2004; Venrick 1993).

4.1.3.3 Climate and Ecosystem Considerations

Most marine animals are cold-blooded and therefore their metabolic rates are strongly affected by temperature. Increases in temperature may increase growth rates and food conversion efficiency. The thermal tolerance of marine organisms generally exhibits an optimum range, with poorer growth at temperatures that are too high or too low. Takasuka (2004) suggested that differences in optimal temperatures for growth during the early life stages of Japanese anchovy (~22 °C) and Japanese sardine (~16 °C) could explain the shifts between the warm "anchovy" regimes and cool "sardine" regimes in the western North Pacific Ocean.

Similarly, spawning times and locations have evolved to match prevailing physical (such as temperature, currents) and biological (such as food) conditions that maximize the chances for a larva to survive to become a reproducing adult (Heath and Gallego 2010); or at the very least to minimize potential disruptions caused by climate events. For example, climate change is likely to induce strong selection on the date of spawning of Pacific salmon in the Columbia River system and growth of Japanese chum salmon through zooplankton concentration (Kishi et al. 2010). Advances have been achieved in climate forecasts used to provide fisheries management advice (Fluharty 2011).

As mentioned above, observations and model outputs suggest that climate change is likely to lead to increased vertical stratification and water column stability in oceans, thereby reducing nutrient availability to the euphotic zone and a reduction of primary and secondary production (Roemmich and McGowan 1995). However, in high latitude regions the resultant increased stability of the water column and increased growing season can have a positive effect on production.

Hashioka and Yamanaka (2007) modeled the Northwest Pacific region under a global warming scenario and predicted increases in vertical stratification and decreases in nutrient and chlorophyll-*a* concentrations in the surface water by the end of the twenty-first century. Their results suggest that the onset of the diatom (large phytoplankton) spring bloom will take place one half-month earlier than presently, and that the maximum biomass in the spring bloom will decrease significantly. In contrast, the biomass maximum of small phytoplankton at the end of the diatom spring bloom remains unchanged relative to present conditions, because of their ability to adapt to low nutrient conditions.

In the tropical Pacific, models have been developed to understand the links between climate, primary and secondary production, forage fish and tuna, such as skipjack and yellowfin. Key to these models is the definition of a suitable tuna habitat that is linked to varying regimes of the principal El Niño-La Niña climate indices, and the Pacific Decadal Oscillation. Coupled biogeochemical models (Lehodey et al. 2006) indicate the decrease of equatorial upwelling, which may have caused primary production and biomass decreases of approximately 10 % over the last three decades.

For secondary production Richardson (2008) provides a general review of the potential climate warming impacts to zooplankton. Shifts in biomass have been observed in the North Atlantic (Beaugrand and Reid 2003), the North Pacific (Karl 1999) and the Indian Ocean (Hirawake et al. 2005). In the California Current system McGowan et al. (2003) show a large, decadal decline in zooplankton biomass, along with a rise in upper-ocean temperature, as well as changes associated with the abrupt 1976–1977 regime shift.

Globally planktonic records as related to distributional and phonological shifts, show strong shifts of communities associated with regional oceanic climate regime shifts, as well as poleward range shifts and changes in timing of peak biomass. Some copepod communities in the eastern North Atlantic Ocean and European shelf seas have shifted poleward by roughly 1,000 km (Beaugrand et al. 2002). The northern limit of California sardines in Canadian waters is related to sea surface temperature, expanding north during summer months and returning south when sea temperatures cool (McFarlane et al. 2005). Skipjack tuna in the western Pacific alter their distribution to follow the convergence zone between the tropical Pacific warm pool and the eastern Pacific cold tongue as it moves in response to ENSO cycles (Lehodey et al. 1997).

The demographic timing of zooplankton in Northeast Pacific (Mackas et al. 2007) is strongly correlated with the temperature that the juvenile zooplankton encounters during early spring. The researchers found that in the sub-Arctic North East Pacific the strongest responses observed are poleward shifts in centers of abundance of many species and changes in the life cycle timing. Transfer of primary and secondary production to higher trophic levels such as commercially important fish species depends on the temporal synchrony between trophic production peaks in temperate systems. Shifts in the timing of blooms of primary or secondary producers as well as higher trophic levels can cause a mismatch with their predators (Cushing 1995).

Climate change is likely to affect ecosystems and their species both directly and indirectly through food web processes. Whether direct or indirect, processes are dominant and likely to vary between systems, and depend on whether the ecosystems are structured from the top down, from the bottom up or from the middle out (Cury et al. 2003). For example, increases in the frequency of blooms of gelatinous zooplankton have been observed (in the Bering Sea: Brodeur et al. 2008). In the tropical Pacific, it appears that direct effects on the dominant pelagic fish species predominate, whereas food web processes are more significant in the western Gulf of Alaska and even more so in the Barents Sea (Ciannelli et al. 2005). Takasuka et al. (2008) showed that multi-species comparisons extracted a simple and direct pathway: if viewed at large scales, direct temperature impacts on vital parameters may explain the regime shifts of multi-species of small pelagic fish in the western North Pacific (see Fig. 4.4).

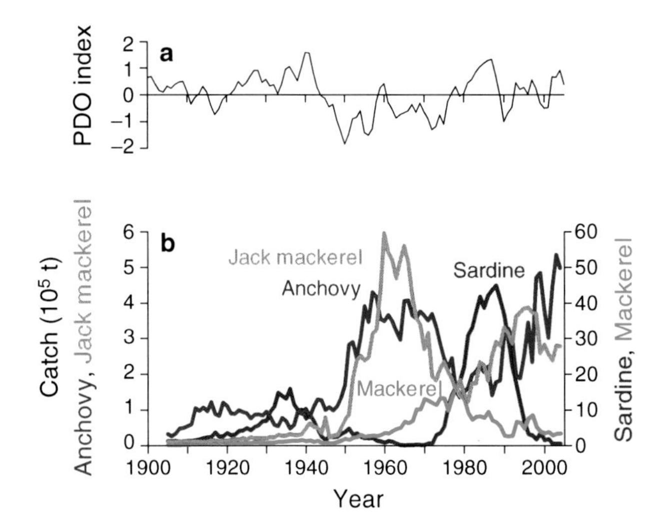

Fig. 4.4 Image of climate-related regime shift (Source: Takasuka et al. 2008). Reprinted with permission from "Multi-species regime shifts reflected in spawning temperature optima of small pelagic fish in the western North Pacific," by A. Takasuka, Y. Oozeki, and H. Kubota, 2008, Marine Ecology Progress Series, 6(2), p. 3. Copyright 2008 by Inter-Research Science Center

4.1.3.4 Examples of Fish Ecology Change and Climate Change

Salinger (2013) has reviewed the effects of ENSO, PDO, Indian Ocean Dipole (IOD), and the Southern Annular Mode (SAM) on variability of oceanic fisheries. ENSO events affect the distribution of tuna species in the equatorial Pacific, especially skipjack tuna as well as the abundance and distribution of fish along the western coasts of the Americas. The IOD modulates the distribution of tuna populations and catches in the Indian Ocean. The SAM, and its effects on sea surface temperatures influence krill biomass and fisheries catches in the Southern Ocean. The response of oceanic fish stocks to these sources of climatic variability can be used as a guide to the likely effects of climate change on these valuable resources. The decadal change of air temperature together with the periods when Japanese sardine was abundant was shown in by Klashtorin (2001), suggesting that climate change is related to pelagic fish abundance. Sakurai et al. (2000) pointed out that during the warm regime (i.e. the period when the sea surface temperature anomaly is positive) there is a good abundance of common squid and this is a consequence of good vertical temperature structure of the spawning ground.

Here, we highlight some examples of changing fish ecology with climate change.

(a) *Pacific tuna*

Oceanic fisheries of the tropical Pacific Ocean are of great importance to many of the economies and people of the region (Gillett et al. 2001; Gillett 2009; Gillett and Cartwright 2010). In the waters surrounding the Pacific Islands, the four main species that underpin these oceanic fisheries, skipjack tuna, yellowfin tuna, bigeye tuna and South Pacific albacore tuna, yield combined harvests well in excess of 1 million tonnes each year, and support fishing operations ranging from industrial fleets to subsistence catches. Across the wider Western and Central Pacific Ocean (WCPO) these four species of tuna provide catches of about 2.5 million tonnes a year (Williams and Terawasi 2010).

The Western and Central Pacific Fisheries Commission (WCPFC) was established for the conservation and management of migratory fish stocks in the Western and Central Pacific. They published time series data "WCPFC Tuna Fishery Yearbook" that presents annual catch estimates in the WCPFC Statistical Area from 1950 to 2010 (http://www.wcpfc.int/statistical-bulletins). They have tag data as well as fisheries and are proposing sustainable fisheries in future.

The most recent stock assessments for skipjack tuna show that this species is currently exploited at a moderate level relative to its biological potential. Similar conclusions have been drawn for albacore (Hoyle 2008) and yellowfin tuna (Langley et al. 2009). For bigeye tuna, assessments are less optimistic with the current estimated fishing mortality rates significantly greater, and now overfishing is occurring for this species (Harley et al. 2010).

Climate variability is important and the large-scale, east–west displacements of skipjack tuna in the equatorial Pacific are correlated with ENSO events (Lehodey et al. 1997). The extension of the warmer water preferred by skipjack tuna to the east during El Niño episodes results in greater catches of this species in the region where the Warm Pool and Pacific Equatorial Divergence (PEQD) converge: this convergence appears to promote the aggregation of the macro-zooplankton and micronekton that are the prey of skipjack.

Preliminary simulations of the potential impact of global warming on tuna populations using the model SEAPODYM are presently available only for skipjack and bigeye tuna. These simulations are based on atmospheric CO_2 concentrations reaching 850 ppm in 2100, and historical data between 1860 and 2000. For the entire population of skipjack tuna, the projections are that density will decrease in the waters of Papua New Guinea and the Solomon Islands, and increase in the eastern equatorial Pacific.

The biology and ecology of yellowfin tuna can be considered to lie between those of skipjack and bigeye tuna. Therefore, the effects of climate change on yellowfin tuna should be similar to those already described for these two species. In particular, there is expected to be a progressive extension of spawning grounds towards mid-latitudes and the central equatorial Pacific, and deterioration of foraging habitat in the WCPO. These changes are projected to result in a decrease in total biomass of yellowfin tuna in the WCPO, and an increase in the EPO.

(b) *Common squid*

In the near future, common squid may become the most dominant fish around Japan (Sakurai, personal communication). The life span of common squid is 1 year, and its abundance is strongly related with the recruitment of juvenile, the success of which depends on the temperature structure of the spawning area. Their food (prey) changes from plankton to small fish within a year. Squid is a key species in the marine ecosystem. Sakurai et al. (2000, 2003, 2007) pointed out that recruitment of common squid, especially winter-spawned groups, become worse during cold regimes. This observation is supported by laboratory experiments (Sakurai et al. 2007; Miyanaga and Sakurai 2006).

The most important condition for the survival of juveniles is the vertical stratification of sea water on the continental shelf, which is a main spawning ground for Japanese common squid. Based on this hypothesis, Kishi et al. (2009, 2010) using physical outputs by Kawamiya et al. (2007) suggested the spawning area of common squid would not shrink but that its spawning season would shift towards spring with global warming.

(c) *Walleye pollack*

Based on the same predicted physical data, Kishi et al. (2010) described the possible habitat area of walleye pollack around Japan. The stock of the Tohoku area (off the northeastern Honshu Island), which is decreasing even now, is estimated to decrease further by 2050. And by 2100, all stocks except the Pacific coast and Sea of Okhotsk stocks around Hokkaido are estimated to collapse.

(d) *Chum salmon*

Japanese chum salmon (*Oncorhynchus* spp.) released in Hokkaido migrate towards the Bering Sea through Sea of Okhotsk and the northwestern Pacific. After 3 or 4 years, they return to their birth place (i.e., the river where they were released) to spawn. Based on the ecological data from Kaeriyama (1986) and Fukuwaka et al. (2007), Kishi et al. (2010) pointed out that the area of optimum temperature for feeding (8–12 °C) will shrink especially in the Bering Sea in summer and also the optimum area for juvenile in the Sea of Okhotsk will disappear by 2050. This means released chum salmon will not be able to survive unless they change their ecology; for example, changes in migration depth.

4.1.3.5 Management Based on Ecosystem Model

The previous examples are based on the present knowledge of ecology and our models are constructed based on present conditions. We cannot accurately predict future changes of fish behavior. GLOBEC (Global Ocean Ecosystem Dynamics, which was a core project of IGBP) and other international projects like PICES (intergovernmental North Pacific Marine Science Organization) have proposed different models that can be used to predict the future status of marine ecosystems and the management of marine resources (GLOBEC 2010).

Small pelagic fish (sardines and anchovies, etc.) constitute a great portion of the fisheries catch in Japan, Korea, the East Coast of China and the Pacific coast of Russia. Young fish recruit in the coastal areas of the Yellow Sea, the Sea of Japan (Korean East Sea) and the Pacific coast of Japan, and the adults expand to occupy the region from Sakhalin (Russia) to the South Coast of China. Dramatic and synchronic fluctuations in their catches have occurred in the above four countries in the last few decades. Research suggests that these stock fluctuations are caused by changes in ocean climate. Understanding how the productivity of small pelagic fish stocks is linked to ocean climate and predicting the productivity cycles of the ocean is essential to avoid overexploitation when the ocean shifts to a period of low productivity (Nakata and APN 2001).

Predicting and understanding effects of global climate change on ecosystems and fish production in oceanic systems is essential to develop quantitative approaches to managing sustainable marine resources. Werner and APN (2004) used a combination of existing data sets, trophodynamic models and a multi-decadal model hindcast scenario to assess changes in ecosystem structure and function of certain regions in the North Pacific. The project's overall hypothesis is that global climate change can alter both the structure and function of the marine ecosystem, causing changes in energy cycling, plankton composition and dynamics, and ultimately fish production.

4.1.4 Conclusions

4.1.4.1 Agriculture: Research

A number of conclusions can be gleaned specifically for research-based activities. These include a number of methods such as:

- Improving the integration of climate, meteorology, and agro-climatology tools with livelihood zone maps and baselines to improve information on vulnerability to weather and climate phenomena;
- Studying the sensitivity of farm incomes to climate variability and change and the accumulated effects on livelihoods of repeated climate-related impacts over time;
- Including farmer livelihood assets in climate impact assessments and development of response strategies;
- Transferring scientific advances into operational and practical applications;
- Interacting with humanitarian, disaster risk reduction, development and other key stakeholders to develop and apply tools for practical applications of risk management strategies including disaster risk reduction and preparedness; and
- Improving climate models, GCMs, and resultant forecasts.

4.1.4.2 Agriculture: Seasonal Forecasts for Farming Community

For farming communities, seasonal forecasting adaptation strategies can be incorporated that may be useful in addressing issues faced by community-based farmers. These include:

- Mainstreaming climate information, including seasonal forecasting, into agricultural research and development strategy;
- Developing the capacity to effectively use existing climate and demand information to meet emerging needs;

- Ensuring that farmers and the agricultural sector have ownership and an effective voice in the development of climate information products and services; and
- Ensuring that climate services target and foster coordination among an expanded set of applications of seasonal forecast information, (e.g. coordinating input and credit supply, food crisis management, trade, and agricultural insurance).

4.1.4.3 Fisheries: Future Directions

The consumption of fishery products in Northeast Asia has been rapidly increasing as the overall economic situations have improved; however, there are differences among countries within the region. Fish consumption in China is rapidly increasing, but annual consumption (about 25 kg per person since the late 1990s) is still lower than that of Japan and Republic of Korea (Kim 2010).

In recent years, human populations in Asia's East, including China, Japan, and Republic of Korea have comprised about a quarter of the world total, and total fishery yields (aquaculture and capture fisheries) and consumption of fishery products are about 40 % and 33 % of the world total production, respectively (Kim 2010). Fisheries management plans must be implemented not only regulating Total Allowable Catch (TAC) but also via ecosystem approaches, considering both changes in climate and in social systems.

Fisheries effects as well as interannual to decadal variability may mask any effects of climate change on the distribution and abundance of the fish species supporting oceanic fisheries over the next few decades. Over longer periods the effects of climate change are expected to dominate. Therefore, the projected effects of climate change on oceanic fisheries should be evaluated now. The importance of addressing the risks posed to fisheries by other drivers in the near- to mid-term is illustrated well where the main aim is to provide fish for food security. In such situations, the effects of population growth will have a stronger effect on exploitation rates and the availability of fish than climate change.

Although the effects of climate change are expected to take time to emerge, managers must start to integrate the projected implications of climate change on fish distribution and abundance with sustainable harvest practices now. This will help avoid the coincidence of excessive fishing pressure and unfavorable climatic conditions for target species.

Action is needed now because the management measures needed to confer increased resilience to stocks can take many years to become fully effective, depending on the life span of the species. Salinger and Hobday (2013) suggest there are two important and linked preparatory steps that will provide choices for preserving future food harvesting options from oceanic systems (a) preparing management instruments and jurisdictional arrangements, and (b) improving observational capability to support management decisions. It is encouraging to see that several countries are actively moving towards eliminating overfishing, rebuilding overfished stocks and evaluating the vulnerability of their fisheries to climate change.

4.2 Climate and Regional Water Security

4.2.1 Introduction

Water security has become a defining issue of the twenty-first century for Asia and the Pacific. The recent decade-long drought in Australia, for example, resulted in less run-off into rivers, streams and dams, and as a consequence led to severe water restrictions for many regions and cities in Australia. This section examines two jurisdictions in the Asia-Pacific region, namely the Himalaya-Tibetan Plateau (HTP) and the Pacific Island Countries (PICs), which have come under greater scrutiny in recent years in relation to climate change and its impacts on the region's fragile water resources. In the case of the HTP region, cross-border conflicts over international water rights have also led to increased geopolitical tensions between countries equipped with nuclear weapon capabilities.

4.2.2 Water Security in the Himalaya-Tibetan Plateau

The Himalaya-Tibetan Plateau (HTP), also known as the Water Tower of Asia, covers an area of approximately 7 million km^2 and is home to the largest and highest mountains and glaciers in the world. The region's glaciers, snow and ice represents the world's third largest store of freshwater after the polar ice caps; it is for this reason the region is also referred to as the Third Pole. Melt water from this huge reserve feeds five major Southeast Asian river basins – Indus, Ganges, Brahmaputra, Yangtze and Yellow, home to an estimated 1.4 billion people or approximately 20 % of the world's population. Each river and associated basin have their own unique hydrological characteristics, which provide a valuable source of water for irrigation, drinking, power generation and industry. Glacial ice melt is a major contributor to stream flow, particularly in dry seasons and is an important source of water for around a quarter of the world's population living in the most rapidly developing region on earth (Behrman 2010).

Increasing population growth and urbanization has led to an increase in demand for electricity generation and food production, which in turn has placed increased pressure on the region's fragile water resources. The HTP also experiences a number of natural disasters of which the most notable of these are earthquakes, droughts and floods. The region is particularly sensitive to climate change, which has exacerbated current stress factors leading to the displacement of people from degraded farmland and fishery habitats. Water security is without doubt the most contentious issue facing the region and is expected to worsen over the coming decades as climate change will continue to have major implications for river flows, the pattern of seasonal flows and discharge to the ocean and associated impacts; for example, salt water intrusion (Behrman 2010).

Apart from precipitation, hydrological processes in this region are largely determined by snow and glacial melt, which in turn are affected by changes in air temperature, as well as changes in precipitation. Regional changes in the hydrological cycle have already been observed in response to a changing climate and further changes have been anticipated in the future (Singh and Bengtsson 2004; Rees and Collins 2006; Asokan and Dutta 2008; Ma et al. 2009; Kehrwald et al. 2008; Gosain et al. 2010; Immerzeel et al. 2010). Given the impacts of climate change on hydrological processes across the HTP, there is an urgent need to develop comprehensive adaptive response strategies to safeguard valuable water resources through improved institutional capacity and cross-border water governance. There is also a need to develop a comprehensive research agenda that facilitates the exchange of information and knowledge to enhance research efforts across the region.

4.2.2.1 Changes in Temperature and Precipitation

The latest Intergovernmental Panel on Climate Change (IPCC 2007a) report states that *global warming is unequivocal* and is due largely to an increase in greenhouse gases, such as carbon dioxide (CO_2), caused by burning fossil fuels. Research has shown that the HTP region is particularly sensitive to global warming. Meteorological observations from across the HTP have shown that atmospheric temperatures have been warming in the Himalayas over the past 100 years (Yao et al. 2006), with an acceleration in temperature that is greater than the global average (Liu and Chen 2000), suggesting there are other mechanisms, such as air pollutants, also impacting localized warming (Lau et al. 2010).

Apart from increases in atmospheric temperature, changes in the region's major climate system, the South Asian summer monsoon, has also been observed. The South Asian monsoon system is responsible for delivering around 75 % of the region's total annual rainfall (Dhar and Nandargi 2003). Over the last 30 years there has been a declining trend in annual mean rainfall of 3.2 mm/year for Central Northeast India (Subash et al. 2011) and, similarly, severe droughts have also become more notable in places such as Pakistan. One of the worst droughts on record occurred in Pakistan during 1998–2002 with water availability falling by as much as 30 % (Manton et al. 2011). Where droughts are a common problem in regions such as Pakistan and India, devastating floods are a major humanitarian problem for low lying countries such as Bangladesh and Thailand. Recent flood analysis work has shown that the frequency of floods in countries such as Bangladesh is increasing (Manton et al. 2011).

Some climate modeling has projected a delay to the start of the monsoon season and suppression in summer precipitation in response to an increase in GHG concentrations (Ashfaq et al. 2009). However, predicting changes in the complex dynamics of the Asian Monsoon climate system in response to global warming is difficult and, therefore, many uncertainties are associated with the ability to predict changes in future precipitation (Bandyopadhyay 2009). Predicting extreme events is also somewhat problematic.

4.2.2.2 Impacts on Hydrology

As expected, the observed rise in atmospheric temperatures has resulted in changes to the region's ice and snow cover, which in turn, has had a direct impact on the region's water resources and will continue to do so as atmospheric temperatures continue to rise. Higher temperatures are contributing to permafrost melt and landslides as well as the rapid retreat of small glaciers at lower elevations with severe implications for seasonal river flows. The Himalayan glaciers are retreating albeit at varying rates, depending upon elevation and snowfall (Fujita and Nuimura 2011; Scherler et al. 2011; Bajracharya et al. 2007). There has been an accelerated increase in the rate of glacial retreat attributed to warming in recent decades (Jain 2008) with estimates of glacial retreat of around 10–60 m per year and with small glaciers at lower elevations melting at a much faster rate (Bajracharya et al. 2007). Even more worrying is the rate at which glacial retreat is taking place, with rates of ice and snow melt exceeding those in other parts of the world (National Snow and Ice Data Centre 1999 up-dated 2009). It has been estimated that around 20 % of the glacial area may shrink by the year 2030 if rates of global warming remain constant (Bajracharya et al. 2007).

Glacial melt is a major contributor to river flow and is the principal source of water during the dry season for much of the region's population. The contribution to river flows from snow and glacial melt varies from between 5 % to 45 % of average flows. For the Indus and Ganges basins, around 40 % of melt water is derived from glaciers, which is much less than the contribution from the other basins (Immerzeel et al. 2010). During the dry season the contribution from glacial melt becomes crucial with around 80 % of dry season flow attributed to glacial melt (Jaitly 2009). The effects of climate change on water security will have its greatest impact on the large population (around 60 million) living along the Brahmaputra and Indus rivers because of their high dependency on irrigated agriculture and melt water (Immerzeel et al. 2010).

Although glacial melt is an important contributor to river flow, it is also responsible for a range of natural disasters such as flooding; especially glacial-lake outburst floods (GLOFs). GLOFs represent a major consequence of glacial melt, which often results in massive loss or disruption to essential infrastructure and livelihoods. Extreme deforestation and unprecedented melting of glacial snow due to higher temperatures has led to major floods in Pakistan with seven of the ten worst floods seen in the last 100 years (Manton et al. 2011). Despite the problems associated with extreme ice melt, the melting of the glaciers has also created economic opportunities for the region. Large areas of the plateau's terrain have become exposed as ice margins have receded allowing easier access and exploration of the region's rich natural resources. However, such opportunities could also lead to greater competition and an increased likelihood of new geopolitical tensions in the region (Morton 2011).

There have been a number of hydrological studies that have looked at the impacts of climate change on snow melt and rainfall contribution to stream flow in the Asian basin (Singh and Bengtsson 2004; Rees and Collins 2006; Asokan and Dutta 2008; Ma et al. 2009; Kehrwald et al. 2008; Gosain et al. 2010). The Himalayan basin is highly sensitive to climate change, which is already having an impact on hydrological processes with flow-on effects to ecosystems and livelihoods (Xu et al. 2007;

Eriksson et al. 2009). Recent modeling by the International Centre for Integrated Mountain Development (ICIMOD) on the hydrological responses to increased warming for the Brahmaputra and Koshi basins has shown a substantial increase in water yield (as much as 40 % from the baseline) attributed to an increase in snow melt and precipitation during the wet season with little or no surface water run-off during the dry months (Gosain et al. 2010). These results have also revealed that there will be an accompanying increase in sedimentation loads from increased run-off, which is likely to create a range of water management problems such as the destruction of valuable wetland habitats. However, it is widely accepted that forest cover can reduce run-off and modify stream flow patterns. Extreme land cover change such as afforestation can compensate for climate change effects at least in the short and middle terms through an increase in actual evaporation and a reduction in overland flows (Ma et al. 2010).

Although it is widely accepted that the impacts of climate change on the region's water resources will be substantial, it is difficult to generalize about the overall impact on the entire region because the hydrological characteristics differ substantially between basins and studies. Stream flow is also influenced by a number of factors other than ice and snow melt. One particular study showed that for a typical "Himalayan Catchment",[2] precipitation is more critical in determining inter-annual variability in run-off than changes in the mass balance of a glacier (Thayyen and Gergan 2010). Many of these studies have been cautious to note that these hydrological studies require further evaluation based on better spatial and temporal data sets.

Another major threat to Asia's water security is closely linked to the increasing demand for energy to power the region's industrial growth. In the last decade, the HTP has experienced an expansion of hydroelectric power generation to meet future energy demands and to sustain the region's rapid economic growth. This rapid expansion in hydroelectric power and associated dam building infrastructure projects is seen by most countries as a viable "green" alternative to using fossil fuels for power generation and has become an integral component of future development plans for many countries in the HTP region. Countries such as India and China, for example, are planning on expanding their hydropower generation capacity to help meet their current and future energy demands. Although hydropower is a low carbon form of power generation, the far reaching ramifications associated with its use are immense. The damming of rivers disrupts natural stream flow and can deprive downstream communities of crucial water supplies, particularly during periods of severe drought. Such disruption to people's livelihoods is likely to lead to an escalation in cross border conflicts between major players in the future (Cronin 2009/2010).

There is an ever increasing threat to water security from agricultural production in regions such as Pakistan, which face periods of prolonged drought. Water shortages from aggressive groundwater extraction are apparent in some regions of India, Pakistan and North China. Data from NASA's Gravity Recovery and Climate Experiment (GRACE) satellite system, which measures small changes in the earth's

[2] *A glacier catchment that experiences snowfall in winter and monsoon precipitation in summer with peak discharge from the glacier contributing to the crest of the annual stream-flow hydrograph (Thayyen and Gergan* 2010*).*

gravitation field, has revealed that water tables have been falling at a rate of 4 cm per year (net loss of 109 km^3 of water) in the Indian states of Rajasthan, Punjab and Haryana: the result of aggressive groundwater extraction practices (Rodell et al. 2009). Such groundwater extraction rates are not sustainable and are likely to result in critical shortages in food and potable water supplies leading to socio-economic instability (Selvaraju et al. 2007). Over extraction of groundwater and a rise in sea level has also led to saline contamination of groundwater supplies through salt water intrusion (Cruz et al. 2007). Further, arsenic contamination is also a major problem in regions where there is a high reliance of groundwater extraction for drinking water, as is the case in Bangladesh.

Water-use associated with agricultural production in much of Asia has also been classed as highly inefficient (Jaitly 2009). However, much can be achieved through a bottom-up approach, particularly at the local level where traditional practices can be implemented to improve water conservation. For example, rainwater harvesting and recharge of depleted groundwater stocks have yielded promising results in the Kathmandu valley (ICIMOD 2009). Policy drivers for groundwater extraction could be implemented to ensure conservation of groundwater supplies.

4.2.2.3 Environmental Monitoring for More Effective Management of Asia's Water Resources

The monitoring of environmental change in response to global warming is an important factor in the management of Asia's water resources. The HTP region has the most inhospitable terrain on Earth, which makes it the most challenging in terms of conducting research. Unfortunately, there has been little attention or resources allocated to studying the impacts of climate change on the hydrological cycle for the HTP region. Data on stream flow, snowfall, rainfall, temperature and groundwater use is either lacking or highly deficient. It is not surprising that the IPCC has defined the HTP region as 'data deficient' requiring a long term strategy to improve long-term data acquisition to address the large knowledge gaps that currently exist.

Without adequate monitoring of environmental change, the ability of downstream communities to effectively respond to natural disasters such as glacial-lake outbursts, and to address critical water shortages at short notice is dramatically reduced. Many downstream communities now have access to an arsenal of remote sensing technologies and early warning systems that alert authorities to impending dangers from glacial-lake outbursts. Of equal priority is the need for decision support systems and processes that adequately incorporate hydrological and social-economic parameters (integrative models).

4.2.2.4 Water Policy and Governance Considerations

Poor water governance and a lack of policy coordination are underlying factors contributing to water security problems in the river basins of Asia. Cross-border cooperation in the area of water governance, institutional reform to better deal with crises when they arise and long term water security measures are needed. To date, a process that allows for greater consultation between countries to enhance cooperative efforts and consider adaptive response options is virtually non-existent (Morton 2011). Likewise, a more rigorous evaluation of policy drivers is needed to avoid unintended consequences associated with the over-exploitation of water resources (Selvaraju et al. 2007). Despite this, there are some examples of successful bilateral and multilateral water treaties such as the Indus Water Treaty, which was established in 1960 to effectively deal with disputes over international water rights and obligations between India and Pakistan (Swain 2002). The emergence of the region's only multilateral response mechanism, the Abu Dhabi Dialogue, has also helped build a sense of trust and cooperation on issues affecting the rivers of the Greater Himalayans. The dialogue between experts and government officials from Afghanistan, Bangladesh, Bhutan, China, India, Nepal and Pakistan has led to a consensus around developing a cooperative and knowledge-based approach to managing Himalayan river systems (Abu Dhabi Environment Agency 2009).

Countries in the HTP region will need to adapt rapidly to climate change. The consequences of no adaptation or a lack of preparedness planning will inevitably lead to long term problems. New approaches to water governance are slowly emerging with a shift towards a more participatory approach involving greater stakeholder engagement (Nikitina et al. 2009b; 2010). However, many legally binding agreements, such as the Mekong River Basin Agreement of 1995, have failed to provide provisions for encouraging greater stakeholder involvement (Nikitina et al. 2009b). To support adaptation, the establishment of a cross-border governance framework (i.e. Multinational Water Commission) to allow for greater cooperation and knowledge sharing may be a feasible strategy.

A cornerstone to effective adaptation is the need to build resilience and to improve adaptive response strategies through better integration of knowledge and local experiences. Further, policy makers need to become aware of the broader consequences that some very specific policies might have, such as subsidizing electricity costs for pumping groundwater. The quantification of social hardship originating from extreme events such as landslides, floods, long-term drought and desertification must be strengthened and natural risk assessments and early warning systems should be adopted where possible. However, the capacity for disaster response in the HTP region is limited and there is a need to enhance the capacity of the national and local disaster management authorities. There are still many issues remaining around improving cooperation in areas such as cross-border early warming of impending disasters.

4.2.3 Water Security in the Pacific Islands

The Pacific Ocean has around 30,000 islands with population of densities ranging from 20 to 500 people per km^2 with higher population densities in urban regions (>10,000 people per km^2). There are fourteen Pacific Island Countries (PICs) situated in the tropical and sub-tropical zone. Many island communities have critical water supply and sanitation problems resulting from a range of human and natural factors. The main sources of freshwater for island communities are rainwater, surface water and groundwater. Apart from domestic use, fresh water derived from either surface or groundwater sources is also used for small-scale agricultural production and hydroelectric generation in some countries such as Fiji.

Groundwater represents the main source of freshwater, which is often limited and its quality is frequently compromised (White and Falkland 2010). Therefore, the effective management of water supply systems and processes is a major requirement for many PICs. Continued population growth in association with a changing climate will exacerbate these existing stress factors. Interim climate projections indicate that Fiji, Niue, Tonga and Vanuatu are likely to become drier by 2030 placing further stress on water resources, while Kiribati and Nauru will become wetter. Sea level rise is unlikely to have any major impact on freshwater lenses before 2030, unless there are significant physical changes (i.e. loss of land) to atolls and small islands caused by storm events.

4.2.3.1 Current Water Security Issues for Pacific Islands

Water security for PICs is impacted by natural climate variability, geological hazards as well as anthropogenic factors. Remote communities and urban centers are most vulnerable to these impacts. Prolonged droughts associated with El Niño and La Niña episodes are a major issue for many small islands leading to critical water shortages for some island countries. The effective management of water resources is a major issue in many PICs. Human activities and saline intrusion from seawater and associated mixing can affect both the quality and quantity of groundwater. Anthropogenic impacts on groundwater supplies include over extraction and contamination of fresh water supplies leading to sanitation problems regarded as being 'among the most difficult in the World' (White et al. 2008). Natural threats to freshwater lenses include drought (White et al. 2007), which leads to a reduction in the amount of freshwater available for extraction. Over-wash from storm surges and cyclones, and from extreme geological events such as tsunamis and earthquakes, are also natural factors that can compromise water security. The intrusion of salt water into freshwater lenses from storm surge events has occurred on a number of atolls such as the northern Cook Islands, Republic of the Marshall Islands (RMI) and Tuvalu. It can take up to a year or more for freshwater lenses to recover and restore wells to a potable condition (Terry and Falkland 2010).

Emergency measures such as desalination and the transportation of freshwater by boat are often deployed by some of the smaller islands during periods of extensive drought. Vandalism to valuable water supply infrastructure has also been a major problem in some islands countries (Falkland 2007) as well as significant leakages (as high as 70 %) from freshwater reticulation systems (White and Falkland 2010).

Population growth is likely to have an impact on water demand and supply in the near future. Annual population growth varies from approximately 2.3 % for Niue,

0.3 % for Cook Islands to 2.6 % for Vanuatu and 2.7 % for Solomon Islands (SPC 2011). The population in Kiribati by 2030 is likely to increase from the present 100,000 to 147,000 based on census data and growth rates (White 2011b). The current water supply situation for South (urban) Tarawa, Kiribati is already regarded as critical with existing supplies unable to meet current per capita demands (White 2011a, b). Even future expansion of groundwater extraction or rainwater harvesting would not meet future demands in 20 years, making desalination a more likely alternative to securing future water supply. Majuro atoll is also facing similar problems with existing freshwater water supplies (surface water collected from the airport, groundwater from a small freshwater lens and rainwater catchments and storages) able to meet the high level of demand during periods of drought (USAID 2009).

Although desalination is often viewed as a panacea to securing future water supply needs, historically the use of this technology to ensure adequate potable water supply has had a poor track record for various reasons (White and Falkland 2010). Desalination plants are expensive to operate due to the huge energy requirements and operating expenses. Some plants on Mauro atoll are no longer operational due to on-going costs and maintenance issues (USAID 2009). In Nauru, where demand for fresh water is often greater than supply and rainwater harvesting capacity is limited or is not available in extended droughts, there is reliance on desalination particularly during droughts. However, estimated demand for drinkable water is far greater than current production capacity of existing plants (SOPAC 2010).

4.2.3.2 Impacts of Climate Change on Water Security

This section discusses the various impacts of climatic change to water security in the Pacific and measures to deal with them, as depicted in sub-sections (a), (b) and (c); with information gleaned from various sources including the recent programme on pacific climate change science (Australian Bureau of Meteorology and CSIRO 2011).

(a) *Modeling climate change in the Pacific*
From a water security perspective, the central Pacific Island Countries are extremely vulnerable. The Pacific Climate Change Science Programme (PCCSP) has released a series of interim climate change projections for the PICs (Australian Bureau of Meteorology and CSIRO 2011), which is also addressed in Chap. 2, providing estimates of rainfall and temperature (PCCSP 2011b) through climate models.
(b) *Changes in rainfall and evaporation affecting water supply*
Based on PCCSP projections there are five countries (Fiji, Niue, Tonga, and Vanuatu) that are projected to become drier. All PICs will experience an increase in temperature and there is an expectation that higher evaporative demand will follow (PCCSP 2011a, b). The incidence of extremely hot days is also likely to increase and, as a consequence, there is likely to be a corresponding impact on water demand. The ratio of annual average rainfall to potential evapotranspiration is expected to decrease in most regions leading to increased aridity except

near the equator where there is likely to be an increase in rainfall that exceeds the smaller changes in potential evapotranspiration.

Based on these interim climate projections it is possible to make some assumptions with respect to how climate change will impact water security. The most likely impacts on water resources for the PICs would originate from changes in rainfall patterns, which in turn will affect water availability through a change in stream flow, groundwater recharge and above-ground water storage capacity. Based on the interim PCCSP projections, climate change is likely to result in an increase in rainfall with positive flow on effects for water resources in many island countries, and less rainfall with corresponding negative impacts for a small number of islands. Countries that are expected to exhibit a decrease in annual and seasonal rainfall and surface water run-off are Fiji, Niue, Tonga and Vanuatu. A decrease in rainfall is likely to impact groundwater in the case of Niue, which relies entirely on groundwater. Using a water balance approach, annual stream flow reductions in the order of up to 5 % are projected by 2030 for Tonga, Vanuatu and East Timor under the 'most likely' conditions and higher reductions (up to 15 %) are expected under the 'largest change' for these three PICs (Falkland 2011).

In terms of groundwater recharge, White et al. (2009) used a water balance approach for Tonga to determine estimates of recharge for a range of climate change scenarios relating to rainfall. Relative to 1990, it was estimated that there would be a 5–25 % decrease in recharge by the year 2095. By 2030 the recharge of groundwater on a pro rata basis would decrease by as much as 10 %. The results of this study assumed that mean annual rainfall and potential evaporation would increase and mean dry season rainfall would decrease. The PCCSP projections suggest that for Tonga, dry season rainfall will decrease with little change in mean annual rainfall. Under these climate change projections, the decrease in recharge could be even greater than the values presented by White et al. (2009).

Working on the assumption that there will be no change in ENSO activity, estimates of stream flow and recharge based on these projections are reasonable considering the uncertainties in the PCCSP interim rainfall projections and the lack of projections relating to evaporation.

When it comes to impacts relating to projected increases in rainfall intensity, there is likely to be some beneficial outcomes from enhanced groundwater recharge of freshwater lenses on coral sand and limestone islands and coastal aquifers in high islands. The negative impacts include increased likelihood of flooding and potential damage to infrastructure including water infrastructure (Falkland 2011). The likely impacts of climate change on the parameters most likely to affect water security are summarized in Table 4.3.

(c) *Sea-level rise*

There is a perception that low-lying atolls are particularly vulnerable to sea level rise (Woodroffe 2007). PCCSP modeling projections indicate that there is a very high confidence that sea levels will continue to rise in line with the global average. These projections for sea level rise show a potential mean sea level

Table 4.3 Summary of the likely impacts of climate change on the parameters most likely to affect water security (Modified from Australian Bureau of Meteorology and CSIRO 2011)

Country	Surface air temperature and sea surface temperature	Annual and Seasonal Mean Rainfall	Incident of Drought	Further Sea Level Rises	Number of Tropical Cyclones
Cook Islands	Increase	Increase	Decrease	Increase	Decline
Federated States of Micronesia	Increase	Increase	Decrease	Increase	Decline
Fiji Islands	Increase	Increase in wet season rainfall Little change is projected in annual mean rainfall	Little Change	Increase	Decline
Kiribati	Increase	Increase	Decrease	Increase	N/A
Marshall Islands	Increase	Increase	Decrease	Increase	Decline
Niue	Increase	Wet season and annual mean rainfall is projected to increase Little change is projected in dry season rainfall	Little Change	Increase	Decline
Samoa	Increase	Wet season and annual mean rainfall is projected to increase Little change is projected in dry season rainfall	Little Change	Increase	Decline
Palau	Increase	Increase	Decrease	Increase	Decline
Naru	Increase	Increase	Decrease	Increase	N/A
Solomon Islands	Increase	Increase	Decrease	Increase	Decline
Tuvalu	Increase	Increase	Decrease	Increase	Decline
Vanuatu	Increase	Wet season rainfall is projected to increase Dry season rainfall is projected to decrease Little change is projected in annual mean rainfall	Little Change	Increase	Decline
Tonga	Increase	Wet season rainfall is projected to increase Dry season rainfall is projected to decrease Little change in annual mean rainfall	Little Change	Increase	Decline
Colour legend:	Very high confidence	High confidence	Moderate confidence	Low confidence	

(MSL) rise within the range of 0.03–0.17 m by 2030 or around 0.7–4.1 mm/year from 1990 levels. Based on analysis work commissioned by the World Bank (2000), which looked at the response of MSL rise to a series of climate change scenarios for Bonriki island, Tarawa atoll, and Kiribati, a MSL of 0.2 m (near the upper limits of the PCCSP projections of 0.17 m) would have little impact on the Bonriki freshwater lens. However, a MSL rise of 0.4 m, which is over twice the upper range of the PCCSP projection showed a slight *increase* in the thickness of the lens. This increase in thickness is due to the slight upwards movement of freshwater from more permeable Pleistocene sediments to the overlying less permeable Holocene sediments (Oberdorfer and Buddemeier 1988; Alam and Falkland 1997). These projections are also based on the assumption that there is little or no change in rainfall.

A worrying impact from rising sea levels on freshwater lenses would occur from inundation, which may lead to a peripheral loss of land and/or excessive erosion from storm surge events. For example, a 20 % loss in land from inundation due to a 0.4 m rise in MSL would result in a 30 % reduction in the thickness and volume of the Bonriki freshwater lens. Assuming there is no inundation due to sea level rise, coastal aquifers and freshwater lenses will face no detrimental effects to 2030. However, beyond 2030 there is significant risk to freshwater supplies due to higher MSL rise projections. By 2100 sea level rise may be in the order of at least 0.18–0.67 m when polar ice sheet dynamics are included (IPCC 2007a). However, a sea level rise of between 0.5–1.4 m above the 1990 level by 2100 cannot be ruled out (Rahmstorf et al. 2007).

A critical question that remains is whether a MSL rise of 0.17 m by 2030 would lead to significant inundation and loss of land to the detriment of groundwater supplies. There have been a number of studies (NIWA 2008; Webb and Kench 2010; Woodroffe 2007) that have examined the impacts of sea level rise on island geography and the consequences on groundwater resources (White 2010; White and Falkland 2010). These geomorphological studies have yielded approximations that suggested atoll/reef islands may not be as susceptible to sea level rise as previously thought but they have been cautious to note that further investigation is warranted. The first quantitative study to be undertaken on the physical changes to atoll islands in response to sea level rise was by Webb and Kench (2010). Using historical aerial photography and satellite images over a 19–61 year period, their analysis showed that out of a total of 27 atoll islands presented in their study, 86 % have remained stable in response to a sea level rise (2.0 mm per year). Only 14 % of the islands featured in their study showed a net reduction in land area.

In terms of water security, a change in rainfall variability due to ENSO is a more significant factor than gradual changes in mean rainfall. Groundwater availability is particularly sensitive to climate variability associated with ENSO cycles (White and Falkland 2010). The projected impacts of warming on ENSO intensity (the main cause of climate variability in the Pacific) is still uncertain and no conclusions can be reached based on the PCCSP modeling results.

(d) *Ocean acidification*

Ocean acidification has increased and is projected to increase in the future (IPCC 2007b). While this will impact food security; for example, through the decline in coral reefs that threaten the many fish species who feed on them (Wilkinson 2008), ocean acidification is not expected to have a direct impact on water security, at least in the short term.

On water management issues, there have been a number of programmes and initiatives that account for future climate change impacts in addition to other stresses with the focus on providing freshwater supplies to satisfy demand from an increase in population and extreme climate variability. These programmes and initiatives have considered issues such as water governance, water resource management, capacity building, education and training. Such initiatives are essential to cope with the current and future stresses and pressures imposed on water resources due mainly to population expansion and pollution.

When it comes to research gaps there is much work that needs to be done with respect to data collection and management. From a water security and hydrological perspective, climate change projections used for hydrological modeling should be downscaled to account for local climate conditions.

Clearly, there is no monolithic solution to effectively address water security problems in PICs as the countries differ in their water requirements and geographical characteristics. There is a need for a range of effective water management strategies for dealing with water security issues ranging from more effective water governance through to enhanced community participation. Such strategies should consider the local situation and potential changes in population growth and climate. Finally, further investment in water supply technologies and infrastructure is required to meet future water supply demands.

4.2.4 Institutions and Politics of Flood Disaster Management

4.2.4.1 Introduction

In the Asia-Pacific region, human vulnerability to natural disasters, particularly those amplified by global change, are increasing. Flood disasters are the most frequent and devastating and their impacts have grown in spite of our improved ability to monitor and describe them. For the past 30 years the number of flood disasters has increased compared to other forms of disaster (Dutta and Herath 2004; Fig. 4.5). Floods are at the top of disaster reduction agenda in many countries.

The present section looks at the challenging problem of *how to effectively shape human responses* to floods and flood-related disasters. It focuses on the design and influence of institutions. A variety of domestic and regional institutions, including legislation, policies, agencies, coordinating, decision-making and planning procedures relevant to flood disaster risk management.

4.2.4.2 Disaster Management and Societal Vulnerabilities

Institutions – whether built specifically to address floods or disasters, or, for some other primary purpose – may influence vulnerabilities of various social groups through several pathways (Fig. 4.6). For instance loans for investments in

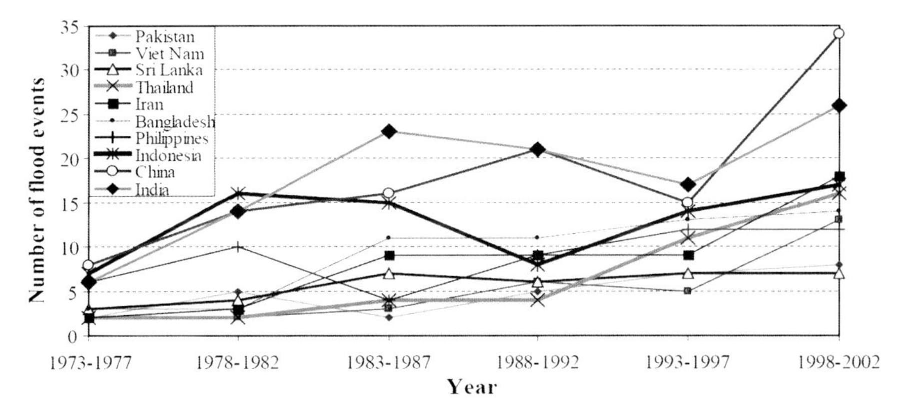

Fig. 4.5 Flood trends in most frequently flood affected countries in Asia (Source: Dutta and Herath 2004)

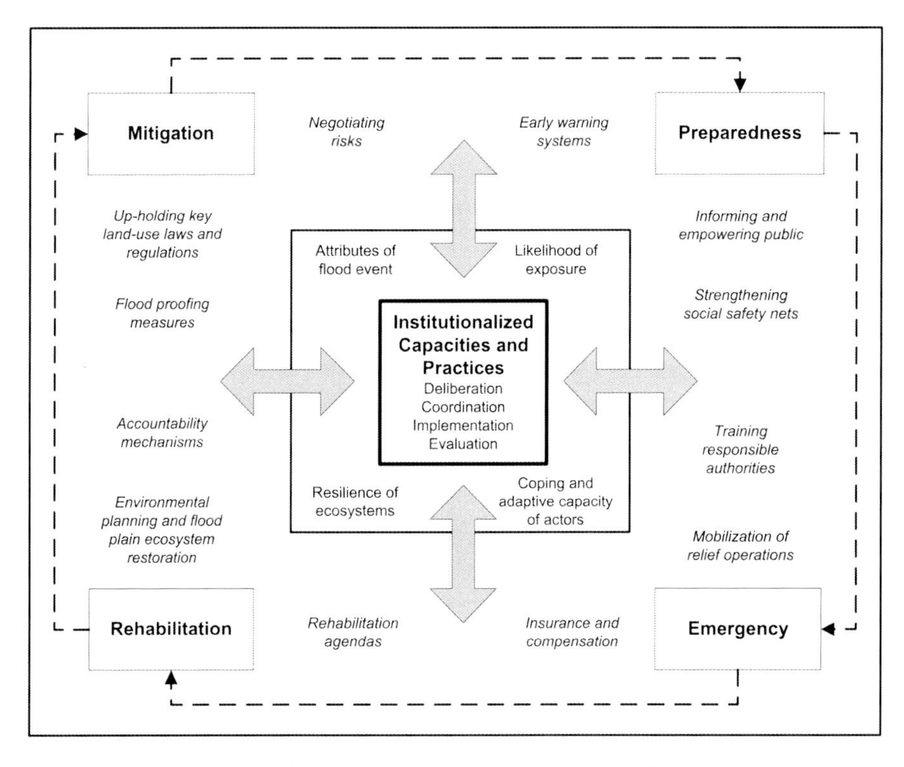

Fig. 4.6 Institutions modify vulnerabilities and hence risks of flood-related disasters through several pathways (Source: Lebel et al. 2006b) (Reprinted with permission from "Assessing institutionalized capacities and practices to reduce the risks of flood disaster," by L. Lebel et al. 2006b. Copyright 2006 by UNU Press)

structural measures and regulatory practices with respect to land use in river basins will alter the attributes of floods in terms of onsets, durations and peak flows by altering run-off, retention times and river-flow regimes. Other pathways alter how involuntary risks are distributed, either by modifying likelihoods of exposure or the capacities of different actors to avoid, cope with or adapt to floods (Lebel and Sinh 2009).

4.2.4.3 Flood Risk Management in Asia: An Institutional and Political Context

This section looks at the institutional and political context of flood risk and associated management practices in Asia and, particularly, addresses the plethora of questions as indicated in sub-sections (a) through (e):

(a) *When is a flood a disaster?*

In the tropical parts of Asia most of the major cities have grown in the deltas literally building on the foundations of a rice-growing civilization. The landscape has been managed for floods for centuries. Communities whose livelihood depends on the productive functions of "normal" seasonal flood cycles have learned to live with floods (Lebel and Sinh 2007; Sinh et al. 2009). Institutions and cultural practices around the "management" of floods are persistent, sometimes, surviving for centuries. Over the last few decades industrialization and the accompanying processes of urbanization have led to very different land-use patterns, economic structure, livelihoods and have modified flood regimes themselves (Lebel et al. 2009a, c). Floods are now perceived as much more threatening events.

As the potential for floods to be a *disaster* has increased, societies have invested more in protective structural measures (Lebel et al. 2009b). Decades of economic growth also mean that the domestic resources available to households, firms and state authorities to address "disaster" risks and events have substantially increased.

Disaster is defined in the United Nations International Strategy for Disaster Reduction as a "*serious disruption of the functioning of a community or society causing widespread human, material, economic or environmental losses, which exceed the ability of the affected community or society to cope using its own resources*"(UN-ISDR 2009). In many parts of Asia a declaration of state of emergency signifies a state's recognition of a disastrous event and often is based on loss of property and investments. Not surprisingly, an operational definition of *what constitutes a flood disaster* remains a contentious political issue (Lebel and Sinh 2007).

There are two main discourses on flood disasters (Bankoff 2004; Dixit 2003; Lebel and Sinh 2007). The first, and dominant view, is that flood disasters are inherently a characteristic of natural hazards. Disasters arise inevitably when the magnitude of a hazard is high. This contrasts with the alternative discourse that sees flood disasters as being jointly produced by interaction of the physical hazard and social vulnerabilities. This alternative discourse brings to the fore socio-economic structures and political processes that make individual, families and communities vulnerable.

(b) *Who and what is at risk?*
Framing disaster as solely a technical problem conceals the politics of shifting risk to already vulnerable groups (Lebel and Sinh 2009). Studies showed that alternative dialogues, the mass media and acts of civil disobedience may be critical to raise issues of unfair distribution of involuntary risks into the design of flood and disaster programmes. Without opportunities for deliberation women-headed households, the elderly, ethnic minorities and other marginalized groups are unlikely to benefit and may even be disadvantaged by programmes and policies aimed at reducing risks of flood disasters (Lebel and Sinh 2009). For example minority households affected by landslides and floods in Thailand were ineligible for most kinds of post-disaster assistance because they were poorly informed about correct reporting procedures or did not hold citizenship documents an apathetic state had failed to provide them (Lebel et al. 2011). In urban areas of Asia, the problems of flooding can be severe and almost chronic for slum dwellers forced into high risk zones because of a lack of low-cost housing in more desirable areas (Lebel et al. 2009c).

(c) *Who is responsible?*
In contrast to the question of "*who and what is at risk?*", the "*who will pay?*" question has been the subject of intense debate, mainly among the various levels of administrative hierarchy (Lebel et al. 2006a). Whether funds should come from local, regional or central budgets is a recurrent debate. Local governments often find they need to allocate additional resources to fund recovery and rehabilitation operations. For example, in Russia and Thailand there are clear regulations for budget requests at various levels of hierarchy depending on levels of damage. Existing problems tend to be with accountability and time lines of available funds. In Russia, the vertical division of responsibilities is institutionally fixed by national regulations, but in crisis and emergency situations provinces and locales tend to do their best to bargain with national-level administration for additional resources (Kotov and Nikitina 2005).

Mobilizing adequate funds, both for protection measures before an event and for recovery and rehabilitation of affected areas and livelihoods after, is the core "coordination" and "cooperation" issue for local authorities because it has a large bearing on their ability to implement plans. Significant gaps and problems exist in this field. There is a diversity of options in place across countries. For example, in Viet Nam, Thailand and Russia flood insurance schemes are at a very rudimentary stage so there is a strong reliance on the state to come to the rescue (Lebel et al. 2006c). In more wealthy countries, like Japan, state guarantees have allowed significant entry by the private sector into insuring against flood disasters (Kitamoto et al. 2005). Here damages are compensated by a household's comprehensive insurance provided by private insurance companies. Insurance is optional, but people who build their houses using housing loans are obliged to buy comprehensive insurance.

If local authorities have the capacity and legal framework that enables them to seek loans and private sector cooperation, then they may be able to secure more diverse funds for disaster risk management. For example, after the 2001 Lena River flood in Siberia, Russia the Sakha Republic administration applied for central bank credit for housing renovation; it also formed a partnership with the Alrossa company, which is a leading world diamond producer based in Sakha, to help rehabilitate and restore livelihoods (Kotov and Nikitina 2005). Elsewhere there are examples of non-government organizations venturing into micro-finance, training and mobilization in intervention programmes to reduce disaster risk. For example, in the aftermath of the Indian Ocean Tsunami in 2004 that caused severe coastal flooding in southern Thailand, fishing communities established "*community-shipyards*" with the support of a private firm (the Siam Cement Group) and an NGO (Save Andaman Network) (Lebel et al. 2006a). A community banking and revolving fund system were established for recovering people's livelihoods.

Being able to count on institutionalized capacities to mobilize and coordinate resources when and where they are needed is crucial in all phases of the disaster cycle, sometimes with very little scope for delay or error (Lebel et al. 2006b). Because there are many uncertainties involved in knowing where disasters will occur and exactly how they will unfold it is important that this "institutionalizing" aspect fosters flexible and coordinated responses.

Coordination of activities across phases of the disaster cycle is necessary because there is often need to link or transfer responsibilities and budgets for programs over time. One approach is through limited-life but clear objective cross-agency and multi-stakeholder task forces that can help guide these transitions.

(d) *How are risks of disaster managed?*

Assessing institutionalized capacities to effectively use resources and execute critical actions requires several different kinds of measures corresponding to different kinds of resources and actions (Lebel et al. 2006b). At the simplest and most conventional level there is a need to look at actual structural and non-structural responses made in preparing for, and responding to, flood disasters.

Good planning and coordination mean little when it comes to disaster risk reduction if there is no follow-through, which can happen as a result of corruption or other institutionalized and ad hoc incapacities preventing appropriate use and allocation of the resources available.

Forecasting and early warning systems sometimes are the weakest element in the chain of purpose-built institutions for reducing risks of flood disasters. First, there are the technical challenges of obtaining critical information and sharing that information in a timely fashion. Second, there are organizational and individual behaviors that undermine otherwise sound information-sharing arrangements. For example, in Russia in 2001, the Hydromet service provided early warning forecasts of dangerous spring thaw conditions in the Lena River basin. Local and provincial administrations in the Sakha Republic were slow in responding; as a result, the population was not well informed and losses were much higher than they needed to be (Kotov and Nikitina 2005).

In most countries national-level institutional frameworks for emergency response are well established (Lebel et al. 2006c). These incorporate a set of administrative structures, governmental programmes and legislation defining the conduct and interactions between specialized task forces that are usually well trained and able to perform effectively in extreme situations. Often the military is involved.

States differ greatly in how they view their own involvement in recovery (Lebel et al. 2006c). In transition economies like Russia and Viet Nam, the state's role remains high. Thus, in the case of the Lena River flood in Russia a combination of tools was applied, including (1) introduction of a programme to resettle populations from the affected areas, (2) allocation of financial resources from federal to provincial budgets for this purpose, (3) allocation of housing certificates from the state Reserve Emergency Fund for those affected, and (4) material compensations for affected livelihoods (although too modest to restore them) (Kotov and Nikitina 2005).

Financial resources matter hugely for plausible action. Studies argue that in Viet Nam and Bangladesh resource constraints have meant that soft measures have, by default, been pursued more frequently than costlier hard measures. Some case studies suggest that community-based initiatives can be more effective than state-led efforts but that they are not without shortcomings. Investment costs in flood protection measures clearly rise with development, for example in Japan (Takeuchi 2001).

For the most part, implementation always lags far behind promises when it comes to addressing the underlying causes of disasters (Lebel et al. 2009b). Consider, for example, issues related to housing and road construction both in mountain areas and in floodplains. Economic imperatives would argue for taking structural measures to protect these investments before disaster strikes, rather than exploring their role as contributors to disasters after the fact. Poorly constructed roads destabilize slopes or act as channels for debris in mountain areas, while in deltas and wetland areas they can prevent and alter natural drainage, thus increasing the duration and height of floods (Ziegler et al. 2004).

During post-disaster periods there is often a flurry of programmes, investments and regulation change (Lebel et al. 2009b). All such actions are far more likely to be followed-up and implemented if there is a significant group of stakeholders involved, who have a sense of ownership and responsibility. This means going beyond the project-bound logic of "implementation" until the allocated budget of the initial action is spent up, towards integrating projects and programmes into local development. In a real sense it is about *creating a sense of stewardship for disaster risk management*.

(e) *How is performance evaluated?*

The performance of institutions and organizations should be monitored and evaluated both to hold authorities accountable and to provide opportunities for organizations to learn. Studies confirm that a number of problems exist in this area (Lebel et al. 2011). For example, as a result of upland flash flood events in northern Thailand, conflicts arose with respect to irregularities, lack of transparency and accountability in compensation payouts involving the village head (Manuta et al. 2006). A mobilization by villagers was able to oust corrupt officials, but delayed compensation. Similar problems have plagued recovery processes in small fisher villages in southern Thailand after the tsunami disaster in December 2004 (Lebel et al. 2006a; Manuta et al. 2005).

An assessment framework could itself be part of an institutionalized learning process by key disaster organizations. Assessments could consult independent expert advice. Thorough and well communicated research could contribute to such evaluations. Prior to reforms in October 2002, the Thai approach to disaster was explicitly reactive, focusing on readiness and response. Since then a more proactive rhetoric has been adopted, which aims to minimize the risks and impacts by using both structural and non-structural measures that include preparedness by mobilizing the resources of the government offices, private sector and community (Lebel et al. 2009a). This development might be evidence of policy learning.

4.2.4.4 Institutional Capacities for Disaster Management

Institutional capacity of societies in the Asia-Pacific region to respond to floods is an important vulnerability indicator. In order to improve institutional capacities for flood risk reduction it is important to distinguish between the *design* and *action* of institutions. Studies show the presence of significant gaps between challenging goals of existing institutions and implementation in flood risk reduction.

(a) *Institutional practices*

Studies suggest that not only well structured and developed institutional design matters for measuring the coping capacity of a country. Equally important is an assessment of *implementation* process and outcomes of how institutions perform at various stages of a flood event. Very often gaps exist between stated policy goals and practice, or between design and action, and good intentions might turn into "dead letters", thus, contributing to increased vulnerability.

This can happen for a variety of reasons, and a broad *variety of factors* define patterns of how institutions act in practice. Identifying the impacts of these factors is critical for assessment of institutional vulnerability. Its causes are often rooted not only in the internal design of institutions, but in many cases are attributed to sets of external factors, which affect the implementation process and performance of institutions in different socio-economic and political circumstances.

Studies in countries with different contexts (developed, developing and transition economies) identify a variety of "situational factors", which significantly alter implementation process (Lebel et al. 2006c). For example, specific conditions such as financial deficiencies, administrative barriers and conflicts between organs, weakness of authority at different levels, corruption, poverty, lack of economic incentives, low public participation and awareness, unsustainable development and many others might contribute to institutional vulnerabilities in flood risk reduction. In many cases situational factors might block or alter the performance of institutions or modify already designed 'good' pathways for implementation of policies and measures – what have been called 'institutional traps' (Lebel et al. 2011).

(b) *Success and failure of institutions*
Effectiveness of existing institutions can be judged from tracking the results of their performance during particular flood events. While a variety of good institutional practices have been demonstrated in Asia, shortages still prevail. Among recent success stories about institutional performance during catastrophic flood events have been, for example, well organized emergency actions of Russian Emercom during the Lena River flood in Siberia, Russia when professionals of this agency provided rescue and relief support to 17,500 citizens of Lensk evacuated from the chill waters mixed with ice during the catastrophic spring freshet flood in 2001. Failures during the Lena River flood include the inability of local administration to properly react to early warnings of Hydromet and Emercom about a potentially difficult freshet situation in the basin. As a result, 87 settlements with over 400,000 people were severely affected by this spring flood. The number of victims could have been reduced if early and more timely warning information could have been relayed by local administration bodies (Kotov and Nikitina 2005).

There is a lot of interesting evidence about recent domestic settings and links between government and community action to increase resilience to floods. A unique innovative programme was introduced in Viet Nam. As children appear to be the major victims of floods (up to 90 % of victims), temporary "emergency kindergartens" are opened where parents can leave their children under organized adult supervision during emergencies, while they are preoccupied with securing personal belongings and other resources crucial for continuing their livelihoods after the floods. At the same time, these need to be sufficiently close or they will not (cannot) be used (Lebel and Sinh 2009).

Another success story about local institutional arrangements is the case of the Fukuoka flood event in Japan (Kitamoto et al. 2005). Local community action has been encouraged by Fukuoka city since the mid-1990s. The city supports the formation of voluntary organizations for disaster prevention within each elementary school district, and these were involved in active emergency measures during the Fukuoka flood inundation. Leaders of small communities using broadcasting equipment available in community centers urged residents to evacuate, or to move their cars to hilly areas.

4.2.4.5 Domestic Institutions

For most countries in the Asia-Pacific region, counteracting floods is at the top of the national disaster risk reduction agenda. However, institutional capacities, their designs and responses, vary across countries. Domestic socio-economic contexts and political cultures within which flood risk reduction institutions perform, are different as well (Lebel et al. 2006c, 2009b).

A variety of institutional frameworks to counteract floods are in place in a number of countries in Asia, including legislation and regulations, administrative organs at different levels, action plans and strategies, financial mechanisms, wide range of tools and measures (structural and non-structural). Some of them are positioned within broader schemes of natural disaster management, while others have special mandates and target flood mitigation (Lebel et al. 2006c, 2009b). Although in some countries the formation of flood risk reduction institutions has a very long tradition, in many others contemporary institutional settings with developed supporting infrastructure have been formed only in the last decade.

While national institutional capacity to counteract floods differs significantly across countries, there are a number of common features across countries:

- Governments at the national level are regarded as the primary authority responsible for flood events.
- Similarities exist in the structure of national institutional designs. For example, institutional organization of disaster risk reduction in Russia has many common features with those of Japan. However, the implementation results and performance of institutions might vary in some cases.
- Most government institutions involved in flood management focus attention on emergency rescue and rehabilitation of affected population and territories and reconstruction of livelihoods once the crisis occurs.
- Much less resource and attention are paid to non-crisis comprehensive flood risk reduction, which incorporates hazard risk assessment, monitoring, forecasting and prevention of catastrophic events in the flood prone areas.
- In some countries, emphasis is primarily on reducing damage to economic assets and infrastructure, rather than enhancing human security.
- There is growing attention to develop comprehensive institutional settings that can provide for flood risk reduction through combining emergency and relief actions, on the one hand, with prevention and planning to reduce flood disasters of catastrophic scales, on the other hand.

4.2.4.6 Regional Cooperation

There are a growing number of regional frameworks for cooperation relevant to disaster risk reduction in the Asia-Pacific region both among nations and non-state actors. The perception is becoming common that responses to weather-related hazards and climate change, which do not respect political boundaries, should be cross-national and draw on cooperation between countries that share common geography, history, cultures and economies.

There is no single regional organization or specialized body on disaster management in the Asia-Pacific, but a number of sub-regional arrangements, for example, for Southeast Asia, South Asia and the Pacific are in effect (UNESCAP and UNISDR 2010). Formation of cooperative frameworks at various stages of development is underway, from declarations of intent, to creation of institutions and joint practical actions. Many existing regional bodies have expanded their agendas to include disaster risk reduction. Most of the joint frameworks between the countries of the region are non-binding in character.

In Southeast Asia, a number of coordinated procedures have been adopted for joint disaster relief and emergency response and more recently risk assessment under the Association of Southeast Asian Nations (ASEAN 2010). They were tested and found helpful in practice during the Cyclone Nargis disaster in Myanmar in 2008. An ASEAN Agreement on Disaster Management and Emergency Response was ratified by its all members and came into force in 2009. More recently cooperation has moved to cover financing (ASEAN 2011).

Other sub-regional cooperative arrangements in disaster risk reduction are developing in South Asia and in the Pacific through such bodies as the South Asian Association for Regional Cooperation (SAARC) and the Pacific Islands Applied Geoscience Commission (SOPAC). In 2005 the Dhaka Declaration by SAARC envisaged development of a permanent regional response mechanism on disaster preparedness, emergency relief and rehabilitation (SAARC 2005). SOPAC has a mandate to enhance cooperation in disaster risk reduction and capacities of the Pacific countries in disaster management (see http://www.sopac.org/).

UN international bodies are active in the Asia-Pacific region. For example, the UN International Strategy for Disaster Reduction (UNISDR) cooperates with two regional bodies – Asian Disaster Reduction Center in Kobe, Japan (see http://www.adrc.asia/index.php) and Asian Disaster Preparedness Center in Bangkok (see http://www.adpc.net/2012/) (UNESCAP and UNISDR 2010); and their practices significantly contribute to better governance of disaster risks and to national and local capacity building (for example, see: ASEAN 2010). Other UN specialized agencies also have initiatives, which promote cooperation, like with the World Meteorological Organization (WMO) and its Weather and Climate Information System. This kind of cross-cooperation is growing; particularly now that natural weather events are showing extreme patterns due to climate change-related phenomena (see Chap. 2).

The Hyogo Framework for Action: Building the Resilience of Nations and Communities to Disasters (HFA), 2005–2015 adopted by 168 countries (UNISDR 2007), has been one driver for improved cooperation in the Asia-Pacific region. Coordination at different levels is underway and includes multiple partnerships and networks; from the Regional Platform for Disaster Risk Management (see http://www.unisdr.org/we/coordinate/regional-platforms) to the ISDR Partnership for Disaster Risk Reduction (IAP) (UNISDR 2011), or thematic regional task forces. Cooperation in knowledge and information exchange, in risk and vulnerability assessment, information management and networking is a typical feature of emerging regional frameworks.

A variety of non-governmental international and regional organizations are also involved in different aspects of disaster management and innovative community-based programs. Many local NGOs and civil society organizations take part. Partnerships are being formed, for example, the Asian Disaster Reduction and Response Network is a partnership of 34 national NGOs from 16 countries of the

Asia-Pacific (ADRRN 2011). Another example includes the Pacific Disaster Net, which is a regional information database launched to assist decision-making in disaster risk reduction (reference). Cross-national comparisons and tracking *common* and *country-specific* problems contribute to "what can be done" and "what innovations and reforms of institutions are needed" to increase coping capacities of society to flood risk (Lebel et al. 2006c).

4.2.4.7 Key Messages

For many countries of the Asia-Pacific region, counteracting floods is at the top of the national risk reduction agenda. Institutional capacities and responses, however, vary considerably across them. Identifying best practices and major lessons learned from rich experiences of these countries as well as the possibilities and constraints for effective risk management is an important challenge. Cross-country transfer and adaptation of best practices in flood risk reduction in the region and learning from each other are becoming more and more important.

States no longer just respond to flood disasters, but they also increasingly manage flood-related disaster risks through institutional frameworks. However, domestic institutional capacities are often still inadequate. Despite the better understanding and monitoring of disasters, property damage, and in some instances also losses of life, remain unacceptably high and are increasing. There is a growing understanding that institutions for floods risk reduction need to be based on broader involvement of stakeholders at the local level that are positioned "closer to the risk" than administrators from central government (Nikitina and Kotov 2008). Studies in Asia-Pacific confirm that part of the responsibilities can be decentralized and coordinated by local governments and communities.

There are a number of reasons why institutional changes to reduce risks of flood-related disasters have not been very successful (Lebel et al. 2006c, 2011). First, a misplaced emphasis on emergency relief to the detriment of crafting institutions to reduce vulnerabilities and prevent disasters continues to underline planning and budgeting. Second, there is a self-serving belief that disaster management is a technical problem that calls for expert judgments that systematically exclude interests of the most socially vulnerable groups. Third, there exists over-emphasis on structural measures, which again and again have been shown to be more about re-distributing risks in time and place than reducing them. Fourth, there is an apparent failure to integrate flood disasters into normal development planning in flood-prone regions. Fifth, there is persistent failure to recognize the importance of learning and developing capacity for building social and ecological resilience and for guiding individual and collective behaviors.

Are current institutional efforts leading to reduced risks of flood disasters? In this context there are overarching questions of such as *how* national and regional institutions are designed; *what* policies and measures are undertaken; and *what* can be done to enhance national-level institutional capacity to increase the resilience of local communities to hazards in the coming years. Changing climates and flood regimes make institutional assessments particular salient for policy and practice.

For longer term disaster risk reduction planning and measures, debate, consultation, public participation, representation procedures should be incorporated as integral elements for 'good governance' of floods (Nikitina et al. 2009a). This also includes addressing the underlying causes of vulnerability. State agencies usually find these very difficult to do by themselves because it requires dealing with fundamental issues of governance and social justice that may undermine positions of authority (This and other governance issues are addressed in Chap. 6).

4.3 Climate and Energy Security

4.3.1 Introduction

Carbon dioxide (CO_2) is a major greenhouse gas (GHG) accounting for 77 % of all anthropogenic GHG emissions in 2004 (IPCC 2007c) and it is imperative to focus on CO_2 emissions when considering climate change mitigation. Within the global CO_2 anthropogenic emissions, nearly 18 % of CO_2 emissions are related to deforestation, decay of biomass, etc. A smaller portion is due to industrial production processes like cement manufacture. Remaining emissions are due to the combustion of fossil fuels, such as coal, oil and natural gas, a result of our energy use. Increases in energy consumption are observed with increasing industrial activity and improvements in living standards (Fig. 4.7; Rogner et al. 2007).

Mitigation[3] of CO_2 emissions can be achieved through the introduction of renewable energy sources such as wind and solar power. Nuclear power is also an option to decouple CO_2 from energy use, but nuclear energy faces risks other than climate change, which is a contentious issue. Carbon capture and storage (CCS) is another option that can reduce CO_2 emissions while consuming fossil fuels, but CCS activities also require energy with associated costs and uncertainties relating to its long term storage and potential in situ impacts. The growth of energy consumption can be further linked to living standards. Improvements in energy efficiency can be made through energy efficiency (intensity) and conserving energy by demand side management. To date, neither increased energy efficiency nor the introduction of renewable energy has stabilized the atmospheric concentrations of CO_2 at a level that would prevent dangerous climate change.

The situation in Asia is more serious than at the global level. GHG emissions from many countries in Asia have been growing rapidly due to expanding economic activities since the early 1990s. Rapid economic growth has resulted in an increase in the absolute amount of CO_2 emissions from the region. Emissions from China, India,

[3] IPCC (2007) defines mitigation as an anthropogenic intervention to reduce the sources or enhance the sinks of greenhouse gases.

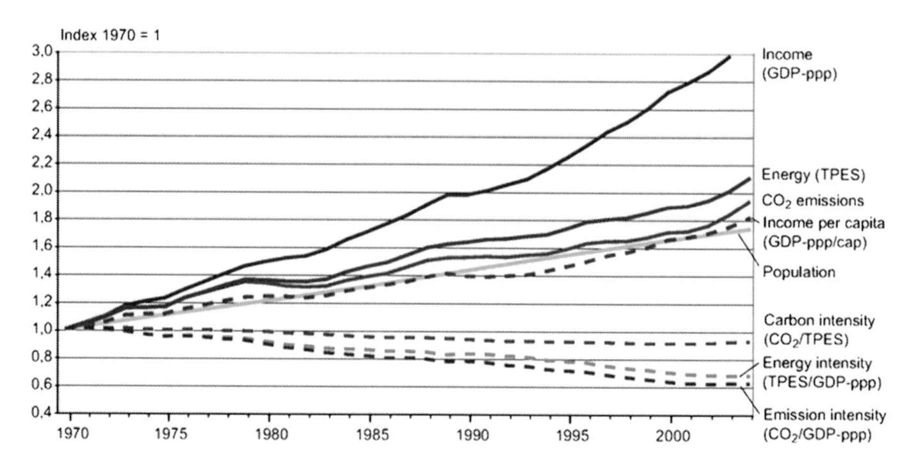

Fig. 4.7 CO_2 emissions and trends in related indicators at the global level (Source: Rogner et al. 2007) (Reprinted with permission from "Climate Change 2007: Mitigation of Climate Change. Working Group III Contribution to the Fourth Assessment Report of the Intergovernmental Panel on Climate Change," Figure 1.5., Cambridge University Press)

Japan and Republic of Korea, the four largest GHG emitters in the region, add up to about one third of all global GHG emissions. Under a business-as-usual scenario, energy demand in developing Asia will almost double by 2030 (IPCC 2007c). Emissions from energy use are projected to increase by 100 % between 2007 and 2030, at which point the region will be responsible for 45 % of all global energy–related emissions, as compared to 31 % in 2007. This means any emission mitigation policies taken in the region could make a significant difference at the global level.

From one perspective, climate change mitigation policies targeted to reduce CO_2 emissions from fossil fuel combustion are criticized because industries would be required to spend more money to replace current infrastructure and improve production processes. In some cases, mitigation policies may need to limit the absolute level of economic production. Such policies may lead to loss in international competitiveness. From another perspective, the same policies could be perceived as a means to improve energy security at the national level, and to save energy and expenditure related to energy costs at a domestic level. Improvements in energy efficiency will lead to cost saving in the long run. The relationship between climate change mitigation, CO_2 emission reduction and energy use could be perceived in a variety of ways, according to different national circumstances.

The present section examines energy security in three major energy-consuming economies in Asia; namely China, India and Japan, and seeks to discuss how climate change mitigation policies are dealt with in relation to energy use in those countries. China and India are chosen because of the sheer size of energy demand and its potential implications. Japan is chosen because of its changes in nuclear policy following the Great Tohoku Earthquake and Tsunami in March, 2011. Data for energy demand and supply in other countries in the Asia and the Pacific region can be obtained from other reference sources (Asian Development Bank 2009).

4.3.2 China

4.3.2.1 Programmes to Promote Energy Efficiency Under the Eleventh 5-Year Plan

China's demand for energy has grown rapidly since the 1990s. To meet the current energy demand, the Chinese government has implemented various policies at the national development planning level. The 16th National Congress of the Chinese Communist Party in 2002 stated that China will achieve the objective of building a society that is prosperous by 2020. Along with further population growth and associated acceleration of urbanization, the rapid development of heavy industry and transportation will lead to a significant increase in the demand for energy. The imbalance between energy constraints and economic development, and the environmental pollution caused by energy utilization could become even more evident.

In November 2004 China announced its medium to long-term Energy Conservation Plan. Encouraging societal change towards energy conservation and energy intensity reduction; the plan aimed to remove energy bottlenecks, build an energy-saving society, and promote sustainable social and economic development in order to achieve it overarching objective of building a prosperous society. The document covers the country's eleventh 5-year plan to 2010; as well as a second target period to 2020 (Chinese Government 2006a). The plan outlines the following:

(a) *Macro energy conservation indicators*
By 2010, energy consumption per 0,000 Yuan GDP (constant price in 1990, the same below) is expected to drop from 2.68 tonnes of coal equivalent (TCE) in 2002 to 2.25 TCE, with an annual average energy conservation rate of 2.2 % from 2003 to 2010. The energy conservation capacity is expected to reach 400 million TCE. Energy consumption per 10,000 Yuan GDP will drop to 1.54 TCE in 2020, with an annual average energy conservation rate of 3 % from 2003 to 2020. The energy conservation capacity is expected to reach 1.4 billion TCE; it is 11 % more than the total planned newly-increased energy production of 1.26 billion TCE during the same period, and correspondingly reduction of 21 million tonnes sulphur dioxide.

(b) *Energy consumption indicators per unit of major products (amount of output)*
By 2010, China's products as a whole was expected to reach or approach the advanced international level of the early 1990s in terms of the indicators, where large and medium sized enterprises are expected to reach the advanced international level at the beginning of the twenty-first century; and by 2020 China is expected to reach or approach the international advanced level.

(c) *Energy efficiency indicators of major energy consuming equipment*
By 2010, the energy efficiency of newly added major energy consuming equipment is expected to reach or approach international advanced level, and some automobiles, motors and household electric appliances are expected to reach the international leading level.

Table 4.4 China's target for renewable energy by 2020 in the Renewable Energy Law in the medium- and long-term Development Plan for Renewable Energy in China (2007) compared with new 2015 targets set in 2013

Energy source	2020 targets (set in 2007)[a]	New 2015 targets (set in 2013)[b]
Wind	30 GW	100 GW
Solar power PV	1.8 GW	21 GW
Solar heater	300 million m^2	400 million m^2
Biomass power	30 GW	13 GW
Biomass diesel	2 Mt	Data not available
Bioethanol	10 Mt	Data not available
Biomass solid fuel	50 million tonne	Data not available
Small hydro	80 GW	Data not available

[a]Medium and Long-Term Development Plan for Renewable Energy in China (2007). Retrieved from http://www.sdpc.gov.cn/zcfb/zcfbtz/2007tongzhi/W020070904607346044110.pdf

[b]Available from State Council http://www.gov.cn/zwgk/2013-01/23/content_2318554.htm

(d) *Objectives of macro-regulation*
In the period from 2010, the plan will establish laws, regulations and a standard system to govern energy conservation. In addition, it will develop policy support, supervision and regulation, and technical service systems suited to a socialist market economy. Key fields for energy conservation are outlined, constituting a framework for national and local governments. In order to reach the targets laid out, ten key projects required implementation during the eleventh 5-year plan:
Coal-fired industrial boiler (kiln) retrofit projects.

- District cogeneration
- Residual heat and pressure utilization
- Petroleum saving and substituting
- Motor system energy saving
- Energy system optimization
- Building energy conservation
- Green lighting
- Government agency energy conservation
- Energy saving monitoring and testing, and technology service system building

In order to support the medium- and long-term energy conservation plan, in 2005, the government set a 20 % energy reduction target between 2005 and 2010 in its eleventh 5-year plan (Chinese Government 2006a). Based on data available in 2010, the energy reduction target of 19.1 % was achieved compared with 2005.

(e) *Development plans for renewable energy and nuclear energy*
In 2005, China announced its Renewable Energy Law (Chinese Government 2005), which sets a target of increasing the share of renewable energy to 15 % by 2020. The renewable energy development plan was further developed and the major targets to be reached by 2020 are shown in Table 4.4.

According to the National Energy Plan, published by the National Development and Reform Commission (NDRC) in 2005, total nuclear power generation should be 40 GW by 2020, while at the same time new plants with 18 GW outputs will be under construction.

4.3.2.2 Recent Energy Policies and the Twelfth 5-Year Plan

China's twelfth 5-year plan (2011–2015) was approved by Congress in March 2012. In addition, Premier Wen's 2006–2010 Work Report (Chinese Government 2006b) includes both an assessment of the previous 5 years and a summary of highlights of the next 5-year plan. The analysis presented in the present section is derived from both the initial twelfth 5-year plan draft and the 2006–2010 Work Report.

Prominent in both documents was the recognition by China of the importance of climate change and energy-use with special emphasis on China's commitment to international cooperation and the UN-led climate negotiation process, including concerns of climate finance and technology transfer. The present 5-year plan also discusses the need to implement more climate adaptation-related policies, such as greater preparedness for extreme weather events.

There are separate targets for energy intensity (16 % reduction by 2015) and CO_2 emissions per unit GDP (17 % reduction by 2015) (Hsu and Seligsohn 2011). These are within the expected range and congruent with the 40–45 % reduction in carbon intensity from 2005 levels that was first announced in the UNFCCC climate change talks in Copenhagen, 2009 and reaffirmed in Cancun, 2010. Clearly defined energy and CO_2 emission targets will help ensure that provinces in China implement their energy policies clearly-articulated goals (Hou and Li 2011). The present plan and work report also include noteworthy policies in the following areas:

(a) *Forest*
China has been steadily increasing forest cover since the founding of the People's Republic in 1949. China's 5-year plan will go a significant distance toward meeting the country's Copenhagen commitment on forests (Su 2010). In the Plan the Chinese government has set a goal to increase the area of forest cover by 12.5 Mha by 2015, while in Premier Wen's Work Report, a forest stock volume goal of 600 million m^3 was established. While the forest cover goal is more or less in line with the already stated 2020 goal to increase forest cover by 40 Mha above 2005 levels, the volume stock target seems more ambitious because it seeks to achieve almost half of the 15-year target, which is of 1.3 billion m^3 by 2020.

(b) *Tracking implementation*
Premier Wen stated that China would put in place "well-equipped statistical and monitoring systems for GHG emissions, energy conservation and emissions reductions" to ensure these policies are tracked and properly implemented.

(c) *Efficiency*
China has had a particularly successful track record on industrial energy efficiency in the previous 5 years. In the new plan, there are policies to promote greater industrial efficiency, and a major push to include all other sectors of the

economy, including both new and existing buildings. For example, the plan introduces a 10,000 Enterprises Program. Following the endorsement of new types of mechanisms in the October Party Plenum Document (Seligsohn 2010), the Plan specifically endorses market approaches like energy service companies (ESCOs) that help to finance energy efficiency.

(d) *Transport*

While China certainly has plans for additional air and road transport, what is striking is its commitment to rail, both for long distance and urban mass transit. The plan includes proposals for the construction of 35,000 km of high-speed rail with the goal of connecting every city with a population greater than 500,000. There are also plans to improve subway and light rail in cities that already have urban transit systems including building new systems in at least nine other cities.

(e) *Non-fossil energy*

The plan incorporates the goal of 11.4 % non-fossil fuels in primary energy consumption by 2015 announced by Zhang Guobao in 2011 (Wang et al. 2011). China continues to exceed earlier targets in non-fossil development. For example, the 5-year target for wind-generated energy is 70 GW, which exceeds the 2020 target set a few years ago. For nuclear power generation, the plan is to install 40 GW of additional capacity by 2015. China currently has around 10 GW of installed nuclear capacity now, which means that if this 5-year target is achieved, China is likely to exceed even the expectation discussed in 2010 of generating 70 GW by 2020 (People's Daily Online 2010). If these goals are realized, China will have the world's highest capacity of nuclear energy by 2020.

While the Plan itself is general on targets, it is much more specific on policies. It assigns specific targets for cities required to reach new motor vehicle emission standards and sets goals for a wide variety of environmental infrastructure, including wastewater and solid waste treatment. There is also a strong emphasis on reuse and recycling, or what the Chinese call "circular economy." There is a clear recognition in these plans of the importance of environmental sustainability in being able to reach not just higher levels of income and but also increased well-being for the Chinese people. The Plan itself is highly specific in some areas but somewhat unclear in other areas such as in relation to target pollutants. Much of the clarity in implementation comes through sectoral plans and later regulations and guidance. In a move that exceeded expectations, China's former Minister in charge of the National Energy Administration, Zhang Guobao, announced in late 2011 that for the twelfth 5-year plan China would cap total energy use at four billion TCE by 2015 (CRIEnglish 2011). There is some speculation that China would adopt a total coal cap in the twelfth 5-year plan, but Zhang's announcement goes beyond just coal to include all energy sources.

The significance of this target is the fact that it is an absolute energy target. The actual number is closely aligned to China's announced 40–45 % 2020 carbon intensity goal (ChinaFAQs 2010) as well as its likely 16–17 % energy intensity goal for 2015. In the past the Chinese government has argued that absolute targets are difficult to project and meet, given the volatile growth of a developing country. The new target suggests more confidence that they can estimate both growth rates and energy use trends, and use these to set absolute goals. It is also significant that this new target is set as part of China's domestic policy-making process. China's domestic imperatives to use resources effectively and control the negative impacts of fuel use, including climate change and other environmental impacts, have been the driver for this decision.

4.3.3 India

4.3.3.1 History of India's Energy Supply Trends

India is the second most populous country in the world with a population of over 1.2 billion in 2011. In the past decade, the size and structure of India's energy system has altered rapidly. India's energy consumption is positively correlated to income and has followed high economic growth patterns in the past two decades, although a marginal decoupling of energy and income was witnessed during the past decade. Fossil fuels have remained the dominant source in India's energy mix, with the share of domestic coal still remaining prominent. In 2009, the consumption of coal reached 586 million tonnes, oil 160 million tonnes and natural gas 47 billion m^3 (GoI-MoPSI 2011). According to the Energy Information Administration (EIA 2011) contributions to India's energy consumption in 2009 consisted of 42 % coal/ peat, 24 % combustible renewables and waste, 24 % oil, 7 % natural gas, 2 % hydroelectric power and other renewables, and 1 % nuclear.

The fastest growth in the past decade has been in the demand for oil and gas from the transport, industry and urban residential sectors. India lacks sufficient domestic energy resources and imports much of its energy requirements. Three quarters of the oil and nearly one third of the total energy consumed in the country are imported. These rising imports are adding to energy security risks. Besides, the sustained high share of fossil fuels in the energy mix is contributing to rising emissions of local pollutants and CO_2.

The share of traditional biofuels in total energy consumption has steadily declined to below 15 %. Due to rural and peri-urban dependence on biofuels as a household energy source, biofuels are a significant portion of in India's energy system. Biofuel-use in households is the main cause of poor indoor air quality, leading to excessive health problems and fatalities especially among women and children. Nearly 300 million people, i.e. one third of India's rural population, do not have access to electricity (IEA 2011) due to inadequate infrastructure and income.

India's energy policy makers are faced with three key challenges: energy security, energy access and energy-related environmental impacts. To meet these challenges government policies have been put into practice that:

- Open up the domestic coal industry to private sector participation;
- Promote oil and gas exploration and production;

- Encourage and facilitate Indian companies engaging in long-term energy contracts and acquire energy assets globally; and
- Step up renewable and nuclear energy shares in the future energy mix.

Over the last decade, India's policy allowing compressed natural gas in public vehicles has helped ease the growing rate of local pollution. However, this has necessitated its stepping up natural gas imports thereby adding to energy security risks. While nuclear power contribution to total energy consumption is presently 1 %, this figure is expected to increase as a result of international civil nuclear energy cooperation deals.

4.3.3.2 Climate Change, Renewable Energy and Energy Efficiency in India

India has participated in global cooperative efforts to combat climate change since the early period of recognition of climate change as a global problem. India has actively worked in global scientific (for example, IPCC) and negotiating (for example, UNFCCC, Kyoto Protocol) forums. India's approach to mitigation and adaption is to align the global climate change policies and actions with the national development plans, policies and actions. By the mid-1990s, the formulation of India's Ninth 5 Year Plan (1997–2002) had recognized climate as an important global environmental issue.

In June 2008, India communicated its National Action Plan on Climate Change (NAPCC) to the UNFCCC (Table 4.5). The plan includes eight goals of which five can be linked to India's energy sector: solar, enhanced energy efficiency, sustainable habitat, water; and sustainable agriculture. The goals are aimed to coordinate policies that overcome technological, financial, and market barriers and align national development policies with global climate change goals.

India has a dedicated ministry (Ministry of New and Renewable Energy: MNRE) for renewable energy and a long history of promoting renewable energy. The two motivations of India's renewable energy policies are to: (i) enhance energy access in rural areas through clean and decentralized energy supply, and (ii) develop a competitive modern renewable energy industry in India to enhance the share of renewable energy in the country's energy mix thus improving energy security and reducing environmental risk.

India's renewable power capacity (in January 2012) was 23,129 MW (Table 4.6). Nearly 70 % of this capacity was from wind power. This capacity was built over nearly two decades as a result of early policy foresight and supportive government programmes. The small hydro power plants amounting to 3 GW are meeting energy demands in the Himalayas and other similar terrains where conventional power is difficult to incorporate. Renewable energy and energy efficiency technologies have benefited from the global carbon market. The power from bio-waste has helped enhance energy access while gaining carbon market benefits through the Clean Development Mechanism (CDM). Solar power, which still generates relatively

Table 4.5 India's National Action Plan on Climate Change (NAPCC)

No	National mission	Targets
1.	National solar mission	Specific targets for increasing use of solar thermal technologies in urban areas, industry, and commercial establishments
2.	National mission for enhanced energy efficiency	Building on the energy conservation Act 2001
3.	National mission on sustainable habitat	Extending the existing energy conservation building code; emphasis on urban waste management and recycling, including power production from waste (3R)
4.	National water mission	20 % improvement in water use efficiency through pricing and other
5.	National mission for sustaining the Himalayan ecosystem	Conservation of biodiversity, forest cover, and other ecological values in the Himalayan region, where glaciers are projected to recede
6.	National mission for a "Green India"	Expanding forest cover from 23 % to 33 %
7.	National mission for sustainable agriculture	Promotion of sustainable agricultural practices
8.	National mission on strategic knowledge for climate change	The plan envisions a new Climate Science Research Fund that supports activities like climate modeling, and increased international collaboration; It also encourage private sector initiatives to develop adaptation and mitigation technologies

Source: GoI, PMCoCC 2008

Table 4.6 India's renewable electric power capacity (MW) (in January 2012)

Power source	Capacity
Wind power	16,179
Small hydro power	3,300
Biomass power	1,143
Bagasse cogeneration	1,953
Waste to power	
Urban	20
Industrial	53
Solar Power (SPV)	481
Total	**23,129**

Source: GoI, MNRE 2012

expensive electricity, has received support through India's National Climate Change Action Plan: NAPCC (GoI, PMCoCC 2008).

Besides its climate change policies and programmes, targeted policies like the New and Renewable Energy Policy 2005 and the Rural Electrification Policy 2006 have promoted the use of renewable energy for enhancing energy access using local renewable and waste resources. These have increased the off-grid and captive power capacities in remote areas: biomass co-generation (348 MW), biomass gasifier (148 MW), waste to energy (93 MW) and solar PV systems (81 MW). In addition, decentralized energy policies have resulted in electricity provision to 9,000 remote villages and the installation of over 2,000 water mills (micro-hydro-units), 45 million family biogas plants and solar water heating units with a collecting area of 5 million m^2.

India initiated its energy efficiency improvement policies even before 1970s oil crisis. The Energy Conservation Act 2001 (TGIE 2001) now provides a legal mandate for implementing energy efficiency measures through its Bureau of Energy Efficiency (BEE). The implementation of the new mandate includes:

- Use of tradable 'energy efficiency certificates' under the 'perform, achieve, and trade (PAT)' scheme wherein large energy-intensive industrial units participate by accepting targets to reduce their energy consumption;
- Promotion of energy efficient appliances; and
- Financial mechanisms to support demand-side management programmes.

4.3.3.3 Future Energy Challenges and Prospects in India

The dynamics of economic growth and the rising population, which will reach nearly 1.5 billion by 2030 (UNDP 2011), are expected to drive India's energy demand. For the development transition stage at which India now stands, economic growth is conventionally accompanied by rapidly rising energy use and associated environmental damage. The future energy challenges for India will include: (i) decoupling economic and energy growth; (ii) decoupling energy use from environmental damage; (iii) making energy accessible to all in the immediate future; and (iv) developing and implementing sustainable energy and technology options on demand and supply sides.

While the forces driving India's long-term energy future must respond to these multiple challenges, in the near-term a business-as-usual scenario is likely to sustain the predominance of fossil fuels, mainly domestic coal. In the next few decades, domestic and imported natural gas will be a likely cleaner energy option (Dhar and Shukla 2010). Strategic support from the international community has opened access for the transfer of advanced technologies to India such as nuclear, solar, next generation bio-fuels, smart grid, coal-bed methane (CBM), industry energy efficiency and carbon capture and storage (CCS). In the long-run, the national sustainable development policies and programmes, the global carbon market and the pace of the development of new energy technologies will likely shift the energy technology mix of India towards a less carbon-intensive and more sustainable energy future (Shukla and Chaturvedi 2013).

4.3.4 Japan

4.3.4.1 History of Japan's Energy Supply Trends

Japan's energy supply is highly dependent on imports from other countries. Especially after the two oil crises in the 1970s, Japan became aware of the importance and significance of energy security. Having a good balance among a wide variety of

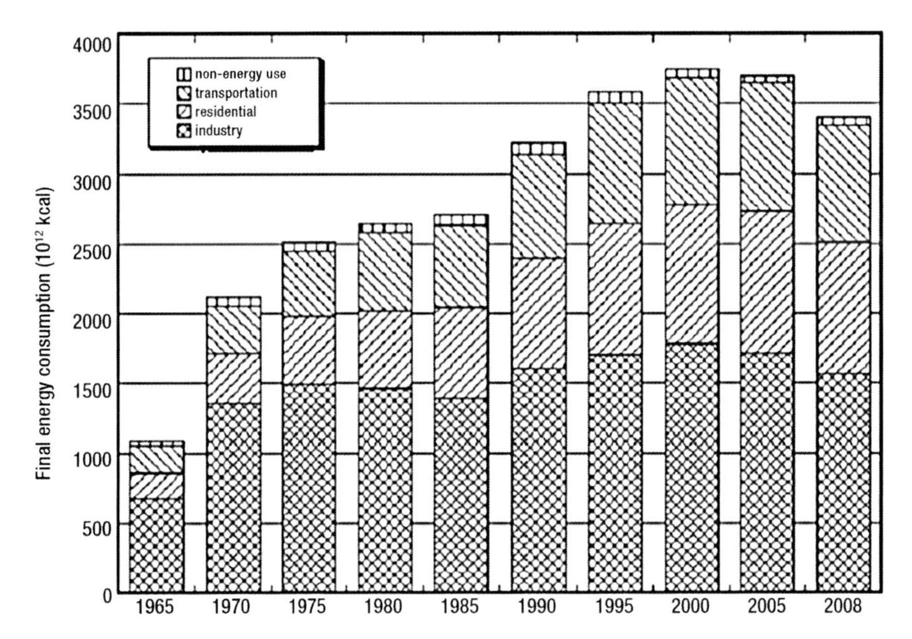

Fig. 4.8 Energy demand trend in Japan (Based on Figures from Energy and Resources Agency, Japan 2011)

energy resources was considered to be the best way for Japan to securitize its energy supply. Since then, the proportion of nuclear power as well as liquefied natural gas (LNG) in the total primary energy supply (TPES) in Japan began to grow to ensure the country's energy future.

Growth in energy demand in Japan can be divided into two periods. The first is the post-world war II period from the 1950s up until the first oil crisis in the early 1970s. The increase in energy demand during this time was mainly due to rapid industrialization of various manufacturing industries including steel, cement, and automobile manufacturing industries. The second energy-demand period was from the early 1980s until the early 2000s, with demand attributed to the household sector. Many Japanese owned their own cars and lived in houses with numerous electrical appliances, such as personal computers and large-size televisions, becoming the norm. As energy consumption in Japan began to swell, CO_2 emissions also continued rising even after 1990.

The growing trend of energy demand in Japan started to change after 2000. First, Japan's manufacturing industries began to shrink as other countries, especially some emerging economies such as China and India, became more industrialized. Japanese industries found it difficult to compete with these emerging economies mainly as a result of cheaper labor costs. Second, Japan's population also started to decrease after 2005. These two factors have led to a gradual decline in energy demand (Fig. 4.8).

4.3.4.2 Japan's Recent Energy Policy and Implications to Climate Change Policy

The increased demand for energy by the Japanese needed to be curtailed in order to reduce the country's carbon emissions and move to a low carbon economy. The year 1990 was chosen as the baseline to compare emissions in the targeted years. While in 2000 the UNFCCC called for emission stabilization to 1990 levels, this was not a legally-binding target and with most Japanese people unaware of this target, emissions from Japan continued to grow.

On the other hand, the Kyoto Protocol set an emissions target reduction of 6 % from 1990 levels between 2008 and 2012. This was perceived to be a legally-binding commitment, which the Japanese government took very seriously. Japan ratified the Kyoto Protocol in 2002, and the Protocol itself entered into force in February 2005. The Japanese government established a "Kyoto Protocol Target Achievement Plan", which contained a set of detailed policies and measures to reduce GHG emissions from various sources in Japan. In order to respond to both energy security and climate change, the Japanese government focused especially on improvements in energy efficiency and greater reliance on new nuclear power plants.

Meanwhile, international negotiations started to discuss long-term targets for 2050. In 2007, then Prime Minister Abe announced the "Beautiful Planet 2050" in which he outlined the idea that the world should aim at cutting its global GHG emissions in half by 2050. Later, Prime Minister Fukuda announced that Japan should aim for a much higher reduction target of 60–80 % by 2050. More recently, Prime Minister Hatoyama, in September 2009, called for a 25 % emission reduction target by 2020 from the 1990 level. Although this target counted on carbon offsetting outside Japan, it was still considered as a hard target to achieve. In order to achieve a 25 % reduction target, it was assumed that 9 new nuclear power plants would need to be commissioned in Japan (Central Council on the Environment 2010). However, this 25 % reduction target was conditional. As Japan was responsible only for 4 % of global GHG emissions, meaningful participation by larger emitting countries was the prerequisite for Japan to commit to a 25 % reduction target. This target was submitted to the UNFCCC Secretariat in January 2010 as a response to what was requested in the Copenhagen Accord, and taken note of at the 15th Conference of the Parties (COP), held in Copenhagen, Denmark, in December 2009.

4.3.4.3 Earthquake and Nuclear Power Plant Incident in March 2011

On 11 March 2011 a very large magnitude (nine on the Richter scale) earthquake struck the Tohoku region in Japan resulting in around 20,000 deaths, the majority of which was caused by the devastating tsunami that followed. Described as the "Great Tohoku Earthquake and Tsunami," this incident also destroyed the Fukushima Daiichi nuclear power plant. People living in the neighborhood had to relocate to safe regions outside the crippled power plant. With this event, the Japanese people became more aware of the dangers of nuclear power and consequently many called

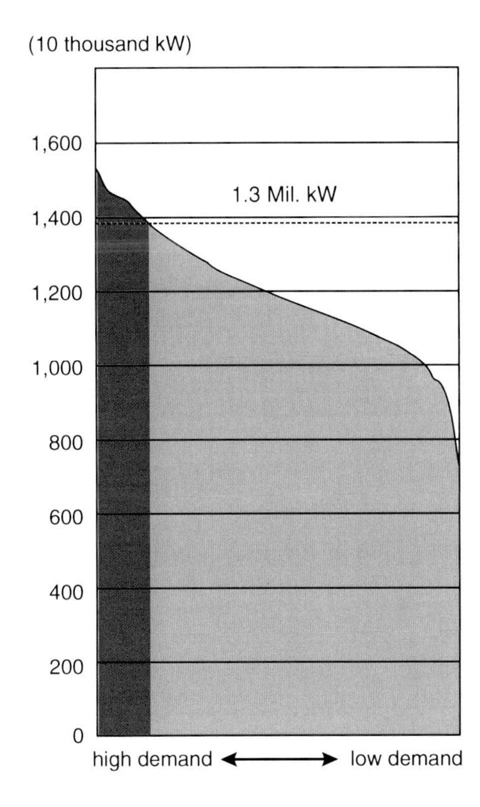

Fig. 4.9 Demand for electricity in Kanto area in 2010, in the order of demand size (Source: Central Council on the Environment 2011)

for a phase out of all nuclear power plants in Japan. In the summer of 2011, coal power plants replaced some nuclear power plants for electricity supply. On the other hand, what was surprising for many was a sudden change in people's behaviour by responding to calls from electricity companies to assist in energy-saving schemes. As a result, demand for electricity was reduced by more than 10 % of expected levels for the summer.

As the debates related to energy supply grew, public support increased for renewable energy technology such as solar, wind, biomass and geothermal sources. In August 2011, new legislation to promote renewable energy was passed in the Diet. The feed-in-tariff (FIT) mechanism will be introduced in the electricity sector to support the introduction of solar and wind power in electricity supply. In addition, in the devastated area of Tohoku, some experts have discussed ways of introducing renewable energy technologies to newly built urban cities and communities.

The Japanese people are also starting to look into the dynamics of electricity demand in more detail. Figure 4.9 shows the daily electricity demand in Japan in 2010. Current electric power plants are built to supply enough electricity to meet the largest demand of the day, but as can be seen in the figure, the maximum demand happens very rarely. Even within daily peak levels, these periods last only for a few hours at a time. Outside peak times, much of the energy potential of power plants is not utilized. The dotted line shows the level of electricity supply without Fukushima Daiichi Nuclear Power Plant. The question is whether to build another power plant to meet the highest demand for electricity, or whether to reduce the demand during those few occasions?

Japan's population started to decrease in 2005, and this decreasing trend is likely to continue in the long-term. Thus, demand for energy in Japan is not expected to grow beyond the present levels. However, a 60–80 % reduction of CO_2 emissions by 2050 is still a difficult target to achieve. A similar situation can be seen for energy security. As Japan's demand for energy decreases, Japan's energy security cannot be managed without consideration of other large energy-consuming countries such as India and China.

4.3.5 Conclusion

What is common among all the three key countries discussed in the present section is that energy security is the important component of mobilizing CO_2 emission mitigation policies. This is an important message, because there is a growing demand for energy in other developing countries in the Asia-Pacific region. Both energy efficiency and renewable energy are important factors in the energy conundrum.

Technological innovation and diffusion is an important component for improving energy efficiency. Energy efficiency of electrical devices such as refrigerators and air conditioners has improved dramatically in the past decades due to energy efficiency regulations. Insulation of houses and buildings are core components to reduce energy consumption in heating and cooling systems. Energy-efficient cars can run for longer distances with less fuel. Tax exemptions and refunds are effective policies to diffuse energy-efficient technologies. Research and development of innovative technologies can be financially supported by governments.

Promotion of renewable energy is a challenge that will require a lot of financial investment and innovation. However, costs of new technologies are likely to decrease as they as become more widely adopted. Prices of solar panels have declined rapidly in the past decade due to higher consumer demand for such products. Subsidies are effective for consumers to purchase renewable energy-related facilities at early stages. In addition, economic instruments such as feed-in-tariffs for electricity from renewable energy sources are an effective way to promote renewable energy in the power sector.

In addition to improvements of energy efficiency and promotion of renewable energy, demand side management is important, especially in relatively more developed countries where energy efficiency has already been achieved. In the summer of 2011, people living in eastern Japan proved they could decrease their demand for electricity by 10–20 % just by changing their habits of electricity usage. School education is also a key means for raising people's awareness on the use of energy and its relation to climate change.

References

Abu Dhabi Environment Agency. (2009, October 29). Abu Dhabi host the 4th dialogue on Southern Asia water cooperation. Retrieved from http://www.ead.ae/en/en/news/dialogue.on.southern.asia.water.cooperation.aspx

ADRRN. (2011, March 31). Civil society taskforce launched at IAP meeting. Asian Disaster Reduction & Response Network. Retrieved from http://www.adrrn.net/recentupdates/earth_concern.html

Alam, K., & Falkland A. (1997, April). *Vulnerability to climate change of the Bonriki freshwater lens, Tarawa* (Rep. No. HWR97/11). Republic of Kiribati: ECOWISE Environmental, ACTEW Corporation, prepared for Ministry of Environment and Social Development.

Antonov, J. I., Levitus, S., & Boyer, T. (2002). Steric sea level variations during 1957–1994: Importance of salinity. *Journal of Geophysical Research: Oceans, 107*(C12), 8013–8021.

ASEAN. (2010). *ASEAN disaster risk management initiative: Synthesis report on ten ASEAN countries disaster risks assessment*. Jakarta: Association for Southeast Asian Nations Secretariat.

ASEAN. (2011). *World bank, GFDRR, ASEAN, and UNISDR cooperate to strengthen fiscal resilience to natural disasters forum kicks-off efforts to develop regional disaster risk financing strategy. Joint news release 8 November 2011*. Jakarta: Association for Southeast Asian Nations Secretariat.

Ashfaq, M., Shi, Y., Tung, W., Trapp, R. J., Gao, X., Pal, J. S., & Diffenbaugh, N. S. (2009). Suppression of south Asian summer monsoon precipitation in the 21st century. *Geophysical Research Letters, 36*, L01704. doi:10.1029/2008GL036500.

Asian Development Bank (ADB). (2009). *Energy outlook for Asia and the Pacific (Report)*. Manila: ADB Publishing.

Asokan, S. M., & Dutta, D. (2008). Analysis of water resources in the Mahanadi River Basin, India under projected climate conditions. *Hydrological Processes, 22*, 3589–3603.

Atlin, G. N., Lafitte, H. R., Tao, D., Laza, M., Amante, M., & Courtois, B. (2006). Developing rice cultivars for high-fertility upland systems in the Asian tropics. *Field Crops Research, 97*, 43–52.

Australian Bureau of Meteorology and CSIRO. (2011). *Climate change in the Pacific: Scientific assessment and new research*. Volume 1: Regional overview. Volume 2: Country reports, Australia.

Bajracharya, S. R., Mool, P. K., & Shrestha, B. R. (2007). *Impact of climate change on Himalayan glaciers and glacial lakes: Case studies on GLOF and associated hazards in Nepal and Bhutan*. Kathmandu: International Centre for Integrated Mountain Development.

Bakun, A., & Weeks, S. J. (2008). The marine ecosystem off Peru: What are the secrets of its fishery productivity and what might its future hold? *Progress in Oceanography, 79*(2–4), 290–299.

Bandyopadhyay, J. (2009). Climate change and Hindu Kush-Himalayan waters – Knowledge gaps and priorities in adaptation. *Sustainable Mountain Development, 56*, 17–20.

Bankoff, G. (2004). In the eye of the storm: The social construction of the forces of nature and the climatic and seismic construction of god in the Philippines. *Journal of Southeast Asian Studies, 35*(1), 91–111.

Barnaby, W. (2009). Do nations go to war over water? *Nature, 458*, 282–283.

Beaugrand, G., Reid, P. C., Ibanez, F., Lindley, J. A., & Edwards, M. (2002). Reorganization of North Atlantic marine copepod biodiversity and climate. *Science, 296*, 1692–1694. doi:10.1126/science.1071329, http://dx.doi.org/10.1126/science.1071329.

Beaugrand, G., & Reid, P. C. (2003). Long-term changes in phytoplankton, zooplankton and salmon linked to climate. *Global Change Biology, 9*, 801–817.

Beddington, J., Asaduzzaman, M., Fernandez, A., Clark, M., Guillou, M., Jahn, M., Wakhungu, J. (2011). *Achieving food security in the face of climate change: Summary for policy makers from the commission on sustainable agriculture and climate change*. Copenhagen: CGIAR Research Programon Climate Change, Agriculture and Food Security (CCAFS). Retrieved from www.ccafs.cgiar.org/commission

Behrman, N. (Ed.). (2010). The waters of the third pole: Sources of threat, sources of survival. London: Aon Benfield UCL Hazard Research Centre, University College London. Retrieved from http://www.chinadialogue.net/UserFiles/File/third_pole_full_report.pdf

Bindoff, N. L., Willebrand, J., Artale, V., Cazenave, A., Gregory, J., Gulev, S., Unnikrishnan, A., et al. (2007). Observations: Oceanic climate change and sea level. In S. Solomon, D. Qin, M. Manning, Z. Chen, M. Marquis, K. B. Averyt, M. Tignor, & H. L. Miller (Eds.), *Climate change 2007: The physical science basis. Contribution of working group I to the fourth assessment report of the intergovernmental panel on climate change*. Cambridge, UK/New York: Cambridge University Press.

Bouman, B. A. M., & van Laar, H. H. (2006). Description and evaluation of the rice growth model ORYZA2000 under nitrogen-limited conditions. *Agricultural Systems, 87*, 249–273.

Bouman, B. A. M., van Laar, G., Zhang, X., Meinke, H., de Voil, P., … & Abawi, Y. (2004, September). Simulating growth and development of lowland rice in APSIM. In T. Fischer, N. Turner, J. Angus, L. McIntyre, M. Robertson, A. Borrell, & D. Lloyd (Eds.), *New directions for a diverse planet: Proceedings for the 4th international crop science congress*. Gosford: The Regional Institute Ltd.

Brodeur, R. D., Decker, M. B., Ciannelli, L., Purcell, J. E., Bond, N. A., Stabeno, P. J., … & Hunt, G. L., Jr. (2008). Rise and fall of jellyfish in the eastern Bering Sea in relation to climate regime shifts. *Progress in Oceanography, 77*(2–3), 103–111.

Central Council on the Environment (Japan). (2010 December). "A Road Map for the Mid-term Target" mid-term report.

Central Council on the Environment (Japan). (2011). Response to low carbon society after the Northeastern Japan's earth quake, a powerpoint slide shown at the meeting on 11 July 2011 (in Japanese).

ChinaFAQs. (2010). China's carbon intensity goal: A guide for the perplexed. World Resources Institute. Retrieved from http://www.chinafaqs.org/files/chinainfo/ChinaFAQs_China's_Carbon_Intensity_Goal_A_Guide_for_the_Perplexed_0.pdf

Chinese Government. (2005). The renewable energy law. Available from http://www.gov.cn/flfg/2009-12/26/content_1497462.htm (in Chinese).

Chinese Government. (2006a). The eleventh five-year plan of the People's Republic of China. http://www.gov.cn/ztzl/2006-03/16/content_228841.htm (in Chinese).

Chinese Government. (2006b, March 5). Premier reports govt work to legislature. Chinese Government Web Portal. Retrieved from http://english.gov.cn/2006-03/05/content_218602.htm

Ciannelli, L., Hjermann, D. Ø., Lehodey, P., Ottersen, G., Duffy-Anderson, J. T., & Stenseth, N. C. (2005). Climate forcing, food web structure, and community dynamics in pelagic marine ecosystems. In A. Belgrano, U. M. Scharler, J. Dunne, & R. E. Ulanowicz (Eds.), *Aquatic food webs: An ecosystem approach* (pp. 143–169). Oxford: Oxford University Press.

CRIEnglish. (2011, March 4). China to cap energy use at 4Bln tonnes of coal equivalent by 2015. CRIEnglish.com. Retrieved from http://english.cri.cn/6909/2011/03/04/1461s624079.htm

Cronin, R. (2009/2010). Mekong dams and the perils of peace. *Survival: Global Politics and Strategy, 51*(6), 147–160.

Cruz, R. V., Harasawa, H., Lal, M., Wu, S., Anokhin, Y., Punsalmaa, B., … & Huu Ninh, N. (2007). Asia. In M. L. Parry, O. F. Canziani, J. P. Palutikof, J. P. van der Linden, & C. E. Hanson (Eds.), *Asia climate change 2007: Impacts, adaptation and vulnerability. Contribution of working group II to the fourth assessment report of the intergovernmental panel on climate change* (pp. 469–506). Cambridge, UK: Cambridge University Press.

Cury, P., Shannon, L. J., & Shin, Y.-J. (2003). The functioning of marine ecosystems: A fisheries perspective. In M. Sinclair & G. Valdimarsson (Eds.), *Responsible fisheries in the marine ecosystem* (pp. 103–123). Wallingford: CAB International.

Cushing, D. (1995). *Population production and regulation in the sea: a fisheries perspective*. Cambridge, UK: Cambridge University Press.

Dhar, O. N., & Nandargi, S. (2003). Hydrometeorological aspects of floods in India. *Natural Hazards, 28*, 1–33.

Dhar, S., & Shukla, P. R. (2010). *Natural gas market in India*. New Delhi: McGraw Hill.

Dingkuhn, M., & Miézan, K. M. (1995). Climatic determinants of irrigated rice performance in the Sahel – II. Validation of photothermal constants and characterization of genotypes. *Agricultural Systems, 48*, 411–434.

Dixit, A. (2003). Floods and vulnerability: Need to rethink flood management. *Natural Hazards, 28*, 155–179.

Dunn, B. W., & Gaydon, D. S. (2011). Rice growth, yield and water productivity responses to irrigation scheduling prior to the delayed application of continuous flooding in south-east Australia. *Agricultural Water Management, 98*(12), 1799–1807.

Dutta, D., & Herath, S. (2004.) Trend of floods in Asia and flood risk management with integrated river basin approach. In 2nd Asian Pacific Association of Hydrology and Water Resources and Conference. Retrieved from http://rwes.dpri.kyoto-u.ac.jp/~tanaka/APHW/APHW2004/proceedings/FWR/56-FWR-A825/56-FWR-A825.pdf

EIA. (2011). International energy statistics. Energy Information Agency, Government of United States of America. Retrieved from http://www.eia.gov/emeu/cabs/India/pdf.pdf

Energy and Resources Agency (Japan). (2011). Japan's energy demand by final consumption sectors (in Japanese).

Eriksson, M., Jianchu, X., Shrestha, A. B., Vaidya, R. A., Nepal, S., & Sandstrom, K. (2009). *The changing Himalayas: Impact of climate change on water resources and livelihoods in the greater Himalayas*. Kathmandu: ICIMOD.

Falkland, A. (2007, March). Report on water investigations, Weno, Chuuk State, Federated States of Micronesia. November 2006–February 2007. Prepared for Asian Development Bank as part of RSC-C60725 (FSM): Loan Water Supply and Sanitation Project, Chuuk Water Supply Follow-up Investigation.

Falkland, T. (2011). Report on water security and vulnerability to climate change and other impacts in Pacific Island countries and East Timor on behalf of GHD Pty. Ltd. for Department of Climate Change and Energy Efficiency. Pacific Adaptation Strategy Assistance Programme.

Feely, A., Feely, A., Byrne, R. H., Betzer, P. R., Gendron, J. F., & Acker, J. G. (1984). Factors influencing the degree of saturation of the surface and intermediate waters of the North Pacific ocean with respect to aragonite. *Journal of Geophysical Research: Oceans, 89*(NC6), 631–640.

Feely, R. A., Sabine, C. L., Lee, K., Berelson, W., Kleypas, J., Fabry, V. J., & Millero, F. J. (2004). Impact of anthropogenic CO_2 on the $CaCO_3$ system in the oceans. *Science, 305*, 362–366.

Fluharty, D. (2011). *Decision-making and action taking: Fisheries management in a changing climate* (OECD food, agriculture and fisheries papers, Vol. 36). Paris: OECD Publishing.

Food and Agriculture Organization of the United Nations (FAO). (2009a). *The state of food insecurity in the world 2009: Economic crises – Impacts and lessons learned*. Rome: FAO.

Food and Agriculture Organization of the United Nations (FAO). (2009b). Crop prospects and food situation No. 3. Retrieved from http://www.fao.org/docrep/012/ai484e/ai484e00.htm

Foresight. (2011). *The future of food and farming (Final project report)*. London: The Government Office for Science.

Fujita, K., & Nuimura, T. (2011). Spatially heterogeneous wastage of Himalayan glaciers. In Hansen, J.E. (Ed.). *Proceedings of the National Academy of Sciences of the United States of America, 108*(34), 14011–14014.

Fujita, D., Santos, R. E., Ebron, L., Yanoria, M. J. T., Kato, H., Kobayashi, S., ... & Kobayashi, N. (2007). Genetic and breeding study on near isogenic lines of IR64 for yield-related traits—QTL analysis for days to heading in early heading lines. *Breeding Science, 9*, 114.

Fukuwaka, M., Sato, S., Takahashi, S., Onuma, T., Sakai, O., Tanimata, N., ... & Moss, J. H. (2007). Winter distribution of chum salmon related to environmental variables in the North Pacific. *North Pacific Anadromous Fish Commission Technical Report, 7*, 29–30.

Ganachaud, A., Sen Gupta, J., Orr, S., Wijffels, K., Ridgway, M., Hemer, C., ... & Kruger, J. (2011). Observed and expected changes to the tropical Pacific Ocean. In J. D. Bell, J. E. Johnson, & A. J. Hobday (Eds.), *Vulnerability of tropical pacific fisheries and aquaculture to climate change* (pp. 101–187). Noumea: Secretariat of the Pacific Community.

Gaydon, D. S., Buresh, R. J., Probert, M. E., & Meinke, H. (2009, July). Simulating rice in farming systems – Modelling transitions between aerobic and ponded soil environments in APSIM. In R.S. Anderssen, R.D. Braddock & L.T.H. Newham (Eds.) *18th world IMACS congress and MODSIM09 international congress on modelling and simulation proceedings*. Paper presented at 18th World IMACS Congress and MODSIM09 International Congress on Modelling and Simulation, The Modelling and Simulation Society of Australia and New Zealand and International Association for Mathematics and Computers in Simulation.

Gaydon, D. S., Meinke, H., & Rodriguez, D. (2011a). The best farm-level irrigation strategy changes seasonally with fluctuating water availability. *Agricultural Water Management, 103*, 33–42.

Gaydon, D. S., Meinke, H., Rodriguez, D., & McGrath, D. J. (2011b). Using modern portfolio theory to compare options for irrigation farmers to invest water. *Agricultural Water Management, 115*, 1–9.

Gaydon, D. S., Probert, M. E., Buresh, R. J., Meinke, H., Suriadi, A., Dobermann, A., … & Timsina, J. (2012a). Rice in cropping systems – Modelling transitions between flooded and non-flooded soil environments. *European Journal of Agronomy, 39*, 9–24.

Gaydon, D. S., Probert, M. E., Buresh, R. J., Meinke, H., & Timsina, J. (2012b). Modelling the role of algae in rice crop production and soil organic carbon maintenance. *European Journal of Agronomy, 39*, 35–43.

Gillett, R. (2009). *Fisheries in the economies of Pacific Island countries and territories* (Pacific studies series). Manila: Asian Development Bank.

Gillett, R., & Cartwright, I. (2010). *The future of Pacific Island fisheries*. Noumea: Secretariat of the Pacific Community.

Gillett, R., McCoy, M., Rodwell, R., & Tamate, J. (2001). *Tuna: A key economic resource in the Pacific Islands* (Pacific studies series). Manila: Asian Development Bank.

GLOBEC. (2010). GLOBEC final report. UK: Global Ocean Ecosystem Dynamics. Retrieved from www.globec.org/downloader.php?id=369

Gosain, A. K., Shrestha. A. B., & Rao, S. (2010). *Modelling climate change impact on the hydrology of the Eastern Himalayas* (Tech. Rep. No. 4. 978 92 9115 151 6.). Kathmandu: International Centre for Integrated Mountain Development (ICIMOD).

Government of India (GoI), Ministry of New and Renewable Energy (MNRE). (2012), Cumulative deployment of various renewable energy systems/devices in the country as on 31/01/2012, Government of India (GoI). Available from http://www.mnre.gov.in

Government of India (GoI), Ministry of Statistics and Programme Implementation (MSPI), MoPSI. (2011). Energy statistics 2011. Ministry of Statistics and Programme Implementation, Government of India (GoI). Available from http://mospi.nic.in/mospi_new/admin/login.aspx

Government of India (GoI), Prime Minister's Council on Climate Change (PMCoCC). (2008). National Action Plan on Climate Change (NAPCC), Prime Minister's Council on Climate Change, Government of India. Retrieved from http://pmindia.nic.in/climate_change_english.pdf

Hammer, G. L., Kropff, M. J., Sinclair, T. R., & Porter, J. R. (2002). Future contributions of crop modelling – From heuristics and supporting decision making to understanding genetic regulation and aiding crop improvement. *European Journal of Agronomy, 18*, 15–31.

Harley, S., Hoyle, S., Williams, P., Hampton, J., & Kleiber, P. (2010, August). *Stock assessment of Bigeye Tuna in the Western and Central Pacific Ocean (WCPFC-SC6-2010/SA-WP-4)*. Nuku'alofa, Kingdom of Tonga: Western and Central Pacific Fisheries Commission Scientific Committee.

Hashioka, T., & Yamanaka, Y. (2007). Ecosystem change in the western North Pacific associated with global warming obtained by 3-D NEMURO. *Ecological Modelling, 202*, 65–104.

Heath, M. R., & Gallego, A. (2010). Ecosystem structure and function. In A. D. Rijnsdorp, M. A. Peck, G. H. Engelhard, C. Möllmann, & J. K. Pinnegar (Eds.), *Resolving climate impacts on fish stocks* (ICES cooperative research report, Vol. 301, pp. 72–79). Copenhagen: International Council for the Exploration of the Sea.

Henson, I. E. (1985). Modification of leaf size in rice (*Oryza sativa* L.) and its effects on water stress-induced abscisic acid accumulation. *Annals of Botany, 56*, 481–487.

Hirawake, T., Odate, T., & Fukuchi, M. (2005). Long-term changes variation of surface phytoplankton chlorophyll α in the Southern Ocean during 1965–2002. *Geophysical Research Letters, 32*. doi 10.1024/2004GL21344

Hou, Y., & Li, J. (2011, March 3). China needs higher targets. chinadialogue. Retrieved from http://www.chinadialogue.net/article/show/single/en/4140--China-needs-higher-targets-

Hsiao, T. C., O'Toole, J. C., Yambao, E. B., & Turner, N. C. (1984). Influence of osmotic adjustment on leaf rolling and tissue death in rice (*Oryza sativa* L.). *Plant Physiology, 75*, 338–341.

Hsu, A., & Seligsohn, D. (2011, March 1). What to look for in China's 12th five-year plan? [Web log comment]. Retrieved from http://www.chinafaqs.org/blog-posts/what-look-chinas-12th-five-year-plan

ICIMOD. (2009). Water storage: A strategy for climate change adaptation. No. 56, Winter 2009. Retrieved from http://www.icimod.org/publications/index.php/search/publication/654

IEA. (2011, October). *World energy outlook 2011: Energy for all*. Paris: International Energy Agency. Retrieved from http://www.iea.org/publications/freepublications/publication/name,4007,en.html

Immerzeel, W. W., van Beek, L. H., & Bierkens, M. F. P. (2010). Climate change will affect the Asian water towers. *Science, 328*, 1382–1385.

IPCC. (2007a). Intergovernmental Panel on Climate Change (IPCC). In S. Solomon, D. Qin, M. Manning, Z. Chen, M. Marquis, K. B. Averyt, M. Tignor, & H. L. Miller (Eds.), *Climate change 2007: The physical science basis. Contribution of working group I to the fourth assessment report of the intergovernmental panel on climate change*. Cambridge, UK/New York: Cambridge University Press.

IPCC. (2007b). Summary for policymakers. In S. Solomon, D. Qin, M. Manning, Z. Chen, M. Marquis, K. B. Averyt, M. Tignor, & H. L. Miller (Eds.), *Climate change 2007: The physical science basis. Contribution of working group I to the fourth assessment report of the intergovernmental panel on climate change*. Cambridge, UK/New York: Cambridge University Press.

IPCC. (2007c). Summary for policymakers. In B. Metz, O. R. Davidson, P. R. Bosch, R. Dave, & L. A. Meyer (Eds.), *Climate change 2007: Mitigation. Contribution of working group III to the fourth assessment report of the intergovernmental panel on climate change*. Cambridge, UK/New York: Cambridge University Press.

IPCC. (2012). Managing the risks of extreme events and disasters to advance climate change adaptation. A special report of working groups I and II of the intergovernmental panel on climate change. In C. B. Field, V. Barros, T. F. Stocker, D. Qin, D. J. Dokken, K. L. Ebi, M. D. Mastrandrea, K. J. Mach, G.-K. Plattner, S. K. Allen, M. Tignor & P. M. Midgley (Eds.). Cambridge, UK/New York: Cambridge University Press. Retrieved from https://www.ipcc.ch/pdf/special.../srex/SREX_Full_Report.pdfShare

Jain, S. K. (2008). Impact of retreat of Gangotri glacier on the flow of Ganga River. *Current Science, 95*, 1012–1014.

Jaitly, A. (2009). South Asian perspectives on climate change and water policy. In D. Michel & A. Pandya (Eds.), *Troubled waters: Climate change, hydropolitics, and transboundary resources*. Washington, DC: The Henry L. Stimson Center.

Kaeriyama, M. (1986). Ecological study on early life of the chum salmon Oncorhynchus keta (Walbaum). *Scientific Reports of the Hokkaido Salmon Hatchery, 40*, 31–92. http://salmon.fra.affrc.go.jp/kankobutu/srhsh/data/srhsh317.pdf.

Karl, D. 1999. A sea of change: biogeochemical variability in the North Pacific Subtropical Gyre. *Ecosystems, 2*, 181–214.

Kawamiya, M., Hasumi, H., Sakamoto, T., & Yoshikawa, C. (2007). Forecast of marine physics with global warming scenario by OGCM. *Monthly Kaiyo, 39*, 285–290 (in Japanese).

Kehrwald, N. M., Thompson, L. G., Tandong, Y., Mosley-Thompson, E., Schotterer, U., Alfimov, V., … & Davis, M. E. (2008). Mass loss on Himalayan glacier endangers water resources. *Geophysical Research Letters, 35*, L22503. doi:10.1029/2008GL035556.

Kim, S. (2010). Fisheries development in northeastern Asia in conjunction with changes in climate and social systems. *Marine Policy, 34*, 803–809.

Kishi, M. J., Nakajima, K., Fujii, M., & Hashioka, T. (2009). Environmental factors which affect growth of Japanese common squid, *Todarodes pacificus*, analyzed by a bioenergetics model coupled with a lower trophic ecosystem model. *Journal of Marine Systems, 78*, 278–287.

Kishi, M. J., Kaeriyama, M., Ueno, H., & Kamezawa, Y. (2010). The effect of climate change on the growth of Japanese chum salmon (*Oncorhynchus keta*) using a bioenergetics model coupled with a three-dimensional lower tropic ecosystem model (NEMURO). *Deep-Sea Research II, 57*, 1257–1265.

Kitamoto, M., Tsunozaki, E., & Teranishi, A. (2005). *Institutional capacity for natural disaster reduction in Japan (USER working paper No 2005–11)*. Chiang Mai: Unit for Social and Environmental Research, Chiang Mai University.

Klyashtorin, L. B. (2001). Cyclic change of climate and main commercial species production in the Pacific. Report of a GLOBEC-SPACC/APN workshop on the causes and consequences of climate-induced changes in Pelagic fish productivity in East Asia.

Kotov, V., & Nikitina, E. (2005). *Institutions, policies and measures towards the Lena river flood risk reduction. USER working paper*. Chiang Mai: Unit for Social and Environmental Research, Chiang Mai University.

Kumar, J., & Abbo, S. (2001). Genetics of flowering time in chickpea and its bearing on productivity in the semi-arid environments. *Advances in Agronomy, 72*, 107–138.

Lagi, M., Bertrand, K. Z., & Bar-Yam, Y. (2011). The food crises and political instability in North Africa and the Middle East. *arXiv* 1108.2455.

Langley A., Harley S., Hoyle S., Davies N., Hampton J., & Kleiber, P. (2009 August 10–21). Stock assessment of yellowfin tuna in the western and central Pacific WCPFC–SC5-2007/SA-WP-3. Port Vila, Vanuatu: Western and Central Pacific Fisheries Commission Scientific Committee, Fifth regular session.

Lau, W., Kim, K. M., Kim, M.-K., & Lee, W.-S. (2010). Enhanced surface warming and accelerated snow melt in the Himalayas and Tibetan Plateau induced by absorbing aerosols. *Environmental Research Letters, 5*(2), doi:10.1088/1748-9326/5/2/025204.

Lebel, L., & Sinh, B. T. (2007). Politics of floods and disasters. In L. Lebel, J. Dore, R. Daniel, & Y. S. Koma (Eds.), *Democratizing water governance in the Mekong region* (pp. 37–54). Chiang Mai: Mekong Press.

Lebel, L., & Sinh, B. T. (2009). Risk reduction or redistribution? Flood management in the Mekong region. *Asian Journal of Environment and Disaster Management, 1*, 23–39.

Lebel, L., Khrutmuang, S., & Manuta, J. (2006a). Tales from the margins: small fishers in post-Tsunami Thailand. *Disaster Prevention and Management, 15*(1), 124–134.

Lebel, L., Nikitina, E., Kotov, V., & Manuta, J. (2006b). Assessing institutionalized capacities and practices to reduce the risks of flood disasters. In J. Birkmann (Ed.), *Measuring vulnerability to natural hazards: Towards disaster resilient societies* (pp. 359–379). Tokyo: United Nations University Press.

Lebel, L., Nikitina, E., & Manuta, J. (2006c). Flood disaster risk management in Asia: An institutional and political perspective. *Science and Culture, 72*(1–2), 2–9.

Lebel, L., Foran, T., Garden, P., & Manuta, B. J. (2009a). Adaptation to climate change and social justice: Challenges for flood and disaster management in Thailand. In F. Ludwig, P. Kabat, H. van Schaik, & M. van der Valk (Eds.), *Climate change adaptation in the water sector*. London: Earthscan.

Lebel, L., Sinh, B. T., Garden, P., Seng, S., Tuan, L. A., & Truc, D. V. (2009b). The promise of flood protection: Dykes and dams, drains and diversions. In F. Molle, T. Foran, & J. Kakonen (Eds.), *Contested waterscapes in the Mekong region* (pp. 283–306). London: Earthscan.

Lebel, L., Perez, R. T., Sukhapunnaphan, T., Hien, B. V., Vinh, N., & Garden, P. (2009c). Reducing vulnerability of urban communities to flooding. In L. Lebel, A. Snidvongs, C.-T. A. Chen, & R. Daniel (Eds.), *Critical states: Environmental challenges to development in Monsoon Asia* (pp. 381–399). Selangor: Strategic Information and Research Development Centre.

Lebel, L., Manuta, B. J., & Garden, P. (2011). Institutional traps and vulnerability to changes in climate and flood regimes in Thailand. *Regional Environmental Change, 11*, 45–58.

Lehodey, P., Bertignac, M., Hampton, J., Lewis, T., & Picaut, J. (1997). El Niño-Southern oscillation and tuna in the western Pacific. *Nature, 389*, 715–718.

Lehodey, P., Alheit, J., Barange, M., Baumgartner, T., Beaugrand, G., Drinkwater, K., … & Werner, F. (2006). Climate variability, fish and fisheries. *Journal of Climate, 19*, 5009–5030.

Levitus, S., Antonov, J. I., & Boyer, T. P. (2005). Warming of the World Ocean, 1955–2003. Geophysical Research Letters, *32*, L02604. doi:10.1029GL021592.

Liu, X., & Chen, B. (2000). Climate warming in the Tibetan Plateau during recent decades. *International Journal of Climatology, 20*, 1729–1742.

Lough, J. M., Meehl, G. A., & Salinger, M. J. (2011). Observed and projected changes in surface climate of the tropical Pacific. In J. D. Bell, J. E. Johnson, & A. J. Hobday (Eds.), *Vulnerability of Tropical Pacific fisheries and aquaculture to climate change* (pp. 49–99). Noumea: Secretariat of the Pacific Community.

Ludlow, M. M., & Muchow, R. C. (1990). A critical evaluation of traits for improving crop yields in water-limited environments. *Advances in Agronomy, 43*, 107–153.

Luquet, D., Clément-Vidal, A., This, D., Fabre, D., Sonderegger, N., & Dingkuhn, M. (2008). Orchestration of transpiration, growth and carbohydrate dynamics in rice during a dry-down cycle. *Functional Plant Biology, 35*, 689–704.

Ma, X., Xu, J. C., Luo, Y., Aggrawal, S. P., & Li, J. T. (2009). Response of hydrological processes to land-cover and climate changes in Kejie watershed, south-west China. *Hydrological Processes, 23*(8), 1179–1191.

Ma, X., Xu, J., & Van Noordwijk, M. (2010). Sensitivity of streamflow from a Himalayan catchment to plausible changes in land cover and climate. *Hydrological Processes, 24*(11), 1379–1390.

Mackas, D.L., Batten, S.D., and Trudel, M., (2007) Effects on zooplankton of a warming ocean: recent evidence from the Northeast Pacific. *Progress in Oceanography, 75*, 223–252.

Manschadi, M., Hammer, G. L., Christopher, J. T., & deVoil, P. (2008). Genotypic variation in seedling root architectural traits and implications for drought adaptation in wheat (*Triticum aestivum* L.). *Plant and Soil, 303*, 115–129. doi:10.1007/s11104-007-9492-1.

Manton, M. J., Heath, L., Salinger, J., & Stevenson, L. A. (2011). *Climate in Asia and the Pacific: A synthesis of APN activities*. Kobe: Asia-Pacific Network for Global Change Research. Retrieved from http://www.apn-gcr.org/resources/items/show/1745

Mantua, N. J., Hare, S. R., Zhang, Y., Wallace, J. M., & Francis, R. C. (1997). A Pacific interdicadal climate oscillation with impacts on salmon production. *Bulletin of the American Meteorological Society, 787*, 1069–1079.

Mantua, N. J., & Hare, S. R. (2002). The Pacific decadal oscillation. *Journal of Oceanography, 58*, 35–44.

Manuta, J., Khrutmuang, S., & Lebel, L. (2005). The politics of recovery: Post-Asian Tsunami reconstruction in southern Thailand. *Tropical Coasts, July*, 30–39.

Manuta, J., Khrutmuang, S., Huaisai, D., & Lebel, L. (2006). Institutionalized incapacities and practice in flood disaster management in Thailand. *Science and Culture, 72*(1–2), 10–22.

McFarlane, N. A., Scinocca, J. F., Lazare, M., Harvey, R., Verseghy, D., & Li, J. (2005). The CCCma third generation atmospheric general circulation model. CCCma Internal Report, 25 pp. http://www.cccma.ec.gc.ca/models/gcm3.shtml

McGowan, J.A., Bograd, S.J., Lynn, R.J., Miller, A., 2003. The biological response to the 1977 regimeshift in the California Current: a tale of two regimes. *Deep-Sea Research II 50*: 2567–2582.

Meehl, G. A., Stocker, T. F., Collins, W. D., Friedlingstein, P., Gaye, A. T., Gregory, J. M., & Zhao, Z.-C. (2007). Global climate projections. In S. Solomon, D. Qin, M. Manning, Z. Chen, M. Marquis, K. B. Averyt, M. Tignor, & H. L. Miller (Eds.), *Climate change 2007: The physical science basis. Contribution of working group I to the fourth assessment report of the intergovernmental panel on climate change*. Cambridge, UK/New York: Cambridge University Press.

Miller, A. J., Chai, F., Chiba, S., Moisan, J. R., & Neilson, D. J. (2004). Decadal-scale climate and ecosystem interactions in the North Pacific Ocean. *Journal of Oceanography, 60*, 163–188.

Minobe, S. (1999). Resonance in bidecadal and pentadecadal climate oscillations over the North Pacific: Role in climatic regime shifts. *Geophysical Research Letters, 26*, 855–858.

Mitchell, P. L., & Sheehy, J. E. (2006). Supercharging rice photosynthesis to increase yield. *New Phytologist, 171*, 688–693.

Miyanaga, S., & Sakurai, Y. (2006). Effect of temperature on the activity and metabolism of Japanese common squid paralarvae. Program and abstract book of "Cophalopod life cycles". CIAC2006, Hobart, Australia.

Morton, K. (2011). Climate change and security at the third pole. *Survival, 53*(1), 121–132.

Mukherji, A., Facon,T., Burke, J., de Fraiture, C., Faures, J. M., Füleki, B., … & Shah, T. (2009). *Revitalizing Asia's irrigation: To sustainably meet tomorrow's food needs*. Colombo: International Water Management Institute; Rome: Food and Agricultural Organization of the United Nations.

Nakata, H. (2001). Report of APN/GLOBEC workshop on the causes and consequences of climate-induced changes in Pelagic fish productivity in East Asia (APN Project Report APN2001-07). APN e-Lib. Retrieved from http://www.apn-gcr.org/resources/items/show/1470

National Snow and Ice Data Center. (1999) Up-dated 2009 world glacier inventory. Available from http://nsidc.org/data/docs/noaa/g01130_glacier_inventory/

Nguyen, H. T., Babu, R. C., & Blum, A. (1997). Breeding for drought resistance in rice: Physiology and molecular genetics considerations. *Crop Science, 37*, 1426–1434.

Nikitina, E., & Kotov, V. (2008, August). *Reducing flood risks through stakeholder participation and partnerships: Lessons learned from river basins in Asia and in Europe (Extended Abstract)*. Davos: International Disaster and Risk Conference, IDRC.

Nikitina, E., Ostrovskaya, E., & Fomenko, M. (2009a, December). Towards better water governance in river basins: Some lessons learned from the Volga. *Regional Environmental Change, 10*, 285–297.

Nikitina, E., Dushmanta, D., Kotov, V., Lebel, L., Sinh, B. T., & Xu, J. (2009b). Reducing water insecurity through stakeholder participation in river basin management in the Asia-Pacific (APN Project Report: ARCP2009-03CMY-Nikitina). APN e-Lib. http://www.apn-gcr.org/resources/items/show/1554

Nikitina, E., Lebel, L., Kotov, V., & Sinh, B. T. (2010). How stakeholder participation and partnerships could reduce water insecurities in shared river basins. In J. Ganoulis, A. Aureli, & J. Fried (Eds.), *Water resources across borders: A multidisciplinary approach* (pp. 280–286). Weinheim: Wiley-VCH Verlag & Co.

NIWA. (2008, September). Sea-levels, waves, run-up and overtopping, Information for climate risk management. Kiribati adaptation programme phase II (Report HAM 2008–22). New Zealand: National Institute for Water and Atmosphere Research Ltd.

Oberdorfer, J. A., & Buddemeier, R. W. (1988). Climate change, effects on reef island resources. *Sixth International Coral Reef Symposium, Townsville, Australia, 3*, 523–527.

Orr, J. C., Fabry, V. J., Aumont, O., Bopp, L., Doney, S. C., Feely, R. A., & Yool, A. (2005). Anthropogenic ocean acidification over the twenty-first century and its impact on calcifying organisms. *Nature, 437*, 681–686. doi:10.1038/nature04095.

PCCSP. (2011a). Interim climate projections, 25 March 2011. CSIRO and Bureau of Meteorology, Pacific Climate Change Science Program (supplied in email from Martin Sharp, Director, International Adaptation Strategies Team, Adaptation, Land and Communications Division, Department of Climate Change and Energy Efficiency on 17th May 2011).

PCCSP. (2011b). Document "PCCSP projections for PASAP v3.xls". CSIRO and Bureau of Meteorology, Pacific Climate Change Science Program (supplied in email from Martin Sharp, Director, International Adaptation Strategies Team, Adaptation, Land and Communications Division, Department of Climate Change and Energy Efficiency on 17th May 2011).

People's Daily Online. (2010, August 27). China's nuclear power capacity to see 7-fold increase in next 10 years. People's Daily Online. Retrieved from http://english.peopledaily.com.cn/90001/90778/90862/7120511

Pielke, R., Jr., Prins, G., Rayner, S., & Sarewitz, D. (2007). Lifting the taboo on adaptation. *Nature, 445*, 597–598.

Pörtner, H. O., Langenbuch, M., & Michaelidis, B. (2005). Synergistic effects of temperature extremes, hypoxia and increases in CO2 on marine animals: from earth history to global change. *Journal of Geophysical Research, 110*, C09S10. doi:10.1029/2004JC002561.

Quadrelli, R., & Wallace, J. M. (2004). A simplified linear framework for interpreting patterns of Northern Hemisphere wintertime climate variability. *Journal of Climate, 17*, 3728–3744.

Rahmstorf, S., Cazenave, A., Church, J. A., Hansen, J. E., Keeling, R. F., Parker, D. E., & Somerville, R. C. J. (2007). Recent climate observations compared to projections. *Science, 316*(5825), 709. doi:10.1126/science.1136843.

Rees, H. G., & Collins, D. N. (2006). Regional differences in response of flow in glacier-fed Himalayan rivers to climatic warming. *Hydrological Processes, 20*, 2157–2169.

Richards, R. A., Rawson, H. M., & Johnson, D. A. (1986). Glaucousness in wheat: Its development, and effect on water-use efficiency, gas exchange and photosynthetic tissue temperatures. *Australian Journal of Plant Physiology, 13*, 465–473.

Richardson, A. J. (2008). In hot water: Zooplankton and climate change. *ICES Journal of Marine Science, 65*, 279–295.

Rodell, M., Velicogna, I., & Famiglietti, J. S. (2009). Satellite-based estimates of groundwater depletion in India. *Nature, 460*, 999–1002. doi:10.1038/nature08238.

Roemmich, D., & McGowan, J. (1995). Climatic warming and the decline of zooplankton in the California current. *Science, 267*(5202), 1324–1326.

Rogner, H.-H., Zhou, D., Bradley, R., Crabbé, P., Edenhofer, O., Hare, B., Kuijpers, L., & Yamaguchi, M. (2007). Introduction. In B. Metz, O. R. Davidson, P. R. Bosch, R. Dave, & L. A. Meyer (Eds.), *Climate change 2007: Mitigation. Contribution of working group III to the fourth assessment report of the intergovernmental panel on climate change*. Cambridge, UK/New York: Cambridge University Press. Data from IEA, Figure 1.5.

Sabine, C. S., Feely, R. A., Gruber, N., Key, R. M., Lee, K., Bullister, J. L., & Rios, A. F. (2004). The oceanic sink for anthropogenic CO2. *Science, 305*, 367–371.

Sakurai, Y., Kiyofuji, H., Saitoh, S., Goto, T., & Hiyama, Y. (2000). Changes in inferred spawning areas of *Todarodes pacificus* (Cephalopoda: Ommastrephidae) due to changing environmental conditions. *ICES Journal of Marine Science, 57*, 24–30.

Sakurai, Y., Bower, J. R., & Ikeda, Y. (2003). Reproductive characteristics of the ommastrephid squid *Todarodes pacificus*. *Fisken og Havet, 12*, 105–115.

Sakurai, Y., Kishi, M. J., & Nakajima, K. (2007). Walleye pollock and common squid. *Monthly Kaiyo, 39*, 323–333.

Salinger, M. J. (2013). A brief introduction to the issue of climate and fisheries. *Climatic Change*. doi:10.1007/s10584-013-0762-z.

Salinger, M. J., & Hobday, A. J. (2013). Safeguarding the future of oceanic fisheries under climate change depends on timely preparation. *Climatic Change*. doi:10.1007/s10584-012-0609-z.

Sandquist, D. R., & Ehleringer, J. R. (2003). Population- and family-level variation of brittlebush (*Encelia farinosa*, Asteraceae) pubescence: Its relation to drought and implications for selection in variable environments. *American Journal of Botany, 90*, 1481–1486.

Sarmiento, J. L., Slater, R. D., Barber, R. T., Bopp, L., Doney, S. C., Hirst, A. C., Stouffer, R. J. (2004). Response of ocean ecosystems to climate warming. *Global Biogeochemical Cycles, 18*(3), doi: 10.1029/2003GB002134.

Scherler, D., Bookhagen, B., & Strecker, M. R. (2011). Hillslope-glacier coupling: The interplay of topography and glacial dynamics in High Asia. *Journal of Geophysical Research, 116*(F2), doi:10.1029/2010JF001751.

Schneider, K. (2009). A rice town's cry. *Circle of Blue Water News*. Retrieved from http://www.circleofblue.org/waternews/2009/world/australia-deniliquin-rice-shortages-drought/

Seligsohn, D. (2010, October 29). China's party plenum recommends climate actions in the 12th five year plan [Web log comment]. Retrieved from http://www.chinafaqs.org/blog-posts/chinas-party-plenum-recommends-climate-actions-12th-five-year-plan

Selvaraju, R., Meinke, H., & Hansen, J. (2007). Climate forecast for better water management in agriculture: A case study for Southern India. In M. V. K. Sivakumar & J. Hansen (Eds.), *Climate prediction and agriculture* (pp. 143–155). Berlin: Springer.

Shimono, H., Okada, M., Yamakawa, Y., Nakamura, H., Kobayashi, K., & Hasegawa, T. (2007). Lodging in rice can be alleviated by atmospheric CO_2 enrichment. *Agriculture, Ecosystems and Environment, 118*, 223–230.

Singh, P., & Bengtsson, L. (2004). Hydrological sensitivity of a large Himalayan basin to climate change. *Hydrological Processes, 18*(13), 2363–2385.
Sinh, B. T., Lebel, L., & Tung, N. T. (2009). Indigenous knowledge and decision making in Vietnam: Living with floods in An Giang Province, Mekong Delta, Vietnam. In R. Shaw (Ed.), *Indigenous knowledge and disaster risk reduction: From practice to policy*. Hauppage: NOVA Science Publishers Inc.
Sivakumar, M. V. K., & Motha, R. P. (Eds.). (2007 October 25–27). Managing weather and climate risks in agriculture. In Proceedings of the International Workshop on Agrometeorological Risk Management held in New Delhi, India, Springer.
Shukla, P. R., & Chaturvedi, V. (2013). Sustainable energy transformations in India under climate policy. *Sustainable Development, 21*, 2013.
SOPAC. (2010, January). Country implementation plan for improving water security in the Republic of Nauru. Prepared by Tony Falkland and John Tagiilima for SOPAC as part of the Multi-Country Project "Support to Disaster Risk Reduction in Eight Pacific ACP States", funded by the European Commission from the EDF9 National B-Envelope.
South Asian Association for Regional Cooperation (SAARC). (2005). Dhaka declaration. Thirteenth SAARC summit, 13 November, 2005. Dhaka, Bangladesh. Retrieved from http://www.saarc-sec.org/userfiles/Summit%20Declarations/13%20-%20Dhaka%20-%2013th%20Summit%2012-13%20Nov%202005.pdf
SPC. (2011). Populations & demographic indicators in spreadsheet "SPCSDP_populations_data_sheet_2011.xls". Statistics for development. Secretariat of the Pacific Community. Available from http://www.spc.int/sdp/
Su, W. (2010, January 28). China's official communication with UNFCCC. Retrieved from http://unfccc.int/files/meetings/cop_15/copenhagen_accord/application/pdf/chinacphaccord_app2.pdf
Subash, N., Sikka, A. K., & Ram Mohan, H. S. (2011). An investigation into observational characteristics of rainfall and temperature in Central Northeast India—A historical perspective 1889–2008. *Theoretical and Applied Climatology, 103*, 305–319.
Swain, A. (2002). Environmental cooperation in South Asia. In K. Conca & G. D. Dabelko (Eds.), *Environmental peacemaking* (pp. 61–85). Washington, DC: Woodrow Wilson Center Press.
Takasuka, A., Oozeki, Y., Kimura, R., Kubota, H., & Aoki, I. (2004). Growth-selective predation hypothesis revisited for larval anchovy in offshore waters: Cannibalism by juveniles versus predation by skipjack tunas. *Marine Ecology Progress Series, 278*, 297–302.
Takasuka, A., Oozeki, Y., & Kubota, H. (2008). Multi-species regime shifts reflected in spawning temperature optima of small pelagic fish in the western North Pacific. *Marine Ecology Progress Series, 360*, 211–217.
Takeuchi, K. (2001). Increasing vulnerability to extreme floods and societal needs of hydrological forecasting. *Hydrological Sciences Journal, 46*(6), 869–881.
Terry, J. P., & Falkland, A. (2010). Responses of atoll freshwater lenses to storm-surge overwash in the Northern Cook Islands. *Hydrogeology Journal, 18*, 749–759.
Thayyen, R. J., & Gergan, J. T. (2010). Role of glaciers in watershed hydrology: A preliminary study of a "Himalayan catchment". *The Cryosphere, 4*, 115–128.
The Gazette of India Extraordinary (TGIE). (2001, October 1). India energy conservation act 2001. The gazette of India extraordinary, part II section 1. Retrieved from http://powermin.nic.in/acts_notification/pdf/ecact2001.pdf
Trenberth, K. E., Jones, P. D., Ambenje, P., Bojariu, R., Easterling, D., Klein Tank, A., … & Zhai, P. (2007). Observations: Surface and atmospheric climate change. In S. Solomon, D. Qin, M. Manning, Z. Chen, M. C. Marquis, K. B. Averyt, M. Tignor, & H. L. Miller (Eds.), *Climate change 2007. The physical science basis*. Contribution of WG 1 to the Fourth Assessment Report of the Intergovernmental Panel on Climate Change (pp. 235–336). Cambridge and New York: Cambridge University Press, plus annex online.
Tripathy, J. N., Zhang, J., Robin, S., Nguyen, T. T., & Nguyen, H. T. (2000). QTLs for cell-membrane stability mapped in rice (*Oryza sativa* L.) under drought stress. *Theoretical and Applied Genetics, 100*, 1197–1202.

UNESCAP & UNISDR. (2010). Protecting development gains: Reducing disaster vulnerability and building resilience in Asia and the Pacific. The Asia-Pacific disaster report, 2010. Retrieved from http://www.unescap.org/idd/pubs/Asia-Pacific-Disaster-Report%20-2010.pdf

UNISDR. (2007). Words into action: A guide for implementing the Hyogo framework. *Hyogo framework for action 2005–2015: Building the resilience of nations and communities to disasters*. Retrieved from http://www.preventionweb.net/files/594_10382.pdf

UNISDR. (2009). 2009 UNIDSR terminology on disaster risk reduction. United Nations International Strategy for Disaster Reduction. Retrieved from www.unisdr.org/files/7817_UNISDRTerminologyEnglish.pdf

UNISDR. (2011). At the crosscroads. Climate change adaptation and disaster risk reduction in Asia and the Pacific. A review of the region's institutional and policy landscape. Retrieved from http://www.unisdr.org/files/21414_21414apregionalmappingdrrcca1.pdf

United Nations Development Programme (UNDP). (2011). *Human development report: s2011. UN development programme*. New York: Oxford University Press.

USAID. (2009, August). *Adaptation to climate change: Case study – Freshwater resources*. Majuro, RMI: United States Agency for International Development. 2009.

Vecchi, G. A., Soden, B. J., Wittenberg, A. T., Held, I. M., Leetmaa, A., & Harrison, M. J. (2006). Weakening of tropical Pacific atmospheric circulation due to anthropogenic forcing. *Nature, 441*, 73–76. doi:10.1038/nature04744.

Venrick, E. L. (1993). Phytoplankton seasonality in the central North Pacific: The endless summer reconsidered. *Limnology and Oceanography, 38*, 1135–1149.

von Grebmer, K., Fritschel, H., Nestorova, B., Olofinbiyi, T., Pandya-Lorch, R., & Yohannes, Y. (2008). *Global hunger index: The challenge of hunger 2008*. Washington, DC/Dublin: Deutsche Welthungerhilfe, International Food Policy Research Institute, and Concern.

Wang, J., Li, T., & Du, Y. (2011, January 6). Zhang Guobao: Towards increasing the share of non-fossil energy to 11.4 percent of the overall primary energy consumption by the end of the twelfth five-year period. People's Daily. Retrieved from http://energy.people.com.cn/GB/13670716.html (in Mandarin).

Wassmann, R., Jagadish, S. V. K., Sumfleth, K., Pathak, H., Howell, G., Ismail, A., … & Heuer, S. (2009). Regional vulnerability of climate change impacts on Asian rice production and scope for adaptation. *Advances in Agronomy, 102*, 91–133.

Watanabe, Y. W., Ishida, H., Nakano, T., & Nagai, N. (2005). Spatiotemporal decreases of nutrients and chlorophyll-a in surface mixed layer of the western North Pacific from 1971 to 2000. *Journal of Oceanography, 61*, 1011–1016.

Webb, A. P., & Kench, P. S. (2010). The dynamic response of reef islands to sea-level rise: Evidence from multi-decadal analysis of island change in the Central Pacific. *Global and Planetary Change, 72*, 234–246.

Werner, F. E., & APN. (2004). Climate interactions and marine ecosystems: Effects of climate on the structure and function of marine food-webs and implications for marine fish production in the North Pacific Ocean Marginal Seas (APN Project Report: ARCP2004-10NSY-Werner). Retrieved from http://www.apn-gcr.org/resources/items/show/1503

White, I. (2010, December). Tarawa water master plan: Te Ran, groundwater, prepared for the Kiribati adaptation programme phase II water component 3.2.1. Australian National University.

White, I. (2011a, February). Tarawa water master plan, 2010–2030. Prepared for the Kiribati adaptation programme phase II water component 3.2.1, Australian National University.

White, I. (2011b, February). Te Ran-Maitira ae Kainanoaki, future water demand, Tarawa water master plan. Prepared for the Kiribati adaptation programme phase II water component 3.2.1, Australian National University.

White, I., & Falkland, A. (2010). Management of freshwater lenses on small Pacific islands. *Hydrogeology Journal, 18*, 227–246.

White, I., Falkland, A., Metutera, T., Metai, E., Overmars, M., Perez, P., & Dray, A. (2007). Climatic and human influences on groundwater. In low atolls. *Vadose Zone Journal, 6*, 581–590.

White, I., Falkland, T., Metutera, T., Katatia, M., Abete-Reema, T., Overmars, M., … & Dray, A. (2008). Safe water for people in low, small Island Pacific Nations: The rural–urban dilemma. *Development, 51*, 282–287.

White, I., Falkland, A., & Fatai, T. (2009). *Groundwater evaluation and monitoring assessment, vulnerability of groundwater in Tongatapu, Kingdom of Tonga. SOPAC/EU EDF8 reducing the vulnerability of Pacific APC States*. Canberra: Australian National University.

Wilkinson, C. (2008). *Status of the coral reefs of the world: 2008*. Townsville: Global Coral Reef Monitoring Network and Rainforest Research Centre.

Williams, P., & Terawasi, P. (2010, August 10–19). *Overview of Tuna fisheries in the Western and Central Pacific Ocean, including economic conditions – 2009*. WCPFC-SC6-2010-GN-WP-01, Western and Central Pacific Fisheries Commission Scientific Committee, Sixth Regular Session, Nuku'alofa, Kingdom of Tonga.

Woodroffe, C. D. (2007). Reef-island topography and the vulnerability of atolls to sea-level rise. *Global and Planetary Change, 62*, 77–96.

World Bank. (2000). *Adapting to climate change. Vol. IV, In cities, seas and Storms, Managing Change in Pacific Island Economies*. Papua New Guinea and Pacific Island Country Unit: The World Bank.

Xu, J., Shrestha, A., Vaidya, R., Eriksson, M., & Hewitt, K. (2007). *The melting Himalayas: Regional challenges and local impacts of climate change on mountain ecosystems and livelihoods* (ICIMOD technical paper). Kathmandu: ICIMOD.

Yao, T. D., Guo, X. J., Lonnie, T., Duan, K. Q., Wang, N. L., Pu, J. C., … & Sun, W. Z. (2006). Record and temperature change over the past 100 years in ice cores on the Tibetan plateau. *Science in China: Series D Earth Science, 49*(1), 1–9.

Yin, X., & Struik, P. C. (2008). Applying modelling experiences from the past to shape crop systems biology: The need to converge crop physiology and functional genomics. *New Phytologist, 179*, 629–642.

Zhang, X., Meinke, H., DeVoil, P., van Laar, G., Bouman, B. A. M., & Abawi, Y. (2004). Simulating growth and development of lowland rice in APSIM. Proceedings 4th International Crop Science Congress, Brisbane. 26 Sept–1 Oct 2004.

Ziegler, A., Giambelluca, T., Sutherland, R., Nullet, M., Yarnasarn, S., Pinthong, J., … & Jaiaree, S. (2004). Towards understanding the cumulative impacts of roads in upland agricultural watersheds of northern Thailand. *Agriculture, Ecosystems and Environment, 104*(1), 145–158.

Chapter 5
Climate and Society

Kanayathu Koshy, Linda Anne Stevenson, Jariya Boonjawat, John R. Campbell, Kristie L. Ebi, Hina Lotia, and Ruben Zondervan,

Abstract The complexity of climate governance at multiple levels leads to fragmented approaches and there is a mismatch between international agreements and commitments that leads to delays in progress. Climate governance requires a great amount of dialogue, action and financing, and strengthening institutional frameworks for

---

K. Koshy (✉)
Centre for Global Sustainability Studies (CGSS), Universiti Sains Malaysia, 11800 Penang, Malaysia
e-mail: kanayathu.koshy@gmail.com

L.A. Stevenson (✉)
Asia-Pacific Network for Global Change Research, APN Secretariat, 4F, East Building, 1-5-2 Wakinohama Kaigan Dori, Chuo-ku, 651-0073 Kobe, Japan
e-mail: lastevenson@apn-gcr.org; clothears_2008@yahoo.co.jp

J. Boonjawat
Southeast Asia START Regional Centre (SEA START RC), Chulalongkorn University, 5th Floor Chulawich Building, 10330 Bangkok, Thailand
e-mail: jariya@start.or.th

J.R. Campbell
The University of Waikato, Te Whare Wānanga o Waikato, Gate 1 Knighton Road, Private Bag 3105, 3240 Hamilton, New Zealand
e-mail: jrc@waikato.ac.nz

K.L. Ebi
ClimAdapt, LLC, 424 Tyndall St., 94022 Los Altos, CA, USA
e-mail: krisebi@essllc.org

H. Lotia
Programme Development Department, Leadership for Environment and Development (LEAD), LEAD House, F-7 markaz, Islamabad, Pakistan
e-mail: hlotia@lead.org.pk

R. Zondervan
Earth System Governance Project, Lund University, P.O. Box 170, SE-221 00 Lund, Sweden
e-mail: ruben.zondervan@esg.lu.se; ipo@earthsystemgovernance.org

M.J. Manton and L.A. Stevenson (eds.), *Climate in Asia and the Pacific: Security, Society and Sustainability*, Advances in Global Change Research 56, DOI 10.1007/978-94-007-7338-7_5,

climate governance is desirable. Developing and widely disseminating technologies and methodologies for securing food and eradicating climate-induced health risks, that especially targets poorer communities, is crucial. Empowering poorer communities in this aspect will help alleviate the problems faced by the most vulnerable. Also crucial to creating a resilient community is the implementation of adaptation measures and poverty reduction approaches, both of which need to be integrated into national development plans. Given the urgency for capacity building to improve climate responses, enhanced efforts in collaboration with national institutions and international partners, including the private sector, is needed. In the longer term, establishing international funding and technology transfer mechanisms based on a comprehensive climate agreement is desirable.

Remote communities are geographical hotspots that suffer the impacts of multiple climate hazards, lacking access to assets, and little adaptive capacity. The Chapter focuses on two very different communities of the Himalaya-Tibetan Plateau (HTP) and the Small Island Developing States (SIDS), situated in both the Pacific and Indian oceans. Capacity building and community-based adaptation is needed for these vulnerable countries. Climate hazards disrupt ecosystem services that support human health and livelihoods, and climate impacts will affect health systems and cause fatalities through various means. Climate-induced threats to food security will lead to increasing malnutrition and consequent nutrition disorders, with implications for child growth and development. Important for small and remote communities is access to information on climate change and its impacts. While awareness-raising activities are becoming more evident, there is still a great need to strengthen climate information networks that clearly outline social vulnerabilities. For this reason, more emphasis on community-based activities to strengthen their adaptive capacity is needed where all sectors are involved and where climate change is integrated and mainstreamed through national adaptation and sustainable development plans.

Keywords Climate governance • Human health • Climate society • Remote communities • Climate resilience • Vulnerability

5.1 Governance

5.1.1 Climate Within Governance Systems of Asia and the Pacific

Human activities have moved numerous natural sub-systems of the planet outside the range of natural variability established over the past 500,000 years (Steffen et al. 2004). The nature of these changes, their magnitude and rates of change are as such already alarming and an immense challenge for society; a society in which at the same time basic human needs are not met in many parts of the world. In addition,

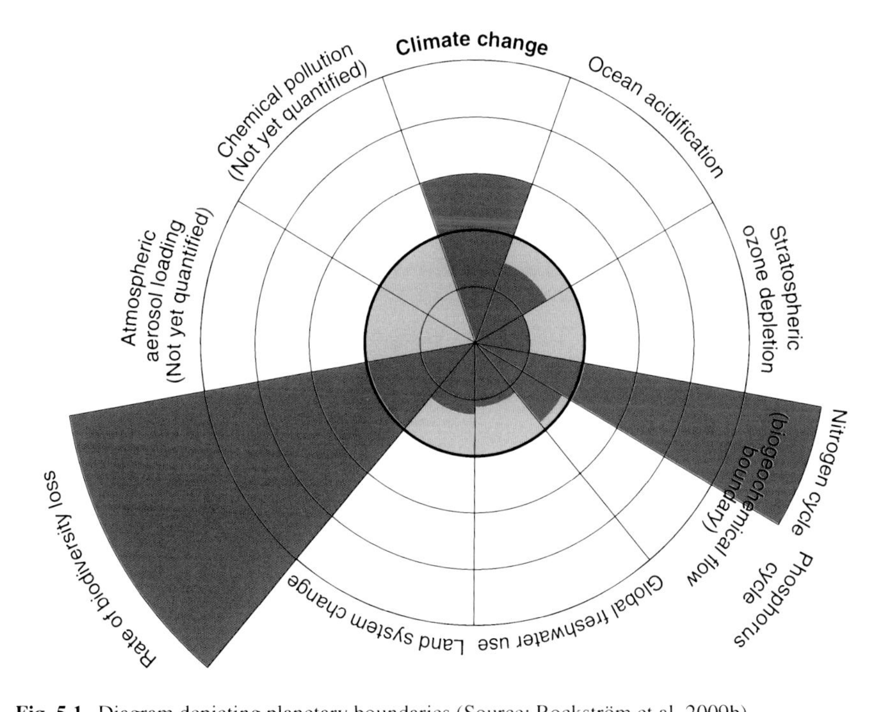

Fig. 5.1 Diagram depicting planetary boundaries (Source: Rockström et al. 2009b)

global environmental change could trigger tipping points in natural systems, causing shifts of equilibrium and new modes of operation (Rockström et al. 2009a). To avoid these potential abrupt and non-linear changes, planetary boundaries, as developed by Rockström et al. (2009b) are indicators of the limits to the safe operating space for humanity. For climate change, the proposed boundary has already been exceeded (Fig. 5.1).

Climate change governance has become one of the key challenges for policy makers and researchers from local to global levels. The establishment of climate governance is not only to mitigate climate change but also to adapt to climatic change already occurring; such as sea level rise, biodiversity loss, and extreme weather events. Response to climate change impacts have to be two-pronged with mitigation addressing the root cause of the issue and adaptation to manage the impacts of those changes already being felt by communities.

Climate change is not just a cumulative and systemic problem at the global level. It has different characteristics causes, and impacts at all levels of governance (Biermann et al. 2010), and is characterized by complexity, interdependency (functional, spatial, and temporal), and uncertainty; both in regards to the problem as well as policy responses (Biermann 2007).

Asia is a key region when it comes to mitigating climate change and adapting to its adverse impacts. While the Asian continent presently accounts for a large share of global carbon dioxide (CO_2) emissions, with some exceptions such as in Japan and Republic of Korea, per capita emissions are still relatively low. However, due to population increase and economic development, both per capita and absolute emissions are likely to increase substantially. Although industrialized countries emit far larger amounts of CO_2 per capita than the world average some rapidly expanding economies in the Asia-Pacific region are significantly increasing their emissions per capita. According to the report released by OECD/IEA (2012), emissions per capita vary even more widely across world regions than GDP per capita.

In addition, there are vast differences in welfare, governance systems, and carbon emission trajectories, both within as well as between states in Asia. Governance systems range from authoritative regimes and closed economies on the one hand, to liberal democracies on the other.

5.1.2 Climate Within Other Socio-Economic Challenges

Because of the immense development challenges still faced in most parts of Asia, climate governance needs to be integrated into broader governance of sustainable development. Climate change can severely hamper social and economic development and vice versa policy responses to climate change could, whether intentional or otherwise, have negative side-effects on social development and economic growth. Therefore, policy makers face interconnected but competing governance challenges related to appropriate allocation of resources (Biermann et al. 2009).

The development of the United Nations Framework Convention on Climate Change (UNFCCC) as part of the Rio 1992 outcome has been the most tangible and concrete measure of success and advancement of international climate law. However, the implementation of its protocol and agreements has been less successful and there has been a growing need to enhance synergies and interlinkages of various Multilateral Environmental Agreements (MEAs) as a means to strengthen UNFCCC framework itself. UNEP's Division of Environmental Law and Conventions has now focused on strengthening interlinkages and synergies of MEAs and the development and implementation of a systematic approach for coordination among MEAs.

Given the far-ranging adverse impacts of climate change, adaptation must be an integral component of an effective strategy to address climate change, along with mitigation. The two are intricately linked; and the more we mitigate, the less we have to adapt. However, even if substantial efforts are undertaken to reduce further GHG emissions, some degree of climate change is unavoidable and will lead to adverse impacts, some of which are already being felt. The world's poor, who have contributed the least to GHGs, will suffer the worst impacts of climate change and have the least capacity to adapt. Elementary principles of justice demand that the world's response strategies and adaptation funds give special priority to the poorest countries (Ricardo and Wirth 2009).

5.1.3 Multi-Level Institutional Frameworks

Climate change challenges the capacity of traditional governance structures at the national level with developing country governments typically dealing with substantial restraints on resources access to information and expertise. In this context non state actors are also beginning to play an important role both nationally and internationally. In such situations issues of participation and inclusiveness, transparency and openness are important in all forms of governance involving public-private partnership. The politics of mitigation and adaptation in the light of growing emissions in the Asia-Pacific region need to address poverty together with ensuring sustainable development. The need for governance structures that innovatively balance these competing goals cannot be overly emphasized.

Pathways of regional development are sequences of interrelated changes in social, economic and governance systems. They vary from place to place and over time, in ways that are likely to have different net consequences for carbon stocks and fluxes, which in turn may constrain or in other ways feedback upon development processes (Dhakal et al. 2011). Thus, the climate problem is not just a cumulative and systemic problem at the global level but has different features causes, and impacts at different levels of governance. These mutual interdependencies and feedbacks place demands on the science community to establish a common, mutually agreed knowledge-base to support policy debate and action, and to develop integrated systems of governance, from the local to the global level, that ensure the sustainable development of the coupled socio-ecological systems.

Apart from government ministries in the Asia-Pacific region, there are a number of other agencies such as university research centers working in the area of climate governance. One such network is the Climate and Environmental Governance Network (CEGNET) of the Australian National University, Canberra (http://regnet.anu.edu.au/cegnet/home), which engages in theoretical, empirical and interdisciplinary research on institutions of regulation and governance (for example, property rights, markets, trade regimes, treaties, national laws) and organizational actors (for example, states, governments, international organizations, corporations, NGOs), and the ways in which relationships and influences amongst them affect the capacity of societies to respond to environmental change and crisis. This network does not underestimate the importance of good climate governance and, with support from the APN; capacity building activities in the region are ongoing (Tienhaara 2012).

Also playing an active role in climate governance is the Energy and Resources Institute of New Delhi, India (http://www.teriin.org/index.php) and their special governance on climate change and migration, development and CDM contributions to sustainable development; and the Pacific Centre for Environment and Sustainable Development (PACE-SD), the University of the South Pacific, Fiji (http://www.usp.ac.fj/index.php?id=570/), with a special role in the development and implementation of PIFACC – Pacific Islands Framework for Action on Climate Change 2006–2015 (SPREP 2006, 2011); focusing especially on community based climate adaptation and capacity building for multiple risk reduction.

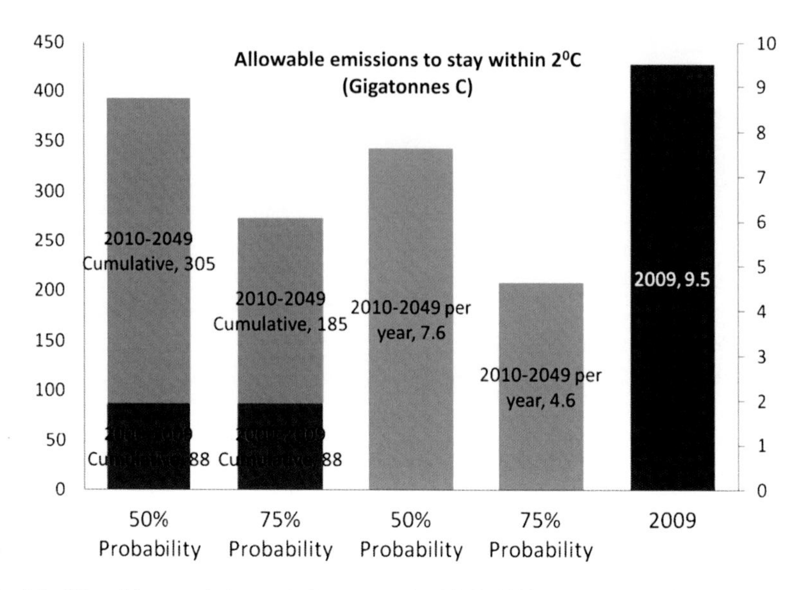

Fig. 5.2 Allowable cumulative emissions quota for 2010–2049 to stay below 2 °C by end of the century (*first two bars* from *left*), per year quotas left (*third* and *fourth bar* from the *left*) and our actual emissions in 2009 (*purple bar*) (Source: Estimation by Dhakal et al. (2011) based on Meinshausen et al. (2009))

Asia is a key region, which is rapidly growing economically. Asian contributions are already dominating the global carbon emissions and Asia will play an even greater role for global carbon management in the foreseeable future. However, within Asia, huge differences in welfare, governance systems, and carbon emission trajectories exist and thus pose a carbon governance challenge. A better understanding of the carbon management challenges across multiple scales is necessary for Asia, which is less understood as of now. Such understanding will provide important insights to design an optimized carbon governance structure.

The Copenhagen Accord, which was adopted at the UNFCCC 15th Conference of the Parties (COP15), 2009, in Denmark and outcomes in the Cancun Framework from the UNFCCC COP16 climate summit in Mexico, 2010 envisions a world below 2 °C temperature rise by the end of the twenty-first century. In order to do that our carbon space, i.e. the amount of anthropogenic CO_2 that we can emit into the atmosphere from now to 2050, is limited. In 2009, the world emitted 9.5 Gigatonnes of Carbon (34.8 Gigatonnes equivalent CO_2) while our annual allowable carbon per year for 2010–2050 for staying below 2 °C (by the end of century) with 50 % probability is 7.6 Gigatonnes of CO_2 (Fig. 5.2). IPCC has shown that world CO_2 emissions need to peak and start declining very soon in order to achieve global 2 °C policy target.

Asia is a key region with a large and growing share of carbon emissions and thus the opportunities as well as challenges for carbon management are expected to be immense. The regional knowledge base on approaches for sound carbon governance is very timely and important for Asia and this capacity building activities aims to share knowledge and expose early career researchers of the Asia-Pacific region to the analyses and improved understanding of carbon governance principles, ways and prevailing experiences. Carbon governance is very timely and important for Asia and this capacity building activities aims to share knowledge and expose early career researchers of the Asia Pacific region to the analyses and improved understanding of carbon governance principles, ways and prevailing experiences.

5.1.4 Architecture and Agents of Climate Governance

5.1.4.1 Global Level

In environmental policy, climate governance is the diplomacy, mechanisms and response measures 'aimed at steering social systems towards preventing, mitigating or adapting to the risks posed by climate change' (Jagers and Stripple 2003). A definitive interpretation is complicated by the wide range of political and social science that are engaged in conceiving and analyzing climate governance at different levels and across different arenas. In the past two decades a paradox has arisen between rising awareness about the causes and consequences of climate change and an increasing concern that the issues that surround it represent an intractable problem. Initially, climate change was approached as a global issue, and climate governance sought to address it on the international stage, taking the form of MEAs and beginning with the UNFCCC in 1992.

While at the global level the architecture of climate governance can be considered as centering around the UNFCCC, it also includes hundreds of organizations, ranging from civil society groups, scientific communities and business associations, to intergovernmental agencies and secretariats (both within and outside the UN System).

5.1.4.2 Regional Level: Asia and the Pacific

The heads of State of the Association of Southeast Asian Nations (ASEAN) are proactively leading ASEAN's efforts to address climate change issues in the region and beyond. Their adopted Road Map for ASEAN Community 2009–2015 (ASEAN 2009) places the ASEAN climate change agenda in the context of sustainable development outlining strategies and actions in the ASEAN Socio-cultural Community Blueprint. Within this context, ASEAN is playing a key role through its declaration to the 2007 Bali UNFCCC COP13, its Joint Statement to the 2009 Copenhagen UNFCCC COP15, and by continuing to be an active player in the UNFCCC COP process. ASEAN is therefore addressing climate change through its framework of ASEAN Community building, with strategies and actions rooted in the various development and sectoral areas.

The Pacific Island countries address climate change within the context of a specific framework, PIFACC (SPREP 2006, 2011), endorsed by the Pacific Islands' leaders in 2005. The vision of PIFACC is 'Pacific island people, their livelihoods and the environment resilient to the risks and impacts of climate change' and the goal of the framework is to ensure Pacific Island people build their capacity to be resilient to the risks and impacts of climate change with the key objective to deliver on the expected outcomes. The measures are to strengthen climate change action in the region, to inform the decisions and actions of national, regional and international partners, and to promote links with, more specific regional and national policies and plans across specific sectors, such as disaster risk management, water, waste management, agriculture, energy, forestry and land use, health, coastal zone management, marine ecosystems, ocean management, tourism, and transport and to secure access to adequate, predictable and sustainable resources to address climate change.

Governance and decision-making is one of the six goals of the framework. Realizing the importance of an effective and efficient enabling environment that ensures administrative feasibility and operational capability, the island nations meet regularly to review progress, to establish institutions needed to link regional, national and local initiatives with the international frameworks for addressing climate change risks, and to represent their interests internationally.

The regional intergovernmental organizations such as the Pacific Islands Forum Secretariat, Suva, and SPREP together with other members of the Council of Regional Organizations in the Pacific agencies organize, coordinate and contribute to the implementation of the climate framework activities at national levels. Multinational donors such as the Global Environmental Facility (GEF) and the European Union and a host of bilateral development partners support climate work in the Pacific.

5.1.4.3 National Level

Three examples are given in this section from Australia, Malaysia and Nepal:

(a) *Australia*

In order to meet the climate-change obligations of Australia, the government introduced (in 2012) a rather controversial carbon tax as part of the Clean Energy Act. The new law requires about 300 of the worst-polluting firms to pay an A$23 (£15; US$24) levy for every tonne of greenhouse gases they produce.

While the government is planning to assist low income earners with any adverse impact of the law, it also hopes that the legislation will force innovation in renewable energy supplies, and free the country from its reliance on fossil fuels. In its Fifth National Communication on Climate Change (2010), Australia reports a number of important steps it has taken to improve climate governance:

- Ratifying the Kyoto protocol in 2007;
- Committing to increased emission reduction targets such as, reducing emissions to 60 % below 2000 levels by 2050 and from 5 % to 15 % and 25 %

below 2000 levels by 2020, depending on the outcome of the current negotiations on a global agreement to limit emissions;
- Developing a Carbon Pollution Reduction Scheme to put a price on carbon and achieve these targets;
- Creating legislation to improve the Renewable Energy contribution to the national electricity production by 20 % by 2020;
- Creating legislation for a national greenhouse gas and energy reporting system; and
- Providing continued and increased investment and cooperation for Asia-Pacific on climate change issues.

Given Australia's considerable background in climate change science and policy, a guide on this theme, *'Climate Change: An Australian Guide to the Science and potential Impacts'* (Pittock 2003) has also been published to explain the science behind climate change, potential impacts at the global and national level, adaptation potential and vulnerability and gaps in our response to climate change.

(b) *Malaysia*

Malaysia's national policy on climate change published by the Ministry of Natural Resources and Environment Malaysia (2010) provides a framework to mobilize and guide government agencies, industry, community groups and other major stakeholders to address the challenges of climate change in a holistic manner that can help the nation navigate towards sustainability. The policy statement to 'ensure climate-resilient development to fulfill national aspirations for sustainability' has three objectives:

- Mainstreaming climate change for strengthened competitiveness and improved quality of life;
- Integrating climate change responses into policies, plans and programmes; and
- Strengthening institutional and implementation capacity.

Malaysia has been actively pursuing ways to integrate climate change impacts and adaptation as part of disaster risk management, public health initiatives, and infrastructure development. There have also been efforts to harmonize existing legislation, policies and plans to develop a balanced approach to adaptation and mitigation taking into account the country's national development aspirations. As a rapidly industrializing economy with relatively higher per capita emissions compared with other developing countries, Malaysia has agreed to play a greater role in the near future in international negotiations for emission reductions. For example, during the UNFCCC COP15 conference in Copenhagen, Malaysia Prime Minster, Mohammad Najib Abdul Razak, delivered the country's proposal to reduce its CO_2 emissions to 40 % by the 2020 compared with 2005 levels, subject to assistance from developed countries.

(c) *Nepal*

Agriculture a crucial component of Nepal's economy and contributes almost 35 % to the national GDP, while engaging more than 65 % of the total labor force (NPC 2010; Gurung et al. 2011). There is predominance of mixed farming system in which crop, livestock and forestry are integrated. Nepal is known to be highly disaster prone and vulnerable to climate change and a study by the Department of Hydrology and Meteorology revealed that the average temperature in Nepal has been increasing at a rate of approximately 0.06 °C per year (Dahal and Khanal 2010; ICIMOD 2009).

Relevant national policies and plans have emerged in the country including Nepal's National Adaptation Plan of Action (Nepal NAPA 2010), Climate Change Adaptation Policy of the Government of Nepal (2010), Nepal Agriculture Research Council Vision 2011–2031. The Asia-Pacific Adaptation Network (APAN) reviewed these plans to assess priority areas at the national and sector levels (Okayama 2011) for training and capacity in relation to climate adaptation governance. The report on Climate Change in the Mid-hills of Nepal (Helvetas, Nepal and Intercooperation 2010) revealed various facts from the farmers' perspectives at grassroots level in the agriculture sector.

Immediate needs were recorded for capacity in climate adaptation governance:

- Build capacity of communities and local governments, including district agriculture offices and farmers' institutions;
- Incorporate climate change adaptation topics into educational curricula at all levels from primary through tertiary education;
- Collaborate with regional and international agencies for human resource capacity building through research, exchange visits, twinning arrangements; and
- Provide short training courses to increase awareness and facilitate the integration of adaptation policies into national development plans and programmes, particularly in thematic areas identified in Nepal's NAPA.

Youth has become a prominent feature in climate governance in Nepal and the younger generation from a broad range of disciplines outside the traditional realms of natural science are taking a proactive stance in the face of climate change. This was visible through an International Graduate Conference on Climate Change and People, the first of its kind, held in Kathmandu (Pradhananga 2010). The conference covered a range of diverse topics of biodiversity, water resources, climate change science, natural hazards, anthropology, biogeography, policy, equity, and ethics among hundreds of the youth community from 17 countries in South Asia, and focusing on multidisciplinary capacity building of graduate students from a range of disciplines, including climate science, hydrology, sociology, journalism, law, etc.

5.1.4.4 Local Level

For climate governance at the local level, women in rural areas have enormous untapped potential to promote sustainable development and poverty eradication. However, 'they lack equal access to opportunities and resources, which will help generate results', noted UN Secretary-General Ban Ki-moon in a message marking the 2011 International Day of Rural Women. According to the UN Food and Agricultural Organization, if rural women had the same access to productive resources as men, then more than 100 million people could be lifted out of poverty (FAO 2011). Mr. Ban also citing FAO stated, 'Productivity on women's farms would increase up to 30 % and the number of hungry people would drop by as much as 17 %, which translates into improvements for up to 150 million individuals'.

Beddington et al. (2012) highlighted a case study in India showing how Indian women benefit from poverty alleviation programmes that also address environmental sustainability objectives, particularly when they involve locally appropriate, bottom-up planning. The Mahatma Gandhi National Rural Employment Guarantee Act (MGNREGA), launched in 2006, now operates similar schemes in all of India's districts.

5.1.4.5 Individual Level

Empowered youth are becoming ever more visible these days, not only thanks to various social media and networks, but through other means that stray from the conventional world of science and academia. For example, the Asia-Pacific Mountain Courier in its special edition (APMC 2011) presents several encouraging cases of young climate champions who express themselves through art forms to build awareness and reach out to their communities and the world. The artwork of Amal Shakeel of Pakistan is one such example in the special edition.

Many other contributions are drawn from Youth such as Empowering Youth with Earth Observation Information for Climate Actions, October 2010 (ICIMOD; Nepal 2010) and the Asia-Pacific Forum: Youth Action on Climate Change Exploration through Cultural Expression, organized by Southeast Asian Ministers of Education Organization (SEAMEO) in Bangkok, January 2012.

The Asia-Pacific Mountain Network (APMN) is growing at a steady pace, with, at the time of writing, 292 organizational and 2,081 individual registered users from 43 countries across the Asia-Pacific region. The work of emerging champions from the youth arena is noteworthy and two examples are highlighted in Box 5.1. There are many other examples of young and old climate champions whose work may not often come to the forefront, but who play an important role in individual climate activism of great local or corporate-sector impact.

5.1.5 Knowledge-Based Climate Governance

In a rapidly globalizing world, progress in climate cooperation and action leading to economic productivity and services will be based on knowledge-intensive activities that rely more on intellectual capabilities than on physical inputs or natural resources. This implies that, in addition to enhancing physical infrastructure, we

Box 5.1: Young Champions for Climate Governance

The Asia-Pacific Mountain Network of Nepal's ICIMOD is raising awareness through various initiatives (such as art mentioned above) and showing how youth is championing climate governance in mountainous areas of Asia (Mahat et al. 2011).

A young bank employee was one international bank's 'climate champion' and is involved in a Climate Partnership programme at the China Climate Centre in Zhejiang. This is a 5-year, US$100 million campaign launched in 2007. The bank hopes an increase in employees' awareness of the business implications of climate change will help make its business more sustainable (SCMP 2011, 6 February).

will also need to urgently invest in knowledge infrastructure aimed at developing human capital to its highest potential. This technical capacity is often the prerequisite for advancing mitigatory and adaptive measures to address climate impacts. This requires sustained investments in governance and informed policies. The present section looks at various approaches to address climate governance, the first through an international science-led top down approach; and the second through local participatory methods that combine science-based approaches.

5.1.5.1 Earth System Governance Project

Responding to climate change is likely to be the defining challenge of this century and the Earth System Governance Project is making steps to advance research and knowledge-base in this area. At the time of writing, for example, a major conference in Tokyo organized by the Earth System Governance project and supported by the APN (Kanie 2012) and many other important stakeholders, was underway. This conference gathered together literally hundreds of key actors from all levels of governance from around the globe; including governments at various levels, academia, practitioners and the private sector to discuss the way forward for Earth System Governance (http://tokyo2013.earthsystemgovernance.org/conference/about/).

Launched in 2009, the Earth System Governance project addresses the problems of environmental governance. The project defines 'earth system governance' as the interrelated system of formal and informal rules, rule-making mechanisms and actor-networks at all levels of human society (from local to global) that are set up to steer societies towards preventing, mitigating, and adapting to global and local environmental change and earth system transformation, within the normative context of sustainable development. The Earth System Governance project science plan (http://www.earthsystemgovernance.org) is organized around five analytical problems:

- *Architecture* relates to the emergence, design and effectiveness of governance arrangements.

- *Agency* addresses questions of who governs the earth system and how.
- *Adaptiveness* research explores the ability of governance systems to change in the face of new knowledge and challenges as well as to enhance adaptiveness of social-ecological systems in the face of major disturbances.
- *Accountability* refers to the democratic quality of environmental governance arrangements.
- *Allocation and access* deal with justice, equity, and fairness. These analytical problems are united by the cross-cutting themes of power, knowledge, norms and scale.

The next generation of scholars that will be at the forefront of mitigation and adaptation research in the pivotal period leading up to 2050 is now entering academia. Young researchers are tackling a wide range of climate issues from a variety of disciplinary perspectives. Although much important work focuses on the development of new technologies and improved understanding of the underlying biogeophysical systems, it is clear that effective and accountable governance systems will also be required to ensure that mitigation and adaptation strategies are implemented in a timely and equitable manner. The Asia-Pacific region will face particularly grave challenges, including significant displacement of human populations, human health issues and loss of valuable ecosystem services, if climate change continues unabated.

The challenge of establishing effective strategies for mediating the relationship between humans and the natural world represents one of the most daunting tasks in the quest for environmental sustainability at all levels, from the local to the global. Environmental problems, such as climate change, biodiversity loss, water quality and access problems, soil erosion and others, call into question the fundamental viability of how humans have organized the relationship between society and nature. There is an urgent need to identify and develop new strategies for steering societies towards a more sustainable relationship with the natural world. Arguably, this is where both formal and informal training become both urgent and important.

5.1.5.2 Participatory Approaches: Combining Science and Local Knowledge

Competence in climate impact assessments is important, particularly among local decision makers who are at the forefront of actions for responding to pressing issues exacerbated by climate change and how it impacts sectors and societies differently. This is due to the variation in impacts among regions, difference in the characteristics of groups and sectors, and difference as well in the extent of their responses (Ionescu et al. 2009). Such creates a need for area-specific responses. Crucial too is the knowledge and experiences of local people considered *together* with scientific or computer-based tools for assessment, to put into context the source of vulnerability. This facilitates adaptation, which should be treated as a process through which measures are carried out, rather than considering it merely as adaptation strategies.

The Adaptation Policy Framework (APF) was used by Pulhin (2009) as a guide in conducting the assessment. This framework represented a holistic approach

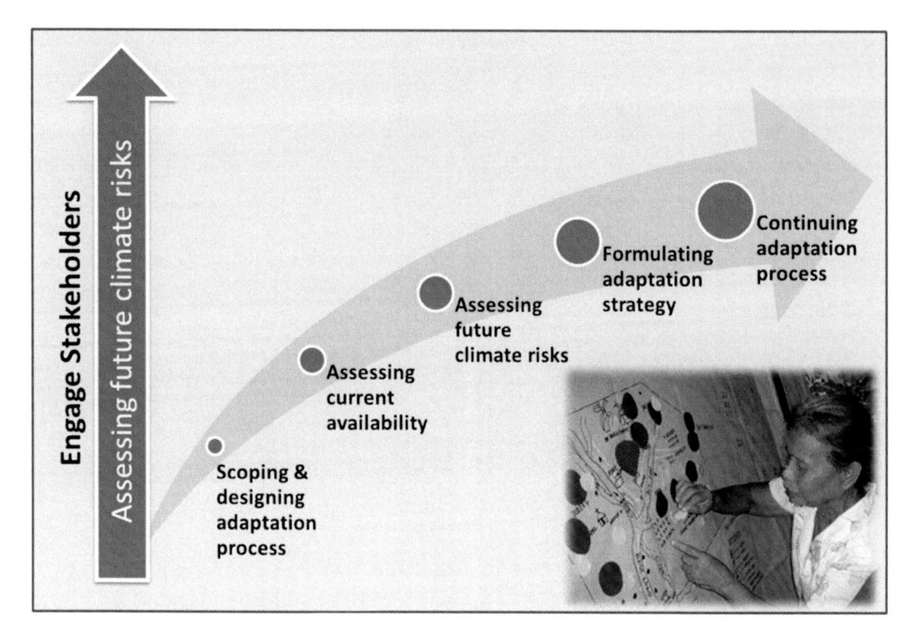

Fig. 5.3 Adaptation policy framework (Modified from Pulhin 2009)

towards impact assessments with a goal of informing adaptation policy (Lim et al. 2004). Central to this is the assessment of current vulnerability, particularly from climate variability and extremes, placing emphasis on understanding current climate risks before moving to the 'uncertain territories of what lies ahead' (Dessai et al. 2005). The APF has five major components linked by two cross-cutting components (adaptive capacity and stakeholder context) (Fig. 5.3). What can be gleaned from the work conducted by Pulhin (2009) is the importance of integrating science-based approaches (for example, climate modeling) and the knowledge and experiences of the local people (for example, through participatory techniques) in assessing impacts, vulnerability and adaptation to climate change. More often than not, the context of vulnerability may not be adequately represented by its generally known defining factors of exposure, sensitivity and adaptive capacity.

Deeper investigation of the intricate interactions in a community reveals non-climatic factors that increase or perpetuate the vulnerability of a certain group to climate- and climate change-related hazards. It is important to be aware of these factors as it is at the community or local level where the locus of the battle against climate change lies.

On the other hand, the future presents an uncertain picture, which is beyond the experience and comprehension of the local communities. This is where the so-called scientific tools such as computer modeling systems could come in to aid in providing a glimpse of what is likely to happen. While uncertainty is prevalent in the climate change arena, planning is about reducing the risk of being vulnerable to uncertainties and surprises. Climate change makes planning for adaptation troublesome because of the many uncertainties involved, but there are (partial) solutions (Dessai 2005). When considering planning ahead, the currently available computer-based tools, which could simulate feasible future scenarios and impacts could facilitate informed decision-making and long-term adaptation planning. Such an approach captures, as well, the "forward-looking aspect" of climate change.

Having the capacity to conduct high quality research that provides underpinning scientific support for decision makers and decision-making processes is vital for least-developed nations in the Asia-Pacific region and is recognized by the APN as crucial for improving the scientific and technical capabilities of these nations. In this context, the importance of capacity building for evidence-based climate decision-making and overall governance cannot be over emphasized. A case in point is the successful climate and extreme events Training Institute in the Pacific in Fiji in 2004; Samoa in 2005; and Kiribati in 2006 (Koshy et al. 2005).

This collaborative training for the Pacific Island Countries involved resource persons mainly from East West Center (EWC), USA; National Institute for Water and Atmospheric Research (NIWA), New Zealand; the Pacific Centre for Environment and Sustainable Development (PACE-SD), Fiji; and was implemented using a well-developed curriculum, which included face-to-face formal presentations and discussions as well as climate scenario development using a SimClim (http://www.climsystems.com/simclim/) training model. Overall about 75 young scientists were trained, all of whom are key climate and sustainable development practitioners at different levels in the Pacific. This APN-funded activity under its CAPaBLE Capacity Development Programme was considered by APN reviewers as an outstanding best practice worth emulating in other places (see Box 5.2).

Significant climate variability and associated societal consequences occur in Oceania resulting from El Niño Southern Oscillation (ENSO), and the Interdecadal Pacific Oscillation (IPO), as described in Chap. 2. Compilation of data on these natural climate variability phenomena has allowed the monitoring, detection and attribution of climate change in the region and also provided useful information for the Intergovernmental Panel on Climate Change.

Scientists from the National Institute for Water and Atmospheric Research (NIWA) conducted three training workshops with funding from the APN for Pacific Island participants in 2000, 2001 and 2003 (Salinger 2003). Arising out of these workshops, the need for an easily-maintained and user friendly climate database,

Box 5.2: Training Institute in the Pacific

APN's Capacity Building Programme, CAPaBLE, Phase 1 In Review (2008) highlights that the Training Institute (Koshy et al. 2005).

> ....has established a modus operandi for dealing with climate change and its impacts across the Pacific island region that has the potential to be applied widely by both the team in this project and by others as well in ways that could have very substantial benefits for the Island States of this region. (APN/CAPaBLE 2008)

digitization of paper records of climate data to allow for contributions to IPCC assessments, enhancement of climate-related education and training in regional institutions, the critical role of media in providing policy makers and the public with timely and relevant information, and the importance of partnership building emerged as the major recommendations.

Similar workshops continue building Pacific Islands' capacity, such as a recent training workshop on improving Pacific Island meteorological data rescue visualization capabilities through involvement in emerging climate research programmes was undertaken (Lorrey 2010).

Another activity conducted under the APN's core research programme built on previous work in India and Pakistan and established a network of research teams with the capacity to apply agricultural systems analysis to evaluate options for managing climatic risk. Building on that foundation, the project documented and delivered benefits from climate information to agricultural decision makers and plotted a course for large-scale, sustained operational support of seasonal climate information and prediction within India, Indonesia and Pakistan. The project also brought scientists in Asia together in a knowledge network for better organizational frameworks for addressing climate impacts on critical sectors (Meinke et al. 2004).

A comprehensive approach to climate mitigation involves enforceable and time bound international policies on the one hand and source reduction and sink enhancement of carbon emission through a variety of innovative and cost-effective mechanisms on the other. Indeed, the current economic crisis can be used to make progress in a new direction, to speed up our efforts to create a new era of low-carbon green economy for sustainable development and poverty eradication. Cost-effectiveness is essential for creating the necessary consensus for effective climate change policies either through carbon tax, 'cap-and-trade' or other market mechanisms such as the UN Reducing Emissions from Deforestation and Forest Degradation Plus (REDD-plus), which goes beyond deforestation and forest degradation, and includes the role of conservation, sustainable management of forests and enhancement of forest carbon stocks.

In this approach, first, the pricing of carbon is critical to secure stakeholder interest. Secondly technological innovation can also bring down the cost over time. Both these approaches need investment and nurturing in the early stages and adequate regulatory mechanisms for sustainability. Studies in the Asia-Pacific region have demonstrated that ecologically sound technology can really make carbon saving in energy, water and sanitation sectors. Training workshops supported by APN has underscored the importance of capacity building to benefit from technology transfer sought by developing countries of the Asia-Pacific region (see, for example, Bambaradeniya 2007; Skole 2011).

Frameworks of international and national environmental law have evolved since the United Nation's Conference on Human Environment held in Stockholm in 1972, providing a sound basis for addressing critical environmental challenges of the present century. Having the necessary legal framework is one thing, but effective implementation has always been a thorny issue at all levels (Karlsson-Vinkhuyzen and Asselt 2009). With this in mind, Chief Justices and senior Judges,

Attorneys-General, Prosecutors, Auditors-General, Senior Legal Advisors and other representatives of the legal community from Asia and the Pacific gathered in Kuala Lumpur in October 2011, in preparation for the World Congress on Justice, Governance and Law for Environmental sustainability, a major part of which is climate and its convention-related legal framework.

The major messages from the participants of that particular meeting were the need for:

- Stronger linkages between social justice and environment for balanced development and social justice – this requires attention to the disproportional distribution of environmental impacts, exposing the poor to a larger share of their impacts while equitable sharing of mitigation of climate change, for example, is still only under consideration.
- Stronger environmental authority to simplify MEA management, and to keep the environment under periodic review, including monitoring and financing.
- Promoting common understanding and implementation of all MEAs, including those for climate change, through further elaborating, clarifying and codifying principles, customary and treaty law. In such an environment, MEA negotiations, ratification, implementation and reporting will be conducted efficiently under broader umbrella of sustainable development and its current institutional frameworks and those recommended in the Rio + 20 outcome (UNSDKP 2012).

5.1.6 Conclusion

While climate governance is not the new kid on the block, the topic is crucial for societies at all levels and is fast becoming a hot topic for international and regional arenas. For this reason, earth system governance, particularly in light of climate change, is receiving more attention these days, with scores of projects being undertaken from global to local levels.

The complexity of climate governance at multiple levels leads to fragmented approaches and there is a mismatch between international agreements and commitments that leads to delays progress. There is a considerable lack in mainstreaming climate into national development strategies and this creates implementation difficulties. With present and severe capacity constraints, there is a growing and urgent need to increase awareness and showcase success stories and best practices to urge accelerated action.

Climate governance requires a great amount of dialogue, action and financing, and strengthening institutional frameworks for climate governance is desirable, with one example being through the engagement in UNFCCC processes for collaborative and win-win outcomes at the national level. Developing and widely disseminating technologies and methodologies for securing food and eradicating climate-induced health risks, that especially targets poorer communities, is crucial. Empowering poorer communities in this aspect will help alleviate the problems faced by the most vulnerable.

Crucial to creating a resilient community is the implementation of adaptation measures and poverty reduction approaches, both of which need to be integrated into national development plans. This also includes promoting systemic and community-based adaptation focused on programmatic approaches. Given the urgency for capacity building to improve climate responses, enhanced efforts in collaboration with national institutions and international partners, including the private sector, is desirable. In the long-term, establishing international funding and technology transfer mechanisms based on a comprehensive climate agreement is desirable.

5.2 Climate and Society: Remote Communities

5.2.1 Introduction

Social vulnerability considers human welfare and livelihood strategies, emphasizing on access to assets, inequity, and social and political determinants. But vulnerability to climate change is described as a function of exposure to change, sensitivity to change and adaptive capacity (IPCC 2007). Based on climate hazards, and access to assets, remote communities, which are described as "geographic hotspots" can be divided into two groups: mountain communities and small islands (Table 5.1).

5.2.2 Mountain Communities

5.2.2.1 Geography

Climate in mountain regions varies with altitude. The foothills can be a tropical climate, whilst the peaks may be covered with ice and snow or just bare rock. In Asia-Pacific, the Himalayas and Tibetan Plateau (HTP) region is the most important mountain ecosystem. Because of the steep terrain and because temperature decreases with altitude, multiple ecological zones are "stacked" upon one another, sometimes ranging from dense tropical forests to glacial ice within a few kilometers. Mountains can affect the climate of nearby lands by blocking rain (rain shadow), so that one side of a mountain range may be rainy and the other side may be a desert. The Gobi desert is located behind the HTP, on the wind protected and dry side; thus it becomes a cold desert.

The Himalayan and Tibetan Plateau (HTP), extends over 2,400 km, marked by Nanga Parbat (8,125 m, 35°N, 74°E) on the west to the Tsankpo-Dihang bend around Namche Bazar (7,755 m, 30°N, 95°E) in the east. The entire HTP is called "the third pole" (Olshak et al. 1987). Numerous earthquakes in the region indicate continuing active orogenic movement by massive tectonic forces and, thus inherently structurally unstable. On the global scale, the greatest value of mountains may arise from their being the source of major rivers, and being centers of biological diversity.

Table 5.1 Geographical hotspots and climate hazards and exposure of people and livelihoods, natural resources, and ecosystems and biodiversity (Modified from USAID Asia 2010)

Item	Hotspots	Hazards							Vulnerabilities								
		Droughts	Floods	Sea Level Rise	Storms	Glacial Lake Outburst Flooding	Ocean Acidification	Warmer Sea Temperatures	Agriculture/Aquaculture/ Food Security	Water Resources	Infrastructure	Ecosystems and Biodiversity	Settlements	Livelihood	Human Health	Transport and communication
a	Pacific Islands	■	■	■	■		■	■	■	■	■	■	■	■	■	■
b	Coral Triangle	■	■	■	■		■	■	■	■	■	■	■	■	■	■
c	Mekong River Basin	■	■	■	■	■	■	■	■	■	■	■	■	■	■	■
d	Greater Himalayan Region/Tibetan		■			■			■	■	■	■	■	■	■	■
e	Asian Coastal Cities	■	■	■	■	■				■	■		■	■	■	■
f	Indian Ocean Islands	■	■	■	■		■	■	■	■	■	■	■	■	■	■
g	Sundarbans	■	■	■	■		■	■	■	■	■	■	■	■	■	■
h	Equatorial South Asian Coral Zone	■	■	■	■		■	■	■	■	■	■	■	■	■	■

The HTP region holds the largest mass of ice outside Polar Regions. It is also known as the Water Tower of Asia (see Sect. 4.2), and it is the source of the ten largest rivers in Asia. The HTP covers approximately seven million km^2 of which glacial ice covers 116,180 km^2 (Owen et al. 2002; Li et al. 2008; cited in Xu et al. 2009). The GIS map of the Water Tower of Asia in Different elevations and inland water sources of the HTP region are shown in Fig. 5.4. *Alpine* refers to all areas above 3,000 m; *mountain* to elevation ranges between 1,000 and 3,000 m, and *lowland* areas below 1,000 m. With higher heterogeneous geography, the HTP region has great climatic variability and it forms a barrier to atmospheric circulation for the summer monsoon and winter winds. Average annual precipitation ranges from less than 50 mm in the Taklimakan Desert (cold desert) in the northwest to about 11,117 mm in Cherapunji, India. The eastern HTP region contains a rich diversity of species and ecosystems that exist along a pronounced humidity gradient. Vegetation changes from the subtropical semi-desert and thorn steppe formation in the northwest to the tropical evergreen rainforest in the southeast (Schickhoff 2005; cited in Xu et al. 2009).

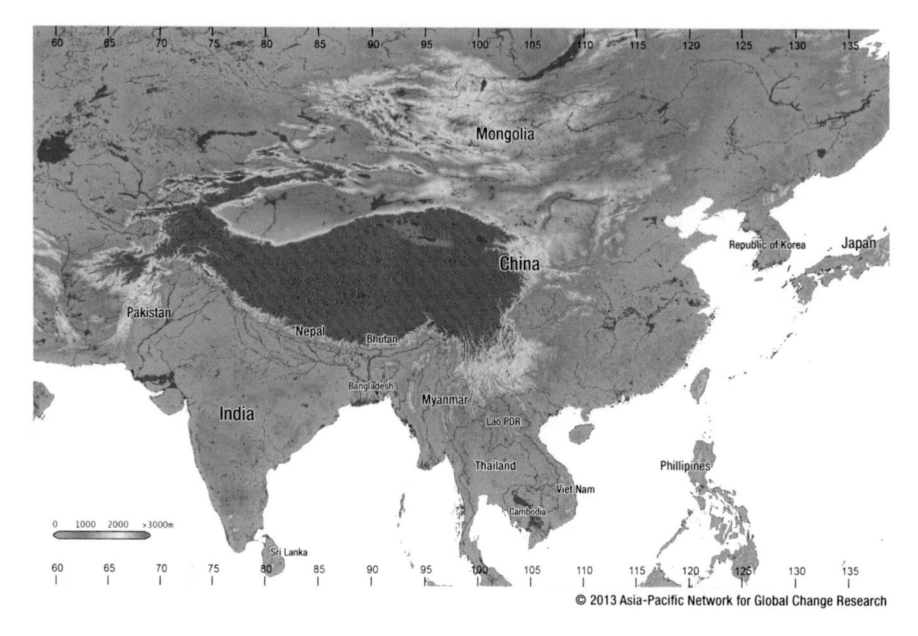

Fig. 5.4 Water tower of Asia showing elevation and inland water sources (Source: APN)

Table 5.2 Remote mountain communities along the highland Himalayan range (Source: Shrestha 2005)

Description	Western Himalayas	Central Himalayas	Eastern Himalayas
Annual average precipitation	700–2,000 mm	3,500 mm	Up to 10,000 mm
Main river system	Indus	Ganges	Brahmaputra
Area	208,359 km^2	113, 162 km^2	137, 839 km^2
Population	18,815,723	9,863,019	2,671,015
Population density	90.3 person/km^2	87.2 person/km^2	19.4 person/km^2
Number of languages	Indo Aryan:11 TibetoBurman: 9	Indo Aryan: 5 TibetoBurman: 11	TibetoBurman: 11
Culture	Caucasoids	Mixed	Mongoloids
Country	Pakistan, India	Nepal, Bhutan, China	China, Myanmar

The HTP and adjacent areas are home to about 100 million people, whereas the highland areas including Bhutan, Indian Himalayan states, and highland Nepal alone have a population of about 31.3 million, with wide variation in population density, language and culture (Shrestha 2005). Table 5.2 shows the spatial variation in annual precipitation, main river system, number of languages, culture and population, and population density of the three most populous mountain zones: Western, Central and Eastern Himalayas.

5.2.2.2 Mountain Ecosystems and Biodiversity

The HTP has unique physical, biological and human systems that are particularly vulnerable to the impacts of various aspects of climate change. Four biodiversity hotspots are located in the Himalayas, including the mountains of Central Asia, Himalaya, south western China, and Indo-Burma. The HTP, especially the eastern Himalayas, is exceptionally rich in biodiversity especially in the alpine environment. In 1995, approximately 10 % of known species in the Himalayas were listed as threatened (IPCC 2001).

Mountain forest ecosystems have multiple functions: they harbor biodiversity, anchor soil and water, provide carbon sinks, and also supply forest products for local livelihoods and economies. In mountain ecosystems it has been projected that a 1 °C increase in mean annual temperature will result in isotherms rising by about 160 m or moving 150 km in latitude. The alpine tree line ecotone is therefore useful for monitoring climate change.

In the Tibetan Plateau, tree lines are expected to shift upward and northward (Song et al. 2004; cited in Xu et al. 2009). In China's northwest Yunnan, a comparison of repeat photographs taken in 1923 and 2003 indicate that tree lines rose by 67 m and tree limits rose by 45 m (Baker and Moseley 2007; cited in Xu et al. 2009). A rise in temperature of 5 °C, which was projected in the region, could result in significant changes in the distribution of life zones. Alpine vegetation is likely to shrink and evergreen forest may also diminish (Xu et al. 2009).

5.2.2.3 Impacts of Glacial Melt and Glacial Lake Outburst Flooding (GLOF)

Progressive increases in warming at high elevations are already occurring as a result of anthropogenic climate change and other factors such as the deposit of black carbon soot (aerosol) on Tibetan glaciers. China's National Report on Climate Change (2007) estimated glacial shrinkage in the Tibetan Plateau to be about 4.5 % over the past 20 years, and about 7 % over the last 40 years. Higher temperatures may also be contributing to a greater proportion of precipitation coming as rainfall, rather than snowfall, which in turn can cause snowmelt to occur sooner and the winter to shorten.

Cascading effects of climate change relating the melting of glaciers and snow can have a destabilizing effect on surrounding slopes and may contribute to increased incidents of glacial lake outburst flooding (GLOF) in areas such as the eastern and central Himalayas, as well as mud-flows and avalanches, with harmful consequences for mountain ecosystems, human lives and settlements, and infrastructure such as bridges and footpaths.

In Nepal and Bhutan, the number of glaciers is 3,252 and 677, and glacial lakes 2,323 and 2,674 with a total area of glacial lakes of 182.47 km^2. Potential GLOFs numbered 44 in 2001 (Mool et al. 2001a, b; cited in Shrestha 2005). An inventory of glaciers and glacial lakes from 1999 to 2004 revealed 15,000 glaciers with total area coverage of 33,340 km^2. There are around 9,000 glacial lakes, 203 of which are potentially dangerous for GLOF incidences throughout Nepal, Bhutan, Pakistan, India and China (Bajracharya and Mool 2009).

The melting of glaciers, ice and snow into runoff and stream flow is a complex process. In the short term increased melting will result in increased discharge. However, in the long term, there will be water shortages and limited supplies for downstream communities, particularly during the dry season. Based on current knowledge, the rivers most likely to experience the greatest loss in water availability due to melting glaciers are the Indus, Tarim, Yangtze, Bramaputra, and Amu Darya (Xu et al. 2009). Landslides, debris flow and flash floods are projected to increase in the mountain areas (from 300 to 3,000 m), with riverine and coastal floods in the lowland areas (less than 300 m).

The HTP region is a crucial water source, and any changes in climatic patterns may seriously impact the Asian continent in terms of water availability for irrigation, drinking water and hydro-electric power. In China, shrinking of glaciers, ice and snow at the source of the Yellow and Yangtze rivers is leading to reduced downstream stream-flow.

In Southeast Asia, the Lancang-Mekong (5,000 km) flows through Yunnan Province in China, Myanmar, Lao PDR, Thailand, Cambodia and Viet Nam; the Salween (2,800 km) flows through China, Myanmar and Thailand; Irrawaddy is Myanmar's largest river; Chao Phraya is the major river in Thailand and it has no direct link with the HTP, but it could be indirectly affected by changes in local weather systems. These rivers feed the rice lands of Asia, the rice-producing countries of which account for about 70 % of the global supply. Thus, declines in river flow could reduce rice production with implications for global food security. Such declines occurred in 2008 and 2010.

Glaciation and snow cover play an important role in the Earth's radiation budget and loss of snow and ice from the HTP region may have climatic effects. These changes may have consequences on precipitation and temperature patterns at regional and global scales. In summer, the vast highlands of Asia heat up more than the Indian Ocean, leading to a pressure gradient and a flow of air and moisture from the ocean intensifying the Indian monsoon (Qiu 2008). Ice and snow loss may increase sea-level rise causing an increase in submerged coastline on the megadeltas of Asia.

5.2.2.4 Community-Based Adaptation in Remote Mountainous Communities

Remote mountain communities have been identified as one of the groups especially vulnerable to climate change. For example, the more than 30 million people living in the HTP region are at risk from natural disasters and from a harsh natural environment. The present section highlights community-based adaptation in three remote mountainous areas in South and Southeast Asia. These studies show how indigenous knowledge on biodiversity preservation, medicinal practices and sustainable 'agropastoral' and 'agroforestry' activities help mountain communities adapt to the impacts of climate change.

Case Study 1: The Eastern Tibetan Himalayas

Supported by villagers, local experts, local government and living Buddha from local temple,

Centre for Tibetan Regional Sustainable Development (CTRSD) of Yunnan Academy of Social Sciences, with funds from the APN, carried out research on Hongpo Watershed and found values of indigenous knowledge and culture of Tibetan people in adapting climate change. The activity also provided very important information for governments relevant to decision-making for climate change in the future.

The activity on community-based adaptation in a remote mountain ecosystem was carried out to increase the scientific capacity of local government, climate scientists and indigenous people in the Eastern Tibetan Himalayas of North-West Yunnan Province, China. Indigenous knowledge and scientific knowledge on current and future climate change were documented to conduct a climate impact and vulnerability assessment and to enable better decision-making in response to future climate change. Climate change data was collected by a Tibetan women's organization known as the 'Sisters Association' on climate-related disasters on traditional subsistence of mixed agriculture and herding, medical herbs, and gynecopathy (Lun 2010).

A climate field school (CFS) was established through the Deqin Tibetan Medical Association organized by local experts with indigenous knowledge of herbs and Tibetan medical knowledge, to protect and research traditional medical treatment. Through training and better understanding of the relationship between climate change and local tradition and culture, traditional crop seeds and seeds of traditional medicinal herbs are grown in Buddhist temples and the Holy Mountain to protect local vegetation and protect the Hongpo Watershed. The Centre for Tibetan Regional Sustainable Development (CTRSD) invited Buddhist monks to carry out the 'Sealing a Mountain' ceremony and build a white pagoda, according to local tradition to establish the "God Mountain" thus forbidding deforestation and hunting practices. Villagers also plant walnut trees in the vicinity of their villages and farmland and protect trees from being eating by domestic animals.

Mud-rock flow, flooding and drought are climate-related disasters already observed in the Hongpo Watershed. Cold-resistant highland crops such as highland barley are cultivated less frequently and are being replaced by wheat, corn and potato. In animal husbandry, the number of cold-resistant yaks and cattle decreased in 2010 compared to 2005, while the number of yellow cattle, pigs and chickens increased (Lun 2010). These days, indigenous knowledge of agropastoralists customs reflects the growing recognition of natural environment change and global warming. Traditional knowledge, while incorporating changes, is being challenged because of the need to change practices in traditional herding and cropping calendars in response to climate change.

For most of Tibetan people living in mountainous area of the Eastern-Tibetan Himalayas, being an 'agropastoralist' is not only a simple way of livelihood, or a method to use natural resources, but also a comprehensive reflection of indigenous knowledge and culture. Agropastoralists maintain local people's existence and biodiversity of crops and domestic animals for inheritance. The most important value of agropastoralists' way of life is to disseminate Tibetan culture. Because of strengthened climate change, species, environment and indigenous knowledge of agropastoralists are affected. Accordingly, alleviating impacts and reducing vulnerability by adapting to climate change becomes very important.

Through the case of the Hongpo Watershed, it was discovered that, although climate change is a global phenomenon, it still can be alleviated at the local level (in this case, the watershed level) through local people's efforts and adaptation based on their indigenous knowledge. In Furthermore, innovation and change of some indigenous knowledge can help local people adapt to climate change to some extent.

Case Study 2: Xihuangbanna (Dai People)

Indigenous people in remote mountain communities have extraordinary cultural diversity. According to historical records the 'Dai people' (also called Tai, Tai Lue), one of the 56 ethnic groups officially recognized in China, lived closely together in modern Yunnan province, while some moved to Laos, Thailand, Viet Nam, and Myanmar. In 1180, the chief of a Dai tribe, Payacheng, conquered other tribes in the Jinghong area and built the Jinghong Golden Temple Kingdom, which was subordinate to the Chinese Song dynasty rulers.

In the Dai language, 'Xishuang' means twelve and 'Banna' means rice field district, hence 'Xishuangbanna' reflects the historical twelve rice field districts (Wu and Ou 1995). The Dai people traditionally practice Theravada Buddhism, and continue to cultivate paddy rice in the lower valleys and basins. Throughout their long history, the Dai people have traditionally adapted in harmony with nature.

Dai philosophy focusses on the intimate relationship between humanity and the natural environment, consisting of mountains, forests, animals, plants and water. Artificial elements, such as buildings, afforested and cultivated lands, are imitations of the former natural elements according to the philosophy. Of all the natural elements, the forest is the most important and occupies the highest position in the hierarchy: forest -> water -> cultivated field -> grain -> humanity. Caring for the environment is a Dai virtue. As a result of prolonged Dai rule in Xishuangbanna these perspectives of the environment have had a great influence on other indigenous people in the region (Qui 2009).

A traditional Dai village was chosen for study near the Holy Hills – a natural virgin forest, which is strictly protected, so that its water resources, animals, plants and the habitats within are inviolable. This traditional approach to environmental conservation has been maintained for more than 2,000 years. An investigation in 1984 identified approximately 400 existing "Holy Hills" in Xishuangbanna, accounting for an area close to 50,000 ha. These Holy Hills are considered "oases of undisturbed biodiversity" among existing cultivated fields and rubber plantations in the region.

Traditional Dai villages cultivate fuel-wood gardens (*Cassia siamea*) near their villages and a Dai family would have small home garden near their house that grows vegetables, medicinal plants, cash crops, fruit trees, fuel tree, and flowers. The Dai in Xishuangbanna were the earliest people in China to grow paddy rice by scattering seeds on the wetland near streams. They trained elephants to reclaim forest and raised oxen to plough the fields. Paddy rice was introduced after traditional shifting cultivation, which was the major mode for upland rice and other crops such as tea, cotton, soybean, peanut, sesame, sugar cane, melon and vegetables.

A rotating system of land management for each village is implemented following design by the village chief who plans three consecutive plantings and then a fallow period that allows land to naturally restore its forest and fertile soil, before being cultivated again after 6–10 years. Traditionally shifting cultivation, when carried out on sufficient land and with enough fallow period, is sustainable for crop production and biodiversity in Xishuangbanna, China (Wu and Ou 1995) and Chiang Mai, Thailand (Rerkasem and Rerkasem 1994; cited in Nakashima and Roué 2002).

Expanding large-scale monocultures such as rubber plantation in Xishuangbanna has become a serious threat to traditional and sustainable forest practices. Yunnan's worst drought for many years in September 2009 has become a compelling example that highlights the devastating impacts on natural habitats through combined impacts of climate change and poor environmental management practices (Wu and Ou 1995).

Case Study 3: Chiang Rai (Karen Community), Thailand

The traditional Pgakenyaw tribe in Northern Thailand practices Deravath Buddhism and is indigenous to the Karen tribes. Their ancestors lived in Mae Chang Kao watershed for centuries, so it was known as the old land of Pgakenyaw. In 1986, the Thai government allowed Chiang Rai Tham Mai logging company to operate in the Khun Jae area, which included the sacred forest (Dae Paw), and the community's cemetery was destroyed in a span of a few years. In 1992, the Hauy Hin Lad area was declared the Khun Jae National Park and villagers were ordered to move out. The villagers joined force with other ethnic groups who faced similar problems and formed the Northern Farmer Network (NFN) to fight for their rights to their land.

As part of the national-level Assembly of the Poor, they conducted a series of protest actions until they were allowed to stay in their own villages. From 1993 to date, the Huay Hin Lad community has established their traditional holy forest and community forest. They have maintained rules and regulations to manage their community forest and natural resources. The Hauy Hin Lad community maintains a sustainable livelihood of agro-forestry that includes sacred forest, community forest, stream, tea garden, shifting cultivation, with upland rice, paddy rice and home garden systems (NDF et al. 2010). The practice of forest sanctification (sacred forest) has rapidly spread among the Pgakenyaw and other indigenous communities in the north (Erni and Nikornuaychai 2004).

The total ecological footprint of the Huay Hin Lad community is composed of three hamlets; the Hin Lad Nai, Pha Yuang, and Hin Lad Nok have only 0.61, 0.54 and 0.4 ha per head, respectively, compared with 1.7 ha per head for the average Thai people, and 9 ha per person in USA. For food security, analysis in 2008 by a research team in Huay Hin Lad community, found that food consumption is mostly from agroforestry products (92 %) obtained from households and natural resources, while the rest 8 % is from markets outside the community.

The Huay Hin Lad community is an example of a "low carbon society" because their farming activities lead to little carbon emission: only 476 tonnes of carbon from cultivation, and 68 tonnes of carbon from urea fertilizer for commercial corn. Paddy rice emits 0.8 tonnes of methane (CH_4), and corn fields release 0.1 tonnes of nitrous oxide (31 tonnes CO_2 equivalent) from the use of urea fertilizer in Hin Lad Nok hamlet. In preserving its natural community forest and by practicing shifting cultivation, the community's total capacity to store carbon is 720,627 tonnes (2,642,299 tonnes CO_2 equivalent); equating to annual emissions of only 0.08 % of their stored carbon. Analysis of carbon emission sources from various activities shows that growing commercial corn, which needs chemical fertilization, tends to increase total emissions from nitrous oxide and the burning of corn fields before planting (NDF et al. 2010).

There is no strong indication that the community has directly felt the impact of climate change. Their farming activities and consumption patterns are lower than average Thai people in the lowland, and they practice sustainable resource management.

5.2.3 *Small Island Remote Communities*

5.2.3.1 Introduction

There are a very large number of small islands in the Asia-Pacific region. In the present section we focus on small island developing states (SIDS) and territories in the Indian and Pacific oceans, but also acknowledge the large numbers of islands in other parts of the region such as the Philippines and Indonesia, the communities of which may also face similar concerns. The majority of countries listed in Table 5.3 are SIDS or colonial territories. This group has been identified as being the most exposed to climate change and extreme events that may result in natural disasters (Barnett and Campbell 2010).

5.2.3.2 What Is a Small Island?

There is no generally accepted definition of what constitutes a small island state and measures of smallness are relative. For example, according to many sources, including the US Census Bureau (http://www.census.gov/), Papua New Guinea is ranked as the 54th largest country in the world and has a land area larger than that of Japan and is massive compared to Tokelau, which is just a few square kilometers (Table 5.3). Similarly, islands range substantially in elevation from having their highest points

Table 5.3 Small island states and territories in Asia and the Pacific (Source: After Barnett and Campbell 2010)

Country	Land area (km^2)	Highest elevation (m)	Main island type
American Samoa	199	964	V
Bahrain	665	122	
Cook Islands	237	652	V & A
Federated States of Micronesia	701	791	V & A
Fiji Islands	18,272	1,324	P-B
French Polynesia	3,521	2,241	V & A
Guam	541	406	V
Kiribati	811	81	A
Maldives	298	2.3	A
Marshall Islands	181	10	A
Nauru	21	61	RL
New Caledonia	18,576	1,628	P-B
Niue	259	68	RL
Northern Mariana Islands	457	965	V
Palau	444	242	V
Papua New Guinea	462,840	4,509	P-B
Samoa	2,935	1,857	V
Singapore	710	163	
Solomon Islands	28,370	2,447	P-B
Tokelau	12	5	A
Tonga	650	1,033	V
Tuvalu	26	5	A
Vanuatu	12,190	1,879	P-B
Wallis and Futuna	142	765	V

A atolls, *P-B* plate boundary islands, *RL* raised limestone, *V* volcanic high islands

only a few meters above sea level compared to those with high mountain ranges; for example, Papua New Guinea's Mount Wilhem stands at 4,905 m.

There are two main categories of island in the PIC region. The first of these are formed by subduction along colliding tectonic plate boundaries and include the larger islands that make up Melanesia in the south-west Pacific. In comparison, 'oceanic' islands (see Fig. 5.5) are formed over 'hotspots' in the earth's mantle and begin as volcanoes and gradually subside as the plate moves away from the hotspot and are eroded by rain and wave action. They range from steep mountainous topographies to atolls, formed by coral growing on the original reefs that surrounded the volcanic islands, and which are typically only a few meters above sea-level. Atolls were formed by corals growing on the original reefs that surrounded the volcanic islands. While most islands are typified by coastal settlement patterns and many communities on the islands may be affected by sea-level rise, the communities on atolls may be particularly exposed.

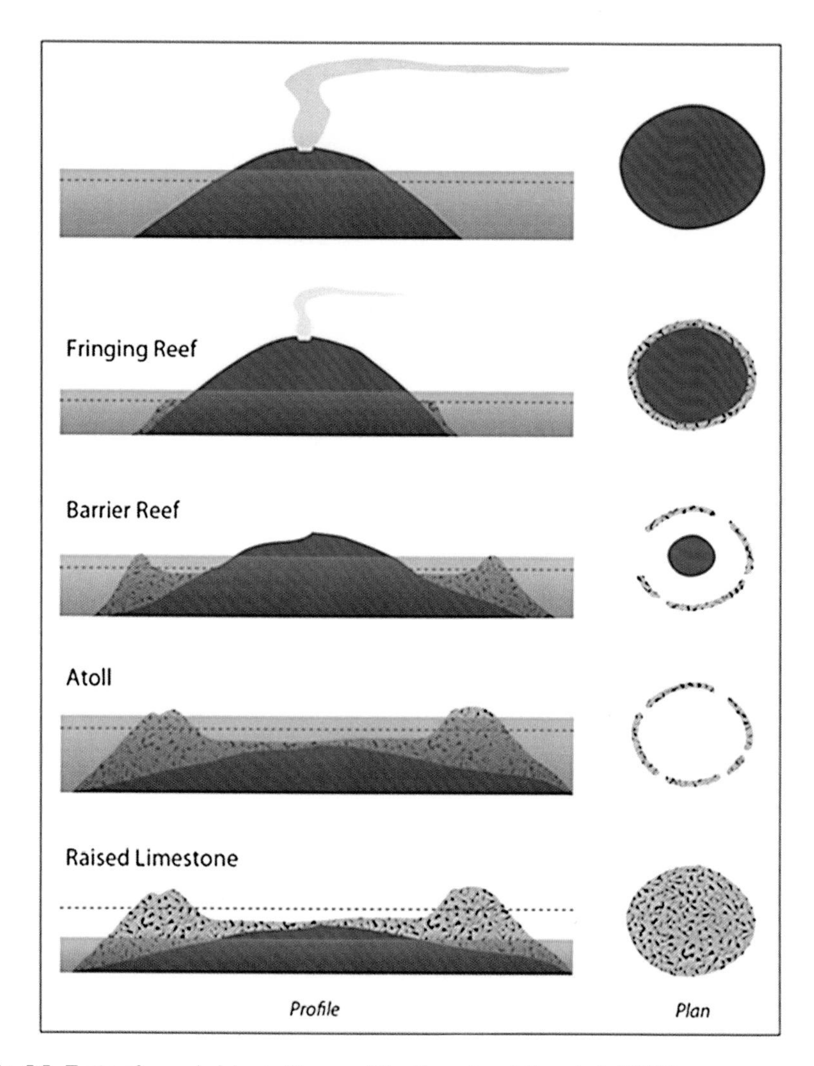

Fig. 5.5 Types of oceanic islands (Source: After Barnett and Campbell (2010))

5.2.3.3 The Significance of Isolation for Small Island Communities

As there is no single definition of a small island, it is also difficult to define or measure isolation. Singapore is located almost adjacent to the Asian mainland while all but three small islands in the Pacific are more than 2,000 km from their nearest metropolitan country. Most Pacific islands are also isolated from each other and there is isolation within countries as well with many outer island communities reliant on irregular and infrequent shipping services. Isolation brings with it many costs, particularly associated with transport. In times of natural disasters, isolation can cause delays in post-disaster assessment, medical assistance and the delivery of relief supplies (Campbell 2010).

There is considerable diversity among the small islands of the Asia-Pacific region in addition to the variations in land area and elevation. These include both environmental characteristics with important implications for resource use and disaster risk management as well as exposure to the effects of climate change. In addition there is great cultural, social, political and economic diversity among small islands, which are likely to result in variable levels and types of vulnerability (Campbell 2009).

5.2.3.4 Small Islands Ecological and Physical Characteristics

The larger plate boundary islands are characterized by higher levels of biodiversity than the smaller islands to the east, which are also more distant from the continental origins of most species. These islands are characterized by river flood plains, fertile soils and delta systems and other than during El Niño events also tend to have greater rainfall.

Volcanic high islands are often characterized by steep slopes although they can be found at different stages of subsidence and erosion, often they have barrier reefs with lagoons separating the reef from the shore, relatively small land areas and less fertile soils than the plate boundary islands. Many volcanic islands have poorly developed river systems though floods may occur during heavy rainfall events. As with the plate boundary islands, volcanic high islands have orographic rainfall, often with distinct wet and dry sides, which affect the types of agriculture that is practiced.

Atolls, by contrast have extremely small land areas, with small islets (the largest of which are on the windward side) encircling a lagoon. Atolls have no surface water although many have an underground fresh water lens, which is replenished by convectional rainfall. Under natural conditions atolls typically have no soil but anthropogenic soils have been developed on many of the islets to enable the cultivation of swamp taro (Cyrtosperma). Raised atolls also have very little soil development and are characterized by steep outer cliffs, a concave interior and karst topography.

In the Pacific region key climatic influences are the trade winds, which blow moisture laden air towards the equator and westward. Typically the western part of the Pacific has much higher rainfall levels. This pattern is disrupted during El Nino events when droughts are more likely to be experienced in the west. The Intertropical Convergence Zone and the South Pacific Convergence Zone are important sources of rainfall in the central Pacific. Most of the Pacific Islands apart from Nauru and Kiribati, which are close to the equator, may be affected by tropical cyclones (Sturman and McGowan 1999).

5.2.3.5 Small Island Social, Political, and Economic Characteristics

In many of the rural areas of small island states there is a significant subsistence component of livelihoods including plant and marine resources. Copra production has become an important commercial activity, especially in outer islands, but the increasing unreliability of shipping and price fluctuations has seen its importance decline in many places.

Of the 22 small island political entities in the Pacific, 14 are independent or self-governing in free association with their former colonizers. With one exception all of these countries have some form of parliamentary democracy. At the local level the most common form of settlement is the village though they vary in size. Typically social relationships are based on kinship and this plays an important role in economic activity, local politics, and land tenure as well as contributing to community resilience through various forms of cooperation (Campbell 2006a).

Most small island developing states have small and open economies with a limited range of exports reflecting narrow resource bases. Terms of trade tend to be adverse. Because of their smallness they often suffer disproportionately high losses, even exceeding annual GDP. Such impacts on economic development are likely to become more disruptive if climate change were to increase the frequency or intensity of climatic extremes.

While the towns and cities in the SIDS are mostly small by global standards there has been a steady growth in urban populations in recent decades. About a quarter of the population of the Pacific region reside in urban areas, with large numbers living in informal settlements, which often include houses that are vulnerable to high winds and in low lying or steeply sloped land. Urban planning is at its infancy in most SIDS; and most small island states have very limited, if any, urban disaster management activities in place. Nearly all of the urban areas in SIDS are in coastal locations (Connell and Lea 2002).

5.2.3.6 Challenges Facing SIDS

SIDS are faced with a number of significant challenges. These include barriers to economic development associated with smallness and isolation – imports cost more and exports earn less. Many of the countries have rapidly growing populations and there is pressure on resources particularly in or near urban areas. While most countries now have some form of environmental ministry or department, they are often understaffed, and struggle to respond to the increasing number of environmental issues both of local and global origin. Waste management and sewerage disposal (especially from urban areas) are areas of difficulty and on small islands such as atolls in particular there is no place for waste to be disposed. Economies of scale make recycling too expensive and the small amounts of waste by international standards make shipping goods for recycling also uneconomic (Finnegan 2011).

5.2.3.7 SIDS and Climate Change

Small islands may be affected by climate change in many ways. These include coastal erosion and inundation caused by sea level rise, increased losses from climatic extremes and impacts upon health. In the Asia-Pacific region, remote small islands communities, described as 'geographical hotspots' as mentioned earlier in Table 5.1 are located in the Indian and Pacific oceans. Hotspots include the 'Coral Triangle' (bordered by Indonesia, Malaysia, Thailand and the Philippines) and the South Asian Coral zones bordered by India.

Coral reefs appear to be among the most vulnerable of ecosystems in the face of climate change. The "Coral Triangle" is one of the World Wild Life's priority areas. Covering only 1 % of the Earth's surface, the Coral Triangle is home to 30 % of the world's coral reefs, 76 % of reef-building coral species and more than 35 % of coral fish species

Observed bleaching occurs when sea surface temperature (SST) increases above 31 °C and remains high for over 3 days during summer (Vivekanandan et al. 2009). Bleaching is a major threat to coral reefs due to expulsion of symbiotic algae, zooxanthellae, which are responsible for the coloration of the corals. SST is also a critical factor in the well-being of the symbiotic association of other host animals like coral reef fish, giant clams and sea anemones. Therefore SST affects the food chain and food security of communities in small islands.

The Andaman Sea and the Nicobar Islands witnessed bleaching events during 1998, 2002 and 2005. Significant bleaching (30–70 %) was observed in 2010 (Krishnan et al. 2011), where branching corals were the worst affected. Coral reefs are attractive dive sites for tourists, and the 90 % coral bleaching in Thailand in 2010 resulted in the closing of 18 famous dive sites in the Andaman Sea and the Gulf of Thailand (Bangkok Post 2011), and leading to a reduction in the income of small-island communities.

5.2.3.8 Sea-Level Rise

Given that most islands have high coastline to width ratios, a large proportion of settlements are coastal: there are numerous atolls, and much economic activity on high islands takes place near the coast. It is not surprising that sea level rise is a cause of much concern. The Fourth Assessment Report of the Intergovernmental Panel on Climate Change (IPCC 2007) identifies accelerated coastal erosion, coastal flooding and salinization of ground water, particularly on atolls, as key effects. There is uncertainty about the response of atolls to sea level rise and Kench et al. (2005) illustrate that uninhabited atoll islets in the Maldives may be resilient to projected sea level changes. Conversely, islands that have been significantly modified may be more vulnerable (Mimura et al. 2007).

5.2.3.9 Changing Extremes

Most small islands in the Asia Pacific region are located in areas that experience tropical cyclones. These events, together with drought, are among the most important environmental extremes for small islands. Under warmer conditions tropical cyclones may have greater wind speeds, increased precipitation intensity and higher storm surges (IPCC 2007). Models of precipitation have considerable variance but even relatively small reductions in rainfall may have major impacts. The severity of tropical cyclones and drought may cause considerable additional disruption to settlements and livelihoods beyond that which is already experienced. Both sea surface and land temperatures of islands are expected to increase with an increase in the frequency of extreme temperatures. While islands tend to have extremes of heat moderated by their maritime locations, heat stress may affect people as well as plants and animals.

5.2.3.10 Human Health Issues

The main health issues linked to climate change in SIDS are possible changes in the distribution of vector borne diseases such as dengue fever, malaria and ciguatera (a form of poisoning caused by affected fish), changes in water quantities and quality (with possible increases in the incidence of diarrhea) in addition to nutritional effects of declining agricultural production and changes in mortality and injury from the increasing frequency and/or incidence of extreme events. Issues for human health are discussed in Sect. 5.3.

5.2.3.11 Resilience and Vulnerability

It is important to recognize that people who live on islands are not intrinsically vulnerable: in the past there is considerable evidence to suggest that many small island societies were remarkably resilient in relation to both ambient and extreme environmental conditions. In the Pacific Islands this was achieved through sustaining food security (in particular by producing surpluses and preserving and storing them), inter- and intra-community cooperation, building and settlement characteristics, and the application of traditional ecological knowledge (Campbell 2006b).

Many of the small islands were among the last places to be colonized but the imposition of a new social order, the introduction of a new religion and exposure to a global economic system all contributed to the decline of many of the traditional practices that enhanced the resilience of these remote communities. For example, traditional rituals that brought different communities together and built networks that could be called upon in times of need (such as a disaster) were often suppressed by missionaries, and many of the 'cash crops' were less resilient in the face of strong winds than traditional cultivars. Disaster relief itself also played a role, for

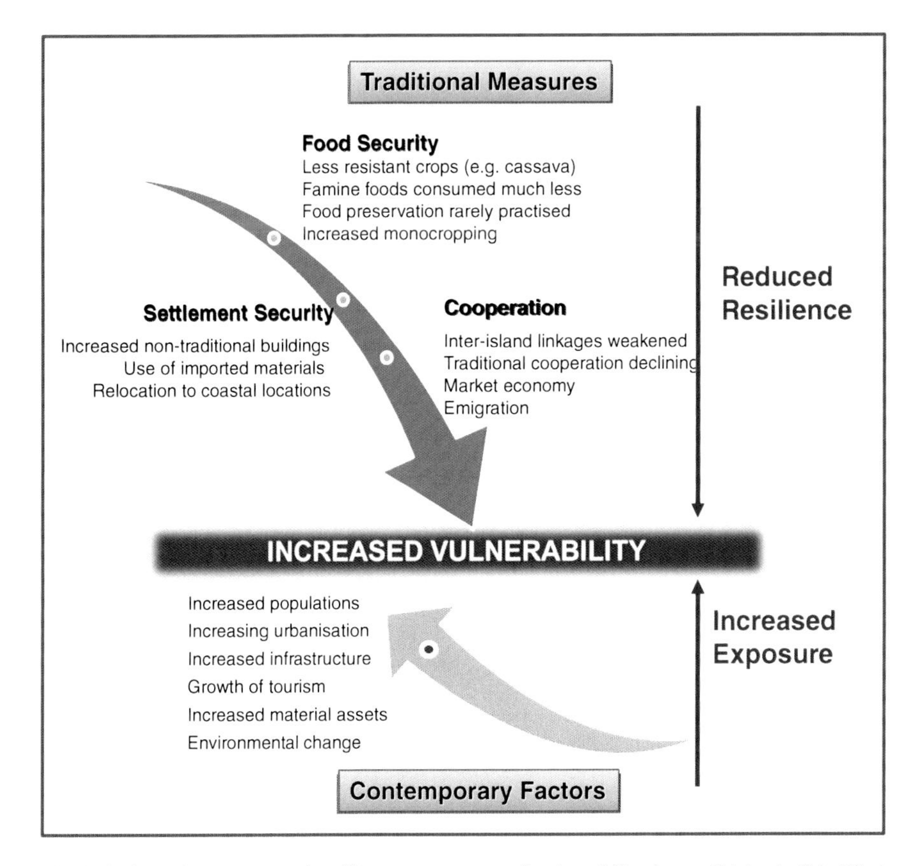

Fig. 5.6 Changing patterns of resilience, exposure and vulnerability in small islands (Modified from Campbell, J.)

example in reducing resilience in that it reduced the need for surplus food production and increased the consumption of famine foods.

In addition to reducing resilience, there have been many changes to small island states that have increased exposure to environmental change and variability. These include population growth, urbanization, the introduction of new livelihood activities that are less well adjusted to local conditions, and the expansion of tourism. Figure 5.6 summarizes these changes. While traditional resilience has been reduced new aspects of island life are increasing exposure to extreme events and sea-level rise.

5.2.3.12 Responding to Climate Change

(a) *Mitigation*

The great majority of small island states and territories in the Asia-Pacific region play a very small role in contributing to the increasing atmospheric concentrations of GHGs both in absolute and per capita terms. Nevertheless, international progress on gaining agreements to increase mitigation initiatives is considered critical. Accordingly, SIDS contribute disproportionately to international negotiations given their small populations. Nevertheless, they are greatly stretched by the demands of the UNFCCC Conferences of the Parties where some national delegations are several times the size of the total from small islands (Barnett and Campbell 2010). One of six principles of the Pacific Islands Framework for Action on Climate Change 2006–2015 (2005) is for all countries as "part of their national policies, to promote cost effective measures to reduce greenhouse gas emissions, including increased energy efficiency and increased use of appropriate low carbon and renewable energy technologies" (SPREP 2005).

(b) *Recent adaptation activities in the region*

The number of adaptation activities undertaken or in progress in the region has increased significantly in recent years but there is much that has not been achieved given the widespread effects of climate change that are likely to be, and in some cases are already being, experienced.

One of the biggest adaptation initiatives has been the Pacific Adaptation to Climate Change Project, which is funded in part by the Global Environmental Facility (GEF). Thirteen SIDS in the Pacific are involved in this $57 million project that has its main focus on improving food security, water resources and coastal management.

There has also been development in community-based approaches, which seek to incorporate local involvement in adaptation planning and implementation. Projects under this umbrella have included the "Capacity Building for the Development of Adaptation Measures in Pacific Island Countries (CBDAMPIC)" project (Nakalevu 2006) and, amongst other activities, was involved in relocating a community on the island of Tegua to a site inland from the coast. The University of the South Pacific (USP) has also been involved in a number of community-based projects funded by the Australian government.

An activity conducted under the APN's capacity building programme CAPaBLE, and undertaken by the University of the South Pacific (USP) and involving Kiribati, Tuvalu and Fiji, focused on the dependence on natural resources of communities in these three countries for their livelihood and, importantly, for maintaining traditions and culture. The project looked at a holistic integrated approach to sustainable development through the integration of climate and variability change with biodiversity conservation and successfully delivered messages and raised awareness by engaging youth (Aalbersberg 2007). This activity emphasized the importance of youth as the future custodians of Pacific Islands' resources, and their contribution to community resource management is vital to ensure security and availability of resources for the next generations. For these reasons, the Pacific Islands' youth were targeted as carriers of the message through lively and culturally appropriate means such as theatre.

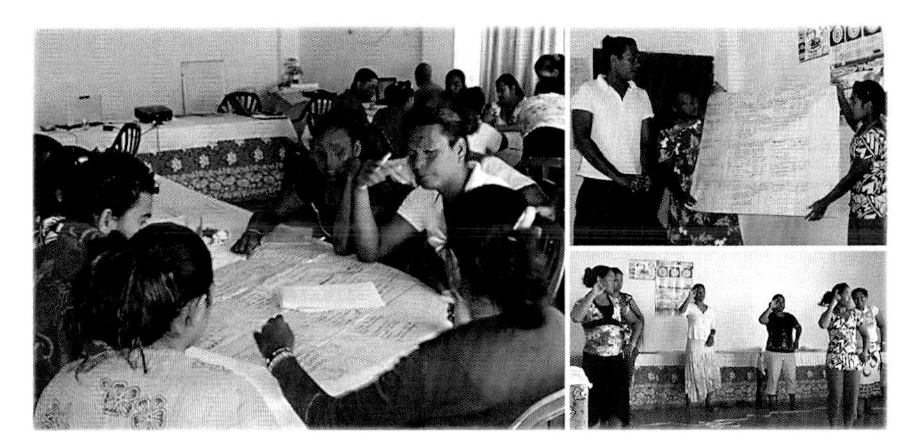

Fig. 5.7 Pacific island youth simulations of climate hazards (Source: APN)

Activities included lessons learned, training and support materials, drama scripts, and songs and dance shared for adaptation (Fig. 5.7).

(c) *Community-based adaptation methodology*

The PACE-SD Integrated Project Cycle: Component 1 depicts a seven-step integrated project cycle, which forms the basis of developing and implementing an adaptation project (Fig. 5.8). Through an integrated and consultative process the vulnerability and adaptive capacity of the community and the adaptation options for the community are identified and assessed. The selected adaptation options are then implemented with community participation. The progress of the project is continuously monitored and evaluated using specific indicators for the time horizons of short-, mid- and long-term.

This method depicted is unique in that it has a strong and equal emphasis on community-based approaches using participatory tools and facilitator-based approaches using technological and scientific tools and methods to assess vulnerability and adaptation options. First, an assessment of the current socio-economic, cultural and environmental problems faced by the community is carried out. Secondly, it addresses climate issues currently faced by the community before assessing the vulnerabilities and implementing cost effective adaptation measures by the community. A similar approach is used in SPREP's CBDAMPIC and Pacific Adaptation and Climate Change (PACC) projects in the Pacific. This methodology is equally suitable for any climate-affected community in other parts of Asia and the Pacific.

(d) *Forced migration*

With the exception of Papua New Guinea, most SIDS in the region are characterized by coastal settlement patterns, and in the case of atoll islands almost all inhabitant live only a few meters above sea level. The effects of sea level rise, increased tropical cyclone intensity and coral bleaching are likely to see many

Fig. 5.8 Steps in the PACE-SD project cycle (Modified from Koshy, K.)

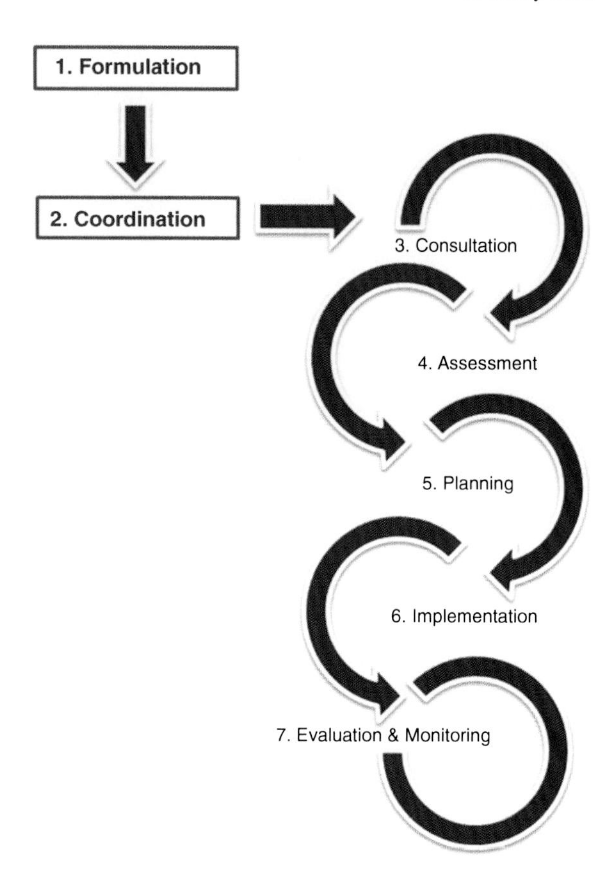

of these settlements rendered increasingly unsuitable for habitation and/or unable to provide adequate sustainable livelihoods. Accordingly, climate change may give rise to increased flow of migrants within and from SIDS. There are already high levels of migration from some Pacific Islands that remain as colonies and some with special arrangements with former colonial countries (for example, Samoa, Tokelau, Cook Islands to New Zealand; American Samoa, Guam, Palau, Federated States of Micronesia and the Marshall Islands to the USA). There are, however, a number with very little access to metropolitan countries, including the Maldives (Indian ocean), Tuvalu and Kiribati, which are all atoll countries, and the larger Melanesian countries of Papua New Guinea, Solomon Islands and Vanuatu.

Climate-induced voluntary migration refers to migrants who choose to migrate from a location that may still be habitable but one that is becoming increasingly environmentally marginal. In such cases a community may remain on the island and its livelihoods supplemented by remittances from the migrants and pressure on local resources reduced by a reduction in numbers. This may be seen as a workable adaptation option (Barnett and Webber 2010).

Interactions between weather and climate and human health are multiple and complex. They could be summarized into two categories: (a) direct effects on morbidity and mortality in relation to temperature and other extreme weather and climate events; and (b) more indirect effects where changing weather patterns cause changes in, for example, ecosystems that then can result in changes in the burden of climate-sensitive health outcomes. For example, respiratory disorders from the air pollution that results from large scale biomass-burning, malnutrition and delays in child growth and development due to decreases in food productivity and water shortages, increases in the geographic range and intensity of transmission of infectious diseases transmitted by mosquito vectors such as malaria, dengue and chikungunya; and diseases transmitted by food and water such as diarrhea and cholera.

The impact on human health from climate change could be significant, if no adaptation and mitigation actions are implemented in the short term. The reduced availability of drinking water could mean more frequent and more severe outbreaks of diarrheal diseases such as cholera. The IPCC Fourth Assessment Report concluded that "*Endemic morbidity and mortality due to diarrhoeal diseases primarily associated with floods and droughts are expected to rise in East, South and South-East Asia due to projected changes in the hydrological cycle associated with global warming*" (IPCC 2007). Further, scarcity of water and food could affect current levels of malnutrition. Water and sanitation programmes in all countries in the region have made significant progress over the past decade towards reaching the Millennium Development Goals, although much still needs to be done, in particular in terms of ensuring safe water in rural areas and sanitation coverage in general. For example, the estimated number of deaths from diarrheal disease dropped from close to 980,000 in 1999 to 504,000 cases in 2005 (WHO 2007). But, many of the achievements can be negated with on-going climate change. There is an urgent need to strengthen these programmes to preserve achieved results and prepare to meet the challenges ahead.

(b) *Extreme weather and climate events*

Extreme events in Asia and the Pacific can significantly affect human health (Climate Institute 2010), such as the October 1999 cyclone in Orissa, India that caused 10,000 deaths and affected 10–15 million. In 2006, Sichuan Province, China experienced its worst drought in modern times, with nearly eight million people and over seven million cattle facing water shortages.

Heat stress can result from exposure to higher than normal ambient temperatures. The frequency, intensity, duration, and spatial extent of heat waves in many parts of the world have increased because of climate change, with further increases projected with on-going climate change (IPCC 2012). As temperatures increase, so do the number of deaths and illnesses occurring from a wide range of health outcomes, including heat stress, heatstroke, cardiovascular disease, respiratory disease, and kidney disease. Older adults, young children, and economically disadvantaged communities are particularly vulnerable to high ambient temperatures.

In 2003, heat waves in Andhra Pradesh, India caused approximately 3,000 deaths; the highest recorded temperature was 51.3 °C (Government of Andhra Pradesh 2004). The 2007 heat wave in Asia (37 °C to 46 °C in March to May) resulted in more than 400 heat-related deaths in Bangladesh, India, Nepal and Pakistan. Russia was also hit by a heat wave at the end of May. Although no deaths were reported, crops over an area of 5,000 km^2 were destroyed. A heat wave also hit Japan in August 2007 with the highest temperature of 40.7 °C recorded in Kumagaya and Tajimi cities. Almost 9,000 people were taken to hospital and at least 13 heat-related deaths occurred.

The Australian State Government of Victoria (2009) reported a 12 % increase in calls for emergency services during the 2009 heat wave in Australia, Melbourne. Saniotis et al. (2011) reviewed the numerous studies in Australia on the impacts of the 2009 heat wave and concluded that immediate policy action is necessary to address the 4,200–15,000 projected heat-related urban deaths by 2100 and the 34 % increase in mortality projected in Sydney by 2050.

Floods can spread bacteria, viruses, and chemical contaminants, foster the growth of fungi, and contribute to the breeding of insects. Prolonged droughts interrupted by heavy rains favor population explosions of insects and rodents. Extreme weather events have been accompanied by new appearances of harmful algal blooms in Asia and North America, and in Latin America and Asia by outbreaks of malaria and various water-borne diseases, such as typhoid, hepatitis A, bacillary dysentery, and cholera (Climate Institute 2010).

(c) *Air Pollution and respiratory disorders*

Higher temperatures increase ground level ozone formation in many areas. Increasing ozone concentrations are associated with increased hospital admission rates and death for people with respiratory diseases such as asthma, and worsen the health of people suffering from cardiac or pulmonary disease (Karl et al. 2009). Particulate matter (PM) is another major pollutant that affects air quality in particularly urban areas. Sources of PM include both natural (wild fire, volcano eruption, wind-blown desert dust, and ocean salt) and anthropogenic sources (aerosols from biomass burning, combustion of fossil fuels from automobiles, power and other industrial plants). The Tata Energy Research Institute (TERI) estimated 18,600 premature deaths per year associated with air quality in the Delhi region alone (TERI 2001). There is limited understanding of how atmospheric concentrations of PM could change with climate change.

Aerosols from biomass burning associated with forest and agricultural land clearing are drivers for regional and global climate change. In the Greater Mekong Sub-region (GMS), approximately 2.4 million tonnes of total particulate matter was emitted from burning forest fire and paddy fields in Cambodia, Laos, Thailand, and Vietnam, estimated using satellite sensors. Forest fires made the greatest contribution, and may lead to trans-boundary air pollution problems in the dry season, from January to April (Towprayoon et al. 2008).

Increased asthma and other respiratory disorders, cardiovascular hospital admissions, and mortality are associated with atmospheric aerosols from forest and bush fires, and from agricultural burning, in Thailand, Malaysia, Indonesia,

Singapore, Australia, and Brazil. Impacts related to dust storm events are found in China and Republic of Korea; increased pulmonary diseases and lung cancer are found in Nepal, India and China from exposure to biomass and coal smoke (Ramanathan et al. 2008).

(d) *Aeroallergens*
Higher temperatures are associated with longer allergenic ragweed pollen seasons, at least in temperate regions. Under current carbon dioxide concentrations, ragweed produces twice as much pollen over historic concentrations; by 2075, that could be four times as much. With increased airborne pollen, those who suffer from seasonal allergies could experience more severe symptoms, including hay fever and asthma (Ziska et al. 2011).

(e) *Malnutrition*
The largest health effect globally is expected to be from malnutrition (IPCC 2007). Malnutrition currently kills 3.7 million people annually, most of them children under 5 years of age. Food production depends not only on water availability but also on ambient temperature. Globally averaged, a 0.5 °C increase in temperature reduces cereal crop yields by 3–5 %. Therefore, a 2 °C higher temperature could imply a 12–20 % fall in global grain production. Overall, projections indicate rice yields in 2080s will have dropped by 14.9 % compared to 2000 (WHO 2007). Further, climate change may worsen floods and droughts, threatening the availability of water for drinking and irrigation. Droughts can harm crops, diminish food variety, nutritional content, and availability; all of which can contribute to malnutrition and the spread of infectious diseases. In addition, warming ocean temperatures can shift the geographic range of fish populations, affecting local food supplies. Climate change's higher temperatures also can increase the risk of food-borne illnesses (NRC 2009).

(f) *Infectious diseases*
Changing weather patterns due to climate change can affect the geographic range and incidence of insect-borne diseases such as dengue fever, West Nile virus, and Lyme disease (Soverow et al. 2009). Projected increases in temperature and changes in precipitation could promote the emergence of more disease-friendly conditions in regions that did not previously host diseases or disease carriers. In addition, hotter temperatures can lead to more rapid development of selected pathogens within insect carriers, allowing these diseases in some regions to expand their range into new, once cooler, regions (NRDC 2011).

Vector-borne diseases (VBD) are infections transmitted by the bite of infected arthropod species, such as mosquitoes, ticks, triatomine bugs, sandflies, and blackflies. Mosquitoes, which can carry many diseases, are very sensitive to temperature changes. Higher temperatures can shorten the viral incubation period in mosquitoes, shortening their breeding cycle, increasing the frequency of mosquito feeding and, in the case of dengue fever, allowing more efficient transmission of dengue virus from mosquitoes to humans. Figure 5.9 shows mean temperature and the incidence of dengue fever in Noumea-New

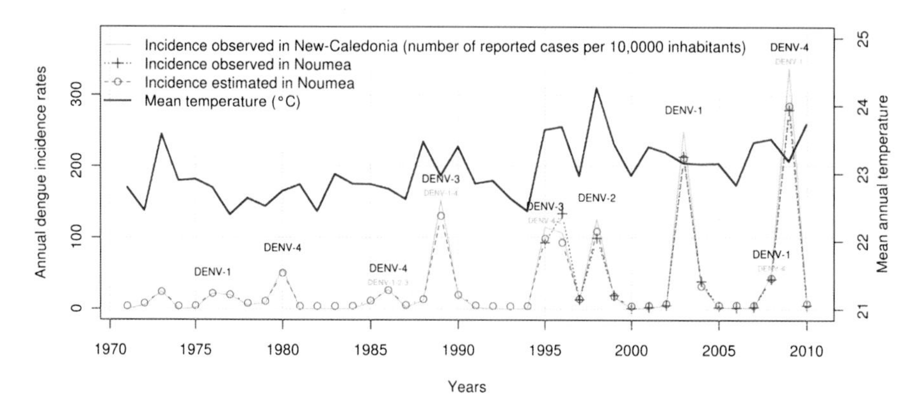

Fig. 5.9 Epidemiology of dengue fever and evolution of annual mean temperature in Noumea-New Caledonia (1971–2010). The predominant circulating serotype (*DENV-1*, *DENV-2*, *DENV-3* or *DENV-4*) is indicated in *black* characters. When other serotypes were detected, they are indicated in *little grey* characters. Annual dengue incidence rates observed in Noumea over the 1995–2010 period are highly correlated with dengue incidence rates observed in New Caledonia (Spearman coefficient rho = 0.99, p-value = 1*10214). Annual dengue incidence rates in Noumea (1971–1994) were estimated (*green dotted line* with *circles*) on the basis of the relationship between incidence rates observed in New Caledonia (*grey line*) and those observed in Noumea (*blue dotted line* with *crosses*) using a linear model. During the 1971–2010 period, dengue incidence rates and annual mean temperatures (from January to December) were significantly correlated in Noumea (Spearman's coefficient rho = 0.426, p-value = 0.007). An increasing trend of dengue outbreaks amplitude and annual mean temperatures were observed during this 40-year study period (Reprinted from "Climate-Based Models for Understanding and Forecasting Dengue Epidemics," by E. Descloux et al., 2012. *PLoS Negl Trop Dis*, 6(2), e1470, p. 3. Copyright 2012 by Descloux et al.)

Caledonia over 1971–2000 (Descloux et al. 2012); maximum temperature and relative humidity were associated with outbreaks.

Dengue cases have been recorded in every season and are widely distributed in many countries in South and Southeast Asia, Central America, and the Western Pacific (Tseng 2008). The number of months with average temperatures higher than 18 °C and the degree of urbanization were found to correlate with increasing risk of dengue fever in some regions (Wu et al. 2009). In Matamoros, Tamaulipas, Mexico, dengue increased by 2.6 % per week for every 1 °C increase in the weekly maximum temperature, and increased 1.9 % for every 1 cm increase in weekly precipitation (Brunkard et al. 2008).

The geographic range and incidence of malaria already kills 1.1 million people each year, may also be affected by changing weather conditions (WHO 2007).

(g) *Contamination of drinking water*

Outbreaks of water-borne diarrheal diseases caused by giardia, cryptosporidium, and other pathogens are associated with heavy rainfall events and flooding, which are likely to become more frequent due to climate change (NRDC 2011).

Although climate change threatens the safety of water supplies worldwide, the impact will be most severe where water infrastructure and treatment are less developed (McMichael et al. 2006).

5.3.3.2 Cross-Cutting Challenges for Climate Change and Health

The availability of relevant hydro-meteorological, socio-economic, and health data is limited. Available data can be inconsistent in low-income countries and are seldom shared in an open and transparent manner. Furthermore, there is insufficient capacity for assessment, research, and communication on climate-sensitive health risks in many countries, as well as insufficient capacity to design and implement mitigation and adaptation programmes.

There is an urgent need to incorporate health concerns into the adaptation and mitigation decisions and actions of other sectors, to ensure these decisions and actions also enhance health. Promoting the use of non-motorized transport systems (for example, bicycles) and fewer private vehicles would reduce greenhouse gas emissions and improve air quality and physical activity.

5.3.3.3 Susceptible, Vulnerable, and Displaced Populations

Certain populations are at increased risk from changing weather patterns, presenting unique concerns when considering the health risks from climate change, such as populations living in poverty, older adults, young children, and pregnant women (Balbus and Malina 2009). Displaced populations also are vulnerable with a higher risk of a number of diseases, including diarrheal and vector-borne diseases resulting from poor sanitation, as well as mental health issues due to increased acute and chronic stress (Myers 2002).

Poverty generally makes people more vulnerable to many of the health effects of climate change. It also makes it increasingly difficult for a population displaced by extreme weather events or environmental degradation to recover, resulting in much higher disease risks. For vulnerable populations, changing weather patterns, including extreme weather and climate events, can disrupt access to public services, such as health care and food assistance programmes. Their pre-existing conditions or situations can magnify stresses. Outdoor workers and people living in coastal and riverine zones also are likely to experience increased vulnerability to climate-induced environmental changes resulting from flooding and extreme weather events.

5.3.4 Public Health and Health Care Infrastructure

The health care delivery infrastructure is more diverse and complicated than the public health infrastructure (though there are multiple overlaps between the two systems). From family doctors in small towns to complex university research hospitals in large cities, health care professionals are the primary source of medical treatment. This infrastructure is vulnerable to climate change in a number of very important ways. Disasters can severely hinder the delivery of health care, with long-term impacts. Changes in the numbers of patients and the spectrum of diseases with which they present could occur in some regions as the climate changes. The types of advice offered to patients with chronic conditions and the infrastructure to support them may need to be adapted to protect against climate-induced changes that may make these individuals more vulnerable. Currently there is limited research to guide these types of decisions.

Storm surges and sea-level rise, combined with coastal erosion and other processes, will likely create challenges for public health and health care infrastructure. Many hospitals are located in low-lying coastal regions, putting them at risk from storm surges. Storm surges can disrupt coastal routes and harbours, and/or affect sewer and water resources, challenging health care delivery and food distribution.

5.3.5 Capacities, Skills and Education Needs

New skills and methods are needed to integrate current and future surveillance activities and retrospective datasets with weather and climate information. Understanding is growing of how to conceptualize and conduct epidemiological analysis using weather and climate as exposures, as are methods and skill in combining spatial epidemiology with ecological approaches. There is a strong need for the ability to translate vulnerability mapping and health impact assessments (HIAs) into behavioral changes and effective public health actions (Portier et al. 2010).

A greater emphasis must be placed on developing and maintaining interdisciplinary and inter-institutional collaborations, as well as on ensuring that established resources and expertise of all of the relevant disciplines, including climatology, modeling, environmental science, risk assessment, public health, and communications and education, are applied to the pressing problems associated with avoiding, preparing for, coping with, and recovering from the health risks of climate change. Additional disciplines including ecology, social science, economics, geography, behavioral psychology, and others also play a vital role in climate and health decision making (Portier et al. 2010).

Public health educators have a strong history of promoting health and wellness through educating individuals and communities about healthy behaviors and disease prevention or management. These skills are critical in raising awareness of the potential impacts of climate change, and translating the scientific research and other technical data into credible and accessible information for the public to use in making informed decisions that will protect their health and environment. Research is needed to determine how to effectively educate and organize the public to respond. Research is also needed to aid climate change communicators and educators in adapting their messages and approaches to most appropriately and effectively reach and be assimilated by each individual audience (Portier et al. 2010).

5.3.6 *Managing the Health Risks of Climate Change*

Effectively responding to the health risks and population vulnerabilities outlined in this chapter necessitates a multidisciplinary and multi-sectoral public health approach that enables coordinated thinking and actions across governments, international agencies, NGOs, and academic institutions. The public recognition of the health implications of climate change should assist the advocacy and political change needed to tackle mitigation and adaptation. Involvement of local communities in monitoring, discussing, advocating, and assisting with the process of adaptation will be crucial (WHO 2008b).

Research and policy action are needed to increase understanding of the relationships between weather/climate and health outcomes, improve modeling of these relationships to develop early warning systems and to project possible future health burdens, identify and evaluate adaptation options to manage current and projected risks to protect the most vulnerable, reduce carbon emissions, and estimate the health co-benefits and co-harms of adaptation and mitigation policies.

Countries that are signatories to the United Nations Framework Convention on Climate Change (UNFCCC) are required to regularly submit National Communications that include national plans for adaptation and mitigation. The extent to which individual plans include the health risks of climate change varies with the capabilities of each country. Actions suggested include new legislation to promote public health, development of disease surveillance systems, and enhancement of emergency response systems.

Current institutional arrangements, private and public, may need to be modified to increase the ability to effectively manage the health risks of climate variability and change. Protecting human health is an issue that crosses institutional, scientific, and political boundaries. No single institution at the local, regional, or national level is able to fully protect public health without cooperation from other institutions. In addition, no single scientific field is capable of accomplishing all aspects of the research needed to understand the human health consequences of global climate change; this requires a broad-based, trans-disciplinary research portfolio. Identifying research needs; mobilizing and creating the expertise, resources, tools, and technologies to address them; and translating these efforts into solutions that will facilitate adaptation to our changing environment while protecting public health will require collaborations on an unprecedented scale (Portier et al. 2010).

5.3.7 *Conclusions*

Climate change is and will further disrupt ecosystem services that support human health and livelihood, and will impact health systems; the magnitude and extent of impacts will depend on the effectiveness of public health and health care infrastructure and organizations. Severe storms and floods can damage critical infrastructure, and can lead to drowning, injuries, drinking water contamination, community displacement, and outbreaks of infectious diseases. Without improvements in current effectiveness of public health policies and programmes, climate change could increase malnutrition, diarrheal disease, and malaria, with implications for child growth and development (IPCC 2007).

Better understanding and management of the health risks of climate change needs is necessary, among other activities, such as:

- Empowering the poor and most vulnerable populations, communities, and countries to understand climate implications and take action. Health professionals and university academics have an important catalytic role.
- Mobilizing and creating the expertise, resources, tools and technologies and translating these efforts into solutions will enable human adaptation to our changing environment while protecting public health. Increased capacity, skills, and education on the adverse health effects of climate change are needed.
- Research to further understand the possible impacts of weather/climate on health, and options to better manage risks, including enhancing infectious disease control programmes, developing vaccines, and implementing early warning systems.
- Improving disaster risk management of extreme weather and climate events.
- Considering climate change and health in multilateral agreements at the international and regional level, while ensuring that policies are not divorced from the agenda for poverty alleviation or for closing the gap on social inequalities and health.
- Reviewing progress towards agreed mitigation targets, to accelerate progress through celebration of success and identification of areas where progress is lagging.
- Advocacy to ensure that the health effects of climate change are placed high on the agenda of relevant organizations and institutions, to promote research and development, education, and action.

Managing the health risks of climate change requires systems-based thinking and approaches that engage other sectors, iteratively managing risks as the climate continues to change. Multidisciplinary and multi-sectoral coordination and collaboration is needed across governments, international agencies, NGOs, and academic institutions.

References

Aalbersberg, W. (2007). *Climate change and variability implications on biodiversity – youth scenario simulations and adaptations*. (APN Project Report CBA2007-02CMY-Aalbersberg). Retrieved from APN E-Lib http://www.apn-gcr.org/resources/items/show/1633

APMC. (2011). *Youth action for climate change through art*. Asia-Pacific Mountain Courier, Vol. 12, No. 1, June 2011. Kathmandu, Nepal. Available from http://lib.icimod.org/record/26964

APN/CAPaBLE. (2008). *'In review' CAPaBLE phase 1 evaluation report*. Kobe: Asia-Pacific Network for Global Change Research. Retrieved from APN E-Lib http://www.apn-gcr.org/resources/items/show/1878

Association of South East Nations (ASEAN). (2009). Annual report 2008–2009. Available from http://www.asean.org/

Australia State Government of Victoria. (2009). *January 2009 heatwave in Victoria: An assessment of health impacts*. Melbourne: Report by Department of Human Services, State Government of Victoria. Available from http://www.health.vic.gov.au

Australia's Fifth national communication on climate change. (2010). *A report under the United Nations Framework Convention on Climate Change*. Department of Climate Change, Commonwealth of Australia. Available from www.climatechange.gov.au

Bajracharya, S. R., & Mool, P. (2009). Glaciers, glacial lakes and glacial lake outburst floods in the Mount Everest region, Nepal. *Annals of Glaciology, 50*(53), 81–86.

Balbus, J., & Malina, C. (2009). Identifying vulnerable subpopulations for climate change health effects in the United States. *Journal of Occupational and Environmental Medicine/American College of Occupational and Environmental Medicine, 51*(1), 33–37. doi:10.1097/JOM.0b013e318193e12e.

Bambaradeniya, C. (2007). *Removing barriers to capacity building in least developed countries – transferring tools and methodologies for managing vulnerability and adaptation to climate change*. Asia-Pacific Network for Global Change Research. Retrieved from APN E-Lib http://www.apn-gcr.org/resources/items/show/1626

Bangkok Post. (2011, 21 January). *18 dive sites closed to save coral reefs*. Bangkok Post. Retrieved from http://ww.bangkokpost.com/news/local/217417/18-dive-sites-closed-to-save-coral-reefs

Barnett, J., & Campbell, J. R. (2010). *Climate change and small island states power, knowledge and the South Pacific*. London: Earthscan.

Barnett, J., & Webber, M. (2010). *Accommodating migration to promote adaptation to climate change* (World Bank Policy Research Working Paper: No. WPS 5270).

Beddington, J., Asaduzzaman, M., Clark, M., Fernandez, A., Guillou, M., Jahn, M., … & Wakhungu, J. (2012). *Achieving food security in the face of climate change: Final report from the commission on sustainable agriculture and climate change*. Copenhagen: CGIAR Research Program on Climate Change, Agriculture and Food Security (CCAFS). Available from www.ccafs.cgiar.org/commission

Biermann, F. (2007). Earth System Governance as a crosscutting theme of global change research. *Global Environmental Change, 17*(3–4), 326–337.

Biermann, F., Betsill, M. M., Gupta, J., Kanie, N., Lebel, L., Liverman, D., … & Siebenhuner, B. (2009). *Earth System Governance: People, places and the planet. Science and implementation plan of the Earth System Governance project* (Earth System Governance report, 1, IHDP Report, 20). Bonn: IHDP, The Earth System Governance Project.

Biermann, F., Betsill, M. M., Gupta, J., Kanie, N., Lebel, L., Liverman, D., … & Zondervan, R. (2010). Earth System Governance: A research framework. *International Environmental Agreements: Politics, Law and Economics, 10*, 277–298. doi:10.1007/s10784-010-9137-3.

Brunkard, J. M., Cifuentes, E., & Rothenberg, S. J. (2008). Assessing the roles of temperature, precipitation, and ENSO in dengue re-emergence on the Texas-Mexico border region. *Salud Pública de México, 50*(3), 227–234.

Campbell, J. R. (2006a). *Traditional disaster reduction in Pacific Island communities, GNS science report 2006/038*. Wellington: Institute of Geological and Nuclear Sciences.

Campbell, J. R. (2006b). *Community relocation as an option for adaptation to the effects of climate change and climate variability in Pacific Island Countries (PICs)*. (APN Project Report ARCP2005-14NSY-Campbell). Retrieved from APN E-Lib http://www.apn-gcr.org/resources/items/show/1519

Campbell, J. R. (2009). Islandness: Vulnerability and resilience in Oceania. Shima. *International Journal of Research into Island Cultures, 3*(1), 85–97.

Campbell, J. R. (2010). An overview of natural hazard planning in the Pacific Island region. *Australasian Journal of Disaster and Trauma Studies, 2010*(1). http://trauma.massey.ac.nz/issues/2010-1/contents.htm

China's National Climate Change Programme. (2007). Prepared under the Auspices of the National Development and Reform Commission. People's Republic of China. Printed in June 2007.

Climate Institute. (2010). *Human health and climate change*. Retrieved from http://www.climate.org/topics/health.html

Connell, J., & Lea, J. P. (2002). *Urbanisation in the Island Pacific: Towards sustainable development*. London: Routledge.

Costello, A., Abbas, M., Allen, A., Ball, S., Bell, S., Bellamy, R., … & Patterson, J. (2009). Managing the health effects of climate change. *The Lancet, 373*(9676), 1693–1733.

Dahal, H., & Khanal, D. R. (2010). Food security and climate change adaptation framework; issues and challenges. Presented in the second stakeholders workshop on NAPA in agriculture sector, Kathmandu, 23 Feb 2010.

Descloux, E., Mangeas, M., Menkes, C. E., Lengaigne, M., Leroy, A., Tehei, T., … & De Lamballerie, X. Climate-based models for understanding and forecasting dengue epidemics. *PLoS Neglected Tropical Diseases, 6*(2), e1470. Epub 2012 Feb 14.

Dessai, S. X. R. (2005). Robust adaptation decisions amid climate change uncertainties. A thesis submitted for the Degree of Doctor of Philosophy in the School of Environmental Sciences, University of East Anglia, Norwich.

Dessai, S., Lu, X., & Risbey, J. S. (2005). The role of climate scenarios for adaptation planning. *Global Environmental Change, 15*, 87–97.

Dhakal, S., Zondervan, R., & Puppim de Oliveira, J. (2011). *Carbon governance in Asia: Bridging scales and disciplines*. (APN Project Report: CBA2010-04NSY-Dhakal). Retrieved from APN E-Lib http://www.apn-gcr.org/resources/items/show/1665

Erni, C., & Nikornuaychai, P. (2004). What kind of environment? Reconciling indigenous people's rights and environmental conservation policies: A case study from Thailand. Paper presented at the expert seminar: *Human right to decent environment; with special reference to the indigenous peoples*, Rovaniemi, 21–22 Aug 2004.

FAO. (2011). *FAO at work – women key to food security*. Rome: Food and Agriculture Organization of the United Nations (FAO).

Finnegan, S. (2011). *Pacific island paradise – wasting away* (IPENZ transactions, Vol. 2011/1). New Zealand: Institution of Professional Engineers New Zealand (IPENZ).

Government of Andhra Pradesh. (2004). Report of the State Level Committee on Heat Wave Conditions in Andhra Pradesh State. Revenue (Disaster Management) Department, Government of Andhra Pradesh, Hyderabad.

Gurung, T. B., Pokhrel, P. K., & Wright, I. (2011). *Climate change; Livestock sector vulnerability and adaptation in Nepal. Proceedings of consultative technical workshop to raise awareness and identify the priority research and development areas in livestock subsector addressing climate change*. Kathmandu: Nepal Agriculture Research Council (NARC).

Helvetas, Nepal & Intercooperation. (2010). *Climate change in the mid hills of Nepal; facts or fictions? from a farmers' perspective*. Helvetas, Nepal and Intercooperation. Helvetas Swiss Intercooperation Nepal, Lalitpur, Nepal.

ICIMOD. (2009). *Climate change impacts and vulnerabilities in the Eastern Himalayas*. Kathmandu: International Center for Integrated Mountain Development.

Ionescu, C., Klein, R. J. T., Hinkel, J., Kavikumar, K. S., & Klein, R. (2009). Towards a formal framework of vulnerability to climate change. *Environmental Modeling and Assessment, 14*, 1–16. doi:10.1007/s10666-008-9179-x.

IPCC. (2001). *Climate change 2001: Impacts, adaptation, and vulnerability. Contribution of working group II to the third assessment report of the Intergovernmental Panel on Climate Change (IPCC)*. Cambridge, UK: Cambridge University Press.

IPCC. (2007). In R. K. Pachauri & A. Reisinger (Eds.), *Climate change 2007: Synthesis report. Contribution of working groups I, II and III to the Fourth Assessment Report of the Intergovernmental Panel on Climate Change*. Cambridge, UK: Cambridge University Press.

IPCC. (2012). Managing the risks of extreme events and disasters to advance climate change adaptation. A special report of Working Groups I and II of the Intergovernmental Panel on Climate Change. In C. B. Field, V. Barros, T. F. Stocker, D. Qin, D. J. Dokken, K. L. Ebi, M. D. Mastrandrea, K. J. Mach, G. -K. Plattner, S. K. Allen, M. Tignor, & P. M. Midgley (Eds.), Cambridge and New York: Cambridge University Press, 582 pp.

Jagers, S. C., & Stripple, J. (2003). Climate governance beyond the state. *Global Governance, 9*(3), 385–400.
Kanie, N. (2012). *Exploring effective architecture for emerging agencies in international environmental governance*. (APN Project Report: CBA2012-04NSY-Kanie). Retrieved from APN E-Lib http://www.apn-gcr.org/resources/items/show/1764
Karl, T. R., Melillo, J. M., & Peterson, T. C. (Eds.). (2009). *Global climate change impacts in the US. US Global Change Research Program (USGCRP)*. Cambridge, UK: Cambridge University Press.
Karlsson-Vinkhuyzen, S., & Asselt, H. (2009). Introduction: Exploring and explaining the Asia-Pacific partnership on clean development and climate. *International Environmental Agreements: Politics, Law and Economics, 9*(3), 195–211.
Kench, P. S., McLean, R. F., & Nichol, S. L. (2005). New model of reef island evolution: Maldives, Indian Ocean. *Geology, 33*(2), 145–148.
Koshy, K., Salinger, J., & Shea, E. (2005). *Training institute on climate and extreme events in the Pacific*. (APN Project Report: CBA2005-CB04-Koshy). Retrieved from http://www.apn-gcr.org/resources/items/show/1605
Krishnan, P., Dam Roy, S., Grinson, G., Srivastava, R. C., Anand, A., Murugesan, S., … & Soundararajan, R. (2011). Elevated sea surface temperature during May 2010 induces mass bleaching of corals in the Andaman. *Current Science*, Research Communications, *100*(1), 111–117.
Lim, B., Spanger-Siegfried, E., Burton, I., Malone, E., & Huq, S. (2004). *Adaptation policy frameworks for climate change: Developing policies, strategies and measures. UNDP/GEF*. Cambridge, UK: Cambridge University Press.
Lorrey, A. (2010). Improving Pacific Island meteorological data rescue and data visualisation capabilities through involvement in emerging climate research programmes. (APN Project Report: CBA2010-10NSY-Lorrey). Retrieved from http://www.apn-gcr.org/resources/items/show/1666
Lun, Y. (2010). *Climate change in the Eastern Himalayas: Advancing community-based scientific capacity to support climate change adaptation*. (APN Project Report: CIA2009-03-Lun). Retrieved from http://www.apn-gcr.org/resources/items/show/1700
Mahat, T. K., Maden, U., Boom, D., Uphadyay, B., Perlis, A., Thapa, R., … & Thaku, A. K. (Eds.). (2011). *Pacific Mountain Courier: Special issue on youth action for climate change through art, 12 (1)*. Kathmandu: ICIMOD.
McMichael, A. J., Woodruff, R. E., & Hales, S. (2006). Climate change and human health: Present and future risks. *The Lancet, 367*, 859–869.
Meinke, H., Hansen, J., Gill, M. A., Gadgil, S., Selvaraju, R., Kumar, K. K., & Boer, R. (2004). Applying climate information to enhance the resilience of farming systems exposed to climatic risk in South and Southeast Asia. (APN Project Report ARCP-04CMY-Meinke). Retrieved from APN E-Lib http://www.apn-gcr.org/resources/items/show/1493
Meinshausen, M., Meinshausen, N., Hare, W., Raper, S. C. B., Frieler, K., Knutti, R., David Frame, J., & Myles, A. R. (2009). Greenhouse-gas emission targets for limiting global warming to 2°C. *Nature, 458*, 1158–1162. doi:10.1038/nature08017.
Mimura, N., Nurse, L., McLean, R. F., Agard, J., Briguglio, L., Lefale, P., Payet, R., & Sem, G. (2007). Small islands. In M. L. Parry, O. F. Canziani, J. P. Palutikof, P. J. van der Linden, & C. E. Hanson (Eds.), *Climate change 2007: Impacts, adaptation and vulnerability. Contribution of working group II to the Fourth Assessment Report of the Intergovernmental Panel on Climate Change* (pp. 687–716). Cambridge, UK: Cambridge University Press.
Ministry of Natural Resources and Environment Malaysia. (2010). Malaysia national policy on climate change. Available from http://www.nre.gov.my/english/Pages/Home.aspx
Myers, N. (2002). Environmental refugees: A growing phenomenon of the 21st century. *Philosophical Transactions of the Royal Society of London. Series B, Biological Sciences, 357*(1420), 609–613. doi:10.1098/rstb.2001.0953.
Nakalevu, T. (2006). *Capacity building for the development of adaptation measures in Pacific Island countries*. Secretariat of the Pacific Regional Environment Programme (SPREP). Retrieved from http://www.sprep.org/publications/capacity-building-for-the-development-of-adaptation-measures-in-pacific-island-countries-cbdampic-project

Nakashima, D., & Roué, M. (2002). Indigenous knowledge, peoples and sustainable practice. In P. Timmerman (Ed.), *Encyclopedia of global environmental change, vol. 5, social and economic dimensions of global environmental change* (pp. 314–324). Chichester: Wiley.

National Research Council (NRC). (2009). *Science and decisions: Advancing risk assessment*. Washington, DC: The National Academies Press.

Natural Resources Defense Council. (2011). Climate and your health: Addressing the most serious effects of climate change. Available from www.nrdc.org/policy

Nepal NAPA. (2010). Government of Nepal Ministry of Science, Technology and Environment. 2010 National adaptation programme of action, Singha Durbar, Kathmandu. Available from http://www.moste.gov.np/

Northern Development Foundation (NDF), Hauy Hin Lad Community, & Oxfam GB. (2010). *A case study on the carbon footprint of a Karen community at Huay Hin Lad*, Chiang Rai. Available from http://www.aippnet.org/home/publication/environment/case-studies/1065-climate-change-trees-and-livelihood-a-case-study-on-the-carbon-footprintof-a-karen-community-in-northern-thailand

NPC. (2010). *Three year plan approach paper*. Kathmandu: National Planning Commission, Government of Nepal, Singh Durbar.

Okayama, T. (2011). *Scientific capacity development of the trainers and policy makers for climate change adaptation planning in Asia and the Pacific*. (APN Project Report CBA2010-09NSY). Retrieved from APN E-Lib http://www.apn-gcr.org/resources/items/show/1670

Olshak, B. C., Ganseer, A., & Gruschke, A. (1987). *Himalayas: Growing mountains, living myths, migrating peoples*. New York: Facts on File.

Pittock, B. (2003). *Climate change: An Australian guide to the science and potential impacts*. Canberra: Australian Greenhouse Office.

Portier, C. J., Thigpen Tart, K., Carter, S. R., Dilworth, C. H., Grambsch, A. E., Gohlke, J., … & Whung, P-Y. (2010). *A human health perspective on climate change: A report outlining the research needs on the human health effects of climate change*. Research Triangle Park: Environmental Health Perspectives/National Institute of Environmental Health Sciences. doi:10.1289/ehp.1002272.

Pradhananga, D. (2010). *Graduate conference on climate change and people*. (APN Project Report CBA2010-12NSY). Retrieved from APN E-Lib http://www.apn-gcr.org/resources/items/show/1673

Pulhin, J. M. (2009). *Capacity development on integration of science and local knowledge for climate change impacts and vulnerability assessments*. (APN Project Report: CIA2009-02NSY-Pulhin). Retrieved from APN E-Lib http://www.apn-gcr.org/resources/items/show/1699

Qiu, J. (2008). The third pole. *Nature, 454*, 393–396.

Qui, J. (2009). Where the rubber meets the garden. *Nature, 457*, 246–247.

Ramanathan, V., Agrawal, M., Akimoto, H., Aufhammer, M., Devotta, S., Emberson, L., … & Zhu, A. (2008). *Atmospheric brown clouds: Regional assessment report with focus on Asia*. Nairobi: United Nations Environment Programme.

Ricardo, L., & Wirth, T. E. (2009). Global leadership for action. Facilitating an international agreement on climate change: Adaptation to climate change. Available from www.globalclimateaction.org

Rockström, J., Steffen, W., Noone, K., Persson, Å., Chapin, F. S., Lambin, E. F., … & Schellnhuber, H. J. (2009a). A safe operating space for humanity. *Nature, 46*, 472–475.

Rockström, J., Steffen, W., Noone, K., Persson, Å., Chapin, III, F. S., Lambin, … & Foley, J. (2009b). Planetary boundaries: Exploring the safe operating space for humanity. *Ecology and Society, 14*(2), 32. Retrieved from http://www.ecologyandsociety.org/vol14/iss2/art32/

Salinger, J. (2003). *APN workshops on climate variability & trends in Oceania*. (APN Project Report: ARCP2003-11). Retrieved from APN E-Lib http://www.apn-gcr.org/resources/items/show/1454

Saniotis, A., Hansen, A., & Bi, P. (2011). Climate change and population health: Possible future scenarios. In J. Blanco & H. Kheradmand (Eds.), *Climate change – socioeconomic effects*. InTech. Available from http://www.intechopen.com/books/climate-change-socioeconomic-effects/climatechange-and-population-health-possible-future-scenarios

Shrestha, K. L. (2005). Global change impact assessment for Himalayan mountain regions for environmental management and sustainable development. *Global Environmental Research, 9*(1), 69–81.

Skole, D. (2011). *Developing an MRV system for REDD+ (Scaling up from project level to a national level REDD+MRV systems for Laos and Vietnam)*. (APN Project Report: EBLU-02CMY(C)-Skole). Retrieved from APN E-Lib http://www.apn-gcr.org/resources/items/show/1708

South China Morning Post (SCMP). (2011, 12 February). 'Climate champions' top league. Retrieved from http://www.scmp.com/article/737928/climate-champions-top-league

Soverow, J. E., Wellenius, G. A., Fisman, D. N., & Mittleman, M. A. (2009). Infectious disease in a warming world: How weather influenced West Nile virus in the United States (2001–2005). *Environmental Health Perspectives, 117*, 1049–1052.

SPREP. (2005). *Secretariat of the Pacific Regional Environment Programme (SPREP); strategic programmes: 2004–2013*. Apia: SPREP.

SPREP. (2006). Secretariat of the Pacific Regional Environment Programme. *Pacific Islands framework for action on climate change 2006–2015*. Second edition. Reprinted (2011). Apia. Available from http://www.sprep.org

SPREP. (2011). *Pacific Islands framework for action on climate change 2006–2015*. Second edition, Reprinted by Secretariat of the Pacific Regional Environment Programme (SPREP), Samoa. Available from http://www.sprep.org

Steffen, W., Sanderson, A., Tyson, P. D., Jager, J., Matson, P. M., Moore, III, B., … & Wasson, R. J. (2004). *Global change and the earth system: A planet under pressure*. New York: Springer.

Sturman, A., & McGowan, H. (1999). Climate. In M. Rapaport (Ed.), *The Pacific islands environment and society*. Honolulu: Bess Press.

TERI. (2001). *Review of past and on-going work on urban air quality in India*. Report submitted to the World Bank, December, Tata Energy Research Institute, 2001EE41.

The Organisation for Economic Co-operation and Development/International Energy Agency IEA (OECD/IEA). (2012). *Highlights of CO_2 emissions from fuel combustion*. International Energy Agency. Paris: IEA. Available from http://www.iea.org

Tienhaara, K. (2012). *Climate change governance in the Asia-Pacific region: Agency, accountability and adaptativeness*. (APN Project Report CBA2011-11NSY-Tienhaara). Retrieved from http://www.apn-gcr.org/resources/items/show/1689

Towprayoon, S., Bonnet, S., Garivait, S., & Chidthaisong, A. (2008). Greenhouse gas and aerosol emissions from rice field and forest in the Mekong River basin sub-region. *GMSARN International Journal, 2*, 163–168.

Tseng, W.-C. (2008). Estimating the economic impacts of climate change on infectious diseases: A case study on dengue fever in Taiwan. *Climatic Change, 92*, 123–140.

United Nations Sustainable Development Knowledge Platform (UNSDKP). (2012). A/RES/66/288 – The future we want (2012). Outcome of the United Nations Conference on sustainable development, Rio de Janeiro from 20 to 22 June 2012. Retrieved from http://www.un.org/ga/search/view_doc.asp?symbol=A/RES/66/288&Lang=E

USAID Asia. (2010). *Asia-Pacific Regional climate change adaptation assessment, final report: Findings and recommendations*. Washington, DC: USAID.

Vivekanandan, E., Ali, M. H., Jasper, B., & Rajaopalan, M. (2009). Vulnerability of corals to warming of the Indian seas: A projection for the 21st century. *Current Science, 97*(11), 1654–1657.

World Health Organisation. (WHO). (2002). *World health report 2002: Reducing risks, promoting healthy life*. Geneva: World Health Organization. http://www.who.int/whr/2002/en/whr02_en.pdf

World Health Organisation. (WHO). (2007). *Climate change and human health in Asia and the Pacific: From evidence to action*. Bali: Report of the WHO regional workshop. 10–12 Dec 2007.

World Health Organisation. (WHO). (2008a). Environmental trends. In health in Asia and the Pacific, World Health Organization regional offices for South-East Asia and the Western Pacific. Available from http://www.wpro.who.int/publications/docs/09_Chapter4Environmentaltrends_5FE0.pdf

World Health Organization. (WHO). (2008b). Regional framework for action to protect human health from effects of climate change in the South East Asia and Pacific Region. Available from http://www.searo.who.int/en/Section260/Section2468_14335.htm

Wu, Z., & Ou, X. (1995). *The Xishuangbanna biosphere reserve* (Working paper, Vol. 2). Paris: UNESCO.

Wu, P.-C., Lay, G.-J., Guo, H.-R., Lin, C.-H., Lung, S.-C., & Su, H.-J. (2009). Higher temperature and urbanization affect the spatial patterns of dengue fever transmission in subtropical Taiwan. *Science of the Total Environment, 407*(7), 2224–2233. doi:10.1016/j.scitotenv.2008.11.034.

Xu, J., Runbine, R. E., Shrestha, E., Erickson, S., Yang, X., & Wilkes, A. (2009). The melting Himalayas: Cascading effects of climate change on water, biodiversity, and livelihoods. *Conservation Biology, 23*(3), 520–530. doi:10.1111/j.1523-1739.2009.01237.x.

Ziska, L. H., Knowlton, K., Rogers, C., Dalan, D., Tierney, N., Elder, M. A., … & Frenz, D. (2011). Recent warming by latitude associated with increased length of ragweed pollen season in central North America. *Proceedings of the National Academy of Sciences, United States of America (PNAS), 108*(10), 4248–4251.

Chapter 6
Climate and Sustainability

Rodel Lasco, Yasuko Kameyama, Kejun Jiang, Linda Peñalba, Juan Pulhin, P.R. Shukla, and Suneetha M. Subramanian

Abstract Projected change in climate in the coming decades adds a layer of complexity in the search for sustainability. Warming temperatures, rising sea levels, changing precipitation patterns and their impacts on natural and human systems could threaten the attainment of development goals. Many countries in Asia and the Pacific are among the most vulnerable to the impacts of climate change and there is growing recognition that climate change adaptation must be tackled as an integral part of the development process, for example in mainstreaming climate change adaptation into national plans and programmes. The aim of Chap. 6 is to explore linkages between sustainable development and efforts to address climate change in

---

R. Lasco (✉)
World Agroforestry Centre (ICRAF), Khush Hall, IRRI, Los Banos 4031,
Laguna, Philippines
e-mail: rlasco@cgiar.org

Y. Kameyama
Centre for Global Environmental Research, National Institute for Environmental Studies,
16-2 Onogawa, Tsukuba-City, Ibaraki 305-8506, Japan
e-mail: ykame@nies.go.jp

K. Jiang
Energy Research Institute, National Development and Reform Commission, B1403,
GuoHong Mansion, Muxidi Beili jia No.11, Xicheng District, Beijing 100038, China
e-mail: kjiang@eri.org.cn

L. Peñalba
Institute of Governance and Rural Development, College of Public Affairs,
University of the Philippines Los Baños, College, Laguna 4031, Philippines
e-mail: lmpenalba@yahoo.com; lmpenalba@gmail.com

J. Pulhin
Department of Social Forestry and Forest Governance, College of Forestry and Natural
Resource, University of the Philippines Los Baños, College, Laguna 4031, Philippines
e-mail: jmpulhin@uplb.edu.ph; jpulhin@yahoo.com

M.J. Manton and L.A. Stevenson (eds.), *Climate in Asia and the Pacific: Security, Society and Sustainability*, Advances in Global Change Research 56,
DOI 10.1007/978-94-007-7338-7_6,

Asia and the Pacific, particularly focussing in two areas of low carbon development (LCD) pathways for the region, and the importance of natural ecosystems in sustaining the delivery of ecosystem services that are essential for climate change adaptation and mitigation. The challenges posed by climate change will be felt in the coming decades in Asia and the Pacific. In parallel, nations in the region will continue to aspire for sustainable development. Policy makers and development workers must find ways to ensure that both these concerns are addressed synergistically while avoiding negative outcomes. One way to mitigate climate change while pursuing sustainable development is through LCD, which will require negotiations across many stakeholders of governments, non-government agencies, industry and broader communities. In Asia and the Pacific natural ecosystems will continue to play a critical role in addressing climate change adaptation and mitigation. Nations in the region will have to find innovative ways to manage and rehabilitate natural ecosystems for a multiplicity of functions and services. This will involve greater collaboration and communication between scientists and policy makers as well as between natural and social scientists. In many developing countries, there is still very limited empirical information and research needs to be ramped up. North-South and South-South partnerships could help fill the gap.

Keywords Climate and ecosystems • Integrated assessment models • Low carbon development • Climate and sustainability

6.1 Introduction

Sustainable development has occupied a place in the global agenda since 1987 when the World Commission of Environment and Development (Brundtland Commission) released its report 'Our Common Future' (WCED 1987).

The Commission defines sustainable development as "*development that meets the needs of the present without compromising the ability of the future generations to meet their own needs*."

Now, 25 years later, sustainable development has become mainstreamed into national and international development discourse. Indeed, it is one of the almost universal aspirations of all nations today. This happened in spite of the proliferation of and

P.R. Shukla
Public Systems Group, Indian Institute of Management,
Vastrapur, Ahmedabad 380015, Gujarat, India
e-mail: shukla@iimahd.ernet.in

S.M. Subramanian
United Nations University Institute of Advanced Studies (UNU-IAS), 6F International
Organizations Center, Pacifico-Yokohama, 1-1-1 Minato Mirai, Nishi-ku,
Yokohama 220-8502, Japan
e-mail: subramanian@ias.unu.edu

disagreements over numerous definitions, frameworks and methods to operationalize it at various scales (Baumgartner 2011; Jabareen 2008; Sneddon et al. 2006).

The essence of sustainable development can be viewed as meeting fundamental human needs while preserving the life support systems of the planet (Kates et al. 2005). It involves efficient management of resources and creation of options for natural ecosystems to support social and economic development. To sustain this capacity requires a full understanding and effective management of feedbacks and interrelations between the system's ecological, social and economic components across temporal and spatial scales (Gunderson and Holling 2002; Kates et al. 2001; cited in Folke et al. 2002).

Ironically, the degree to which sustainable development has been embraced by the global community is not matched by the state of the world's environment. Greenhouse gas (GHG) emissions continue to increase (Friedlingstein et al. 2010). The landmark assessment of the world's ecosystems in 2005 revealed that many of the world's forest, freshwater, coastal and marine resources have been exploited severely (Millennium Ecosystem Assessment 2005). On the other hand, others contend that while the global picture is bleak, there have been many local success stories. In fact, the degree to which the concept of sustainable development has entered mainstream thinking is simply astonishing in itself. In the future, there may be broad international agreement that the goal should be to foster a transition toward development paths that meet human needs while preserving the earth's life support systems and alleviating hunger and poverty (Mexico City Workshop 2002).

There is wide spread concern that the world's development pathway is not sustainable in the long term. This has led to the emergence of what has been called "sustainability science", which seeks to improve on the substantial, but still limited, understanding of nature-society interactions (Statement from Friibergh Workshop on Sustainability Science 2000; de Vries and Petersen 2009). It is premised on the need for a better understanding of the complex and dynamic interactions between society and nature. As such, it will require fundamental advances in our ability to address such issues as the behavior of complex self-organizing systems as well as the responses of the nature-society system to multiple and interacting stresses (Kates et al. 2000). There is also a need to integrate across the full range of scales from local to global and thereby combine different ways of knowing and learning. Increasingly, there is more focus also on examining the contemporary relevance of traditional practices and exploring ways to integrate them with modern approaches for better environmental management (Berkes 2008; Bélair et al. 2010; Arico and Valderrama 2010). This also feeds into the broader issues of equity and the need for participatory planning involving various actors with a stake in the ecosystem (Goma et al. 2001; Kenter et al. 2011).

The projected change in climate in the coming decades adds another layer of complexity in the search for a sustainable development path. Warming temperatures, rising sea levels, changing precipitation patterns and their impacts on natural and human systems could threaten the attainment of development goals, such as those expressed in the Millennium Development Goals (Yohe et al. 2007). Many countries in Asia and the Pacific are among the most vulnerable to the impacts of climate change, such as Bangladesh and the small islands states (Cruz et al. 2007; Mimura et al. 2007). As a result, there is growing recognition that climate change adaptation must be tackled as an integral part of the development process (Munasinghe 2010; Schipper 2007; Robinson et al. 2006; Adger 2003). In many cases this has been expressed in terms of mainstreaming climate change adaptation to national plans and programs (Lasco et al. 2009).

Even climate change mitigation can have strong links with sustainable development. For example, energy utilization typically rises (and thus GHG emissions) with rising economic development as can be seen in fast developing countries in Asia. However, more efficient energy use will lead to lower emissions (i.e. decarburization) while promoting sustainable development (Halsnaes et al. 2011).

The aim of this Chapter is to explore the link between sustainable development and efforts to address climate change in Asia and the Pacific. In Sect. 6.2 we discuss the feasibility and challenges of low carbon development pathways for the region. In Sect. 6.3 we present how natural ecosystems are necessary to sustaining delivery of ecosystem services that are essential for climate change adaptation and mitigation.

6.2 Climate and Economy: Towards Low Carbon Development

6.2.1 Introduction

Low carbon development aims not only to reduce carbon dioxide (CO_2) emissions but also to promote economic development and enhanced community well-being.

Low Carbon Society (LCS) or Low Carbon Development (LCD) has become a familiar notion for anyone involved in climate change policies in the last decade. An LCS or LCD (hereafter LCD) is a comprehensive image that covers both "hard" and "soft" aspects of society that would lead to reduction of CO_2 emissions. The "hard" part includes infrastructure, technology in terms of hardware, buildings and houses, transportation, etc. The "soft" part includes policies, knowledge, people's lifestyles and behavior, institutions, rules, etc.

LCD pathways are not intended just to achieve a certain level of CO_2 emissions reduction only. The notion includes poverty reduction, economic development and fulfillment of people's welfare while reducing greenhouse gas (GHG) emissions.

An LCD society can be defined as one that:

- Takes actions that are compatible with the principles of sustainable development, ensuring that the development needs of all groups within society are met;
- Makes an equitable contribution towards the global effort to stabilize atmospheric concentrations of carbon dioxide and other GHGs at a level that will avoid dangerous climate change through deep cuts in global emissions; and
- Demonstrates high levels of energy efficiency and uses low-carbon energy sources and production technologies (LCS-RNet 2009).

In a way LCD seeks to achieve sustainable development that reaches economic, environmental and social dimensions of development simultaneously. The present section aims at explaining current arguments and research on LCD, especially in the Asia-Pacific region. To reach this aim "integrated assessment" will be considered.

Integrated Assessment Model (IAM) is a tool that contributes to the assessment of complicated policies such as related to climate change mitigation. In the later part of this section, institutional dimensions of integrated assessment are discussed.

6.2.2 Roles of Integrated Assessment Models

6.2.2.1 Integrated Assessment Model (IAM)

Integrated Assessment (IA), when used in conjunction with climate change policies, is a methodology to assess economic development from various policy perspectives. Integrated Assessment Models (IAMs) are large-scale computer simulation models that assimilate a variety of factors and disciplinary inputs to address IA. As CO_2 emissions, energy use and economies are interlinked in a complicated manner, IAMs have been used to assess climate change mitigation policies. Future GHG emissions depend heavily on the development pathways that future societies choose in terms of economic, demographic, technological, land-use, agricultural and energy mix changes. The interactions between these primary driving forces are complex and have profound regional circumstances.

The first trials of IAM model developments can be observed in the early 1970s (Meadows et al. 1972; Mesarovic and Pestel 1974), followed by formal IAM development in the late 1970s for assessment of energy use and economies, and climate change (Nordhaus 1979; Edmonds and Reilly 1985). Full IAM development came of age in the early 1990s, after the Intergovernmental Panel on Climate Change (IPCC) was established in 1988 (Alcamo et al. 1990; Morita et al. 1994).

The IPCC special report on emission scenarios describes four alternative futures that may evolve to change how we view and emit global GHG emissions (SRES 2000). Each scenario is an alternative image of how the future can develop and is an appropriate tool with which to analyze how driving forces such as population and level of economic development may influence future emission outcomes and to assess the associated uncertainties. These scenarios do not include any additional climate initiatives, which mean that no scenario assumes any explicit climate policy intervention by any region or country.

For its Fifth Assessment Report, the IPCC has introduced a slightly modified approach to scenario development. Four representative concentration pathways (RCPs) have been prepared to describe a comprehensive dataset with high spatial and sectoral resolution out to 2100 (van Vuuren et al. 2011).

The IAM studies continue to be developed and the direction of these developments can be categorized into three areas (Kainuma et al. 2003). The first group of IAM models aim at dealing with a wide scope with more detailed data. As IAMs need to support broader audiences, new policy needs and new scientific knowledge, modeling targets and phenomena have become wider and more detailed. The second category involves the application of IAMs to participatory IA processes where stakeholders communicate with each other to determine the priority of information and decisions.

Policy makers are in need of tools to facilitate communication with scientists, industry sectors, and environmental NGOs to reach a consensus on climate change policies. In this case, IAMs are used as a communication tool. The third direction of IAM development is to apply IAMs to regional and local assessment rather than global scale assessment. By focusing on smaller scales, various policy targets can be dealt with simultaneously. Altogether, IAMs have been widely recognized as useful tools to quantitatively assess development of human activities.

6.2.2.2 Case Studies of Countries in Asia by IAM

Studies on future emission scenarios are necessary to support studies of potential anthropogenic impacts on the climate system, to serve as the basis for further analysis, as well as to estimate the consequences of climatic events and define better strategies for adaptation. They provide inputs to climate models and assist in assessing the relative importance of GHGs in changing atmospheric composition and hence climate. Scenarios also have an important role to play as baselines for comparison with stabilization scenarios in order to calculate the required mitigation effort. They have been useful for a multitude of purposes by governments, industry, researchers, and social organizations.

The long-term goal for LCD is to reduce the amount of GHG emissions while maintaining desired lifestyles. The Asia-Pacific Integrated Model (AIM) is one of the IAM models that have been developed for the purpose of assessing climate policies especially focusing on the Asia Pacific region (Kainuma et al. 2003). The AIM model was used to calculate the technological feasibility of achieving 70 % emissions reduction from 1990 levels in Japan (NIES et al. 2008). The figure "70 %" was chosen because, at the time of the study, the G8 summit meeting called for halving global GHG emissions by 2050. If the world was to head for this global target, then industrialized countries would have to reduce their emissions by more than 50 %. Japan announced that it would seek ways to achieve 60–80 % reduction by 2050. As such, two scenarios (A and B, as shown in Figs. 6.1 and 6.2, respectively) were developed to design Japanese LCD by two different approaches.

In both scenarios, it was shown that 70 % GHG emissions reduction was technologically feasible (Figs. 6.1 and 6.2). In the first scenario, Scenario A, development and wide use of fuel cells, photovoltaics, heat pump air-conditioners, etc., is anticipated. Much of the electricity supply will be made by nuclear power plants. Power generation and hydrogen production are combined with CCS technologies.

The second scenario, Scenario B, anticipates that there will be much less demand for energy as a result in changes in societal behavior that will require far less energy consumption. A change in the people's mind would slow down overall consumption,

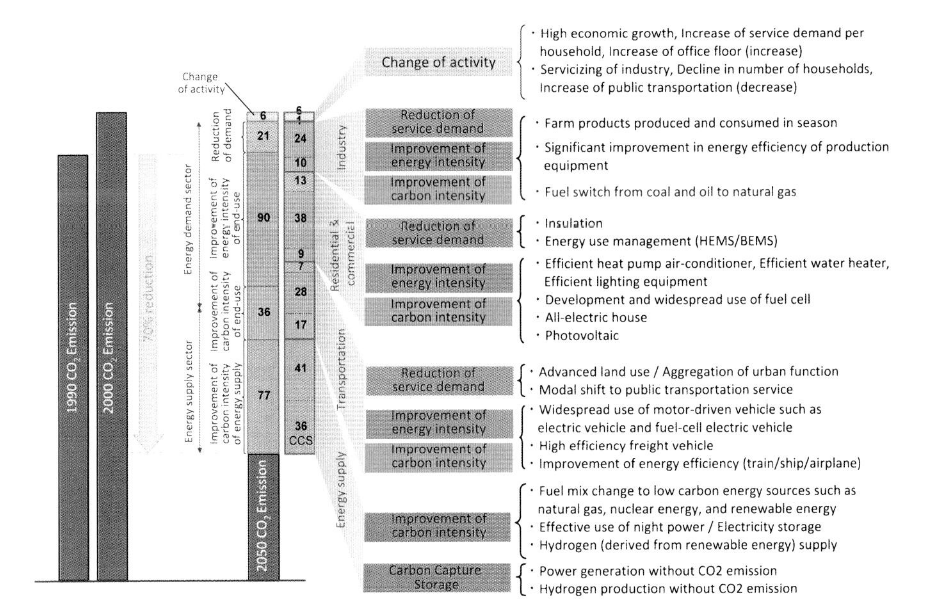

Fig. 6.1 Scenario (A) assumes a high-technology lifestyle where all sorts of energy efficient technologies and non-carbon technologies would be fully installed. This scenario included carbon capture and storage (*CCS*) to play a major role in electricity generation (Source: NIES et al. 2008)

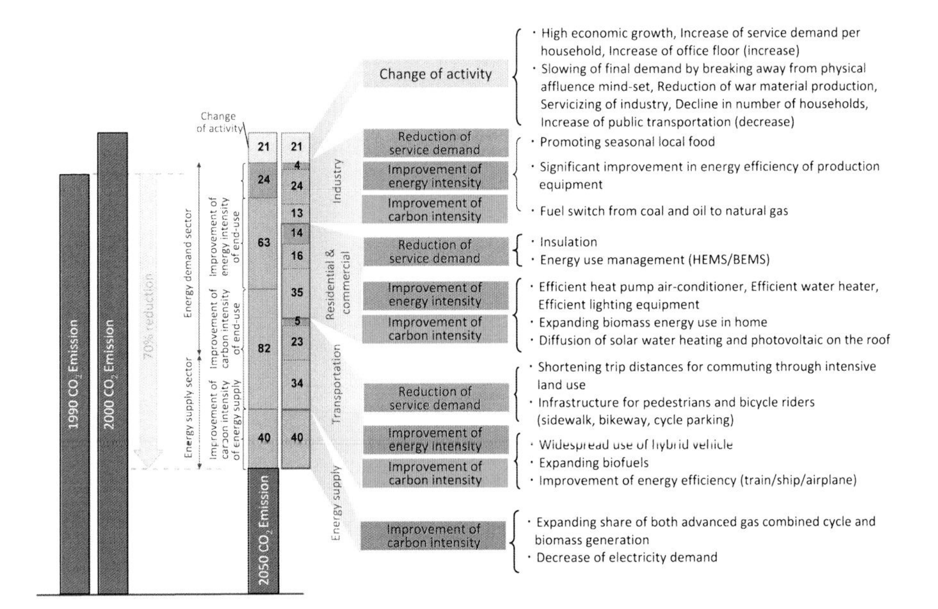

Fig. 6.2 Scenario (B) assumes less high-tech, carbon-intense society where people live in local communities. People would seek a simple and ecological lifestyle, with less eagerness for economic growth. In such a scenario, demand for energy will be less than in Scenario A and society would try to save energy to reduce GHG emissions (Source: NIES et al. 2008)

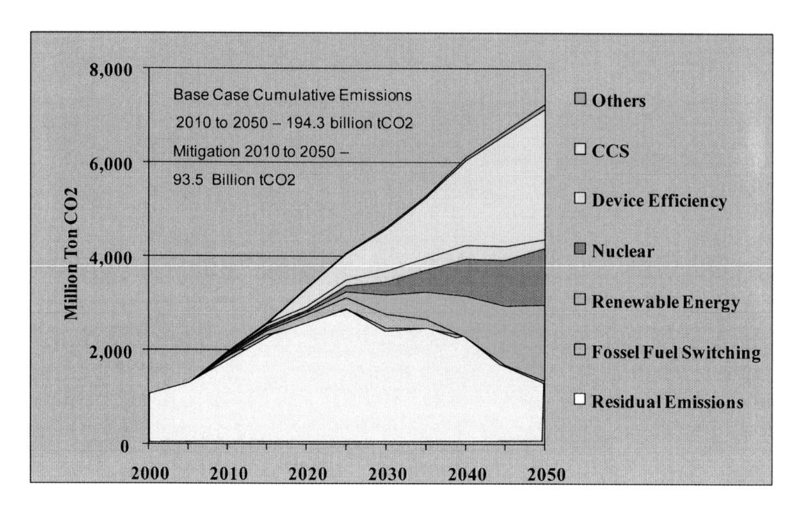

Fig. 6.3 India's carbon tax scenario (Source: IIMA et al. 2009)

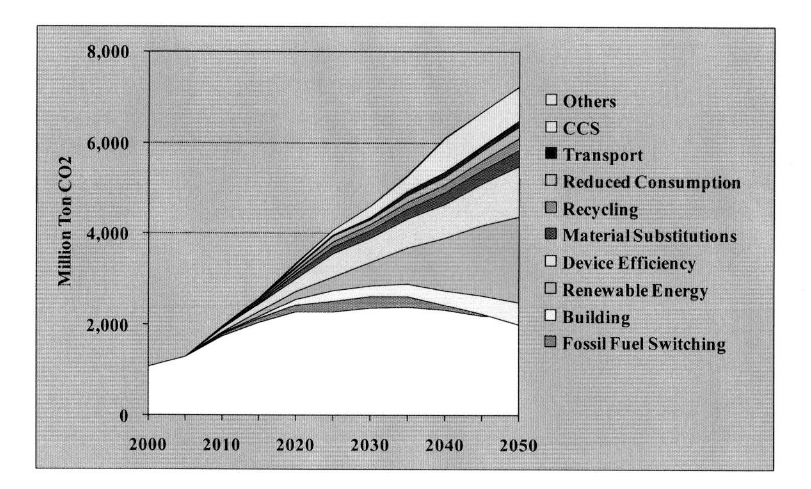

Fig. 6.4 India's sustainability scenario (Source: IIMA et al. 2009)

with the notion of not seeking material affluence but rather mental happiness. Biomass energy will be installed in residential sectors. People will consume food originated in their neighborhood, rather than imported from other countries. The remaining question was whether the people or policy makers are willing to choose such scenarios.

Another study was conducted for India (IIMA et al. 2009). This exercise also assumed two types of scenarios. The first pathway (Fig. 6.3) assumed the conventional development pattern together with a carbon tax that aligns India's emission to an optimal 450 ppmv CO_2 equivalent stabilization global response. The second emissions pathway (Fig. 6.4) assumed an underlying sustainable development pattern characterized by diverse response measures typical of the "sustainability" paradigm.

Under the first scenario (Fig. 6.3), the mitigation target of 93.5 billion tonnes CO_2 is achieved through extensive use of advanced technologies such as CCS and nuclear energy. Meanwhile, the second scenario envisaged the same mitigation target to be achieved by a combination of initiatives on both supply and demand sides. In these scenarios, people in India will benefit not only through climate change mitigation, but also via other co-benefits. For instance, emissions of various gases due to fuel combustion contribute significantly to local air pollution in urban and industrial areas. The control of local air pollutants such as sulphur dioxide (SO_2) has been a major target of environmental policies and programmes in developed countries. Thus, during the low carbon transition, the conjoint policies can deliver benefits of improved air quality through the reduction of costs to achieving air quality targets.

6.2.3 Sustainable Development in Asia and the Pacific

Countries in Asia and the Pacific are facing an expanding future in terms of economy and energy use. Such expansion will affect GHG emissions of these countries, as well as their future for LCD. Especially, energy use in China and India will affect GHG emissions at the global level. This is why we need to look carefully at these two major economies in the region.

6.2.3.1 Sustainable Development in China

As China expects rapid economic development in the years to come, it has been keen on establishing national policies related to energy and environment. After the National Program on Climate Change was released by the Chinese government, China sought ways to deal with GHG emissions mitigation, energy security and economic growth simultaneously (Xiulian and Kejun 2008).

At the same time sustainable development is recognized as an important issue. Agenda 21 for China,[1] whose adoption was announced by the Chinese government in 1994, explicitly states that "*Taking the path of sustainable development is a choice China must make in order to ensure its future development in the century. Because China is a developing country, the goal of increasing social productivity, enhancing overall national strength and improving people's quality of life cannot be realised without giving primacy on economy development. At the same time, it will be necessary to conserve natural resources and to improve the environment, so that the country will see long-term, stable development.*"

Since 1994, Agenda 21s objectives have been translated into other policy plans, including the successive 5-Year plans. Other objectives include reducing large

[1] During the 1992 Environment and Development Summit, the UN launched the Agenda 21 programme to guide sustainable development. As regards China, the government published a specific Agenda 21 for the country in order to implement the broader UN programme.

differences in wealth in different areas (especially rural areas and the regions in the west of the country), and hence to more generally reduce poverty and to control population growth.

Tree-planting and afforestation, together with enhancing ecology restoration and protection, have constituted long-term policies in China since the 1970s. According to the Sixth National Forest Assessment in 2005, the acreage conserved artificial forests in the country was 54 million hectares – ranking top in the world – and the amount of growing stock was 1,505 million cubic meters. The total area covered by forests was 174.91 million hectares, and the percentage of forest cover increased from 13.92 % to 18.21 % from the early 1990s to 2005. In addition to tree-planning and afforestation, China initiated many other policies for ecology restoration and protection, including natural forest protection, converting cultivated land to forest or grassland, pasture restoration and protection, further enhancing the capacity of forests as sinks of GHGs. Meanwhile, urban greening also grew rapidly in China. By the end of 2005, total greenery in built-up urban areas reached 1.06 million hectares with 33 % green coverage and 8.1 m^2 of public green area per capita. These green areas act as carbon sinks and aid the absorption of CO_2 present in the atmosphere.

6.2.3.2 Sustainable Development in India

As electricity becomes available in India, households will be able to enjoy more developed and healthy lifestyles. There are studies that calculate the future of electricity use in India (Shukla et al. 2005; Menon-Choudhury et al. 2006). The electricity power sector consumes about 40 % of the primary energy and nearly 70 % of coal use. The result is that the power sector contributes about half of India's carbon, sulphur, and nitrogen oxide emissions. Electricity consumption in India has more than doubled in the last decade, outpacing economic growth. Without any specific policy to change its trend towards LCD, the power capacity in 2015 will grow 2.5 times the 1995 level. Coal technologies will continue to account for the largest share of new additions to capacity, but will decline from 62 % in 2000 to about 55 % in 2015. CO_2 emissions will more than double from the 1995 level by 2015.

This trend can be changed through several types of policies (Shukla et al. 1999). Market liberalization in India, which has been developed since the 1990s, has led increasing direct foreign investment into India. If such a trend continues, it will lead to minor changes to the power supply profile compared with a "business-as-usual" case, but less power capacity will be needed due to greater energy efficiency and utilization of existing capacity.

One environmentally-aware development scenario is a case in which local governments take stricter action against nitrogen dioxide, sulphur dioxide and particulates. Capacity additions closely resemble the business-as-usual scenario, but fitting coal technologies with sulphur control equipment will cut sulphur dioxide by 40 % in 2015, but CO_2 emissions will remain or even increase in this scenario.

An alternative to the previous development scenario is a combination of progressive policy options, including decentralized governance, environmental conservation, efficiency and renewable energy promotion, and regional cooperation. Requirements for electricity capacity additions fall 22 % from the business-as-usual scenario, but will lead to drastic reduction of both sulphur and CO_2 emissions.

In case studies in any country, there are several paths that lead to a sustainable low carbon future. It is up to the people and decision makers to choose which paths they wish to follow. Some paths require more investment than others. In any case, it is less expensive to start today than to delay investment for the future. It might be good to start from policies that can be considered as no-regrets policies.

6.2.4 International Institutions to Achieve Low Carbon Development

6.2.4.1 Multilateral Institution Set Up Under the UNFCCC: Cancun, Durban and Beyond

Emissions reduction and limitation targets have been negotiated multilaterally under the UN Framework Convention on Climate Change (UNFCCC). As for the GHG emission mitigation, many countries in Asia and the Pacific have pledged their emission reduction targets around the time of the 15th Conference of the Parties to the UNFCCC (COP15), held in Copenhagen in late 2009. Those targets are considered to be voluntary targets and not international commitments, but such target-setting by developing nations is a major initiative that was not observed several years ago.

However, multilateral negotiations under the UNFCCC faced difficulty for many years even after COP15. Countries' emission reduction targets shown in Table 6.1 are voluntary targets which were submitted to the UNFCCC Secretariat to respond to the Copenhagen Accord, noted at COP15. The emissions reduction targets, when summed, are considered to be insufficient to reach the long-term target which had been discussed under the agenda called "shared vision." The Cancun Agreement, agreed at COP16 in 2010, calls for aiming at a maximum 2 °C rise in temperature from pre-industrial levels. It states "deep cuts in global GHG emissions are required according to science, and as documented in the Fourth Assessment Report of the IPCC, with a view to reducing global GHG emissions so as to hold the increase in global average temperature below 2 °C above pre-industrial levels, and that Parties should take urgent action to meet this long-term goal, consistent with science and on the basis of equity." In order to reach the long-term temperature stabilization target, global GHG emission needs to peak as soon as possible. The IPCC report indicates that emissions in the industrialized countries need to be reduced 25–40 % from 1990 levels by 2020 to reach the target. At present the total amount of emissions reduction target set by Annex I countries is not sufficient to achieve the 25–40 % reduction. Further strengthening of emissions reduction by industrialized countries is needed.

COP17, held in Durban, South Africa in late 2011, established a new negotiating process called the "Durban Platform." This process calls for an agreement to be reached by 2015 on "a protocol, another legal instrument or a legal outcome under

Table 6.1 Emission targets of countries in Asia and the Pacific, submitted to the UNFCCC Secretariat by 31 January 2010 (summarized by the author)

Country	Voluntary target
Australia	Reduce GHG emission by 5 % by 2020 compared with 2000 levels unconditionally, and will reduce 20 % by 2020 if the world agrees to a global goal to stabilize GHG concentration at 450 ppm
China	Reduce CO_2 per GDP by 40–45 % by 2020 from 2005
India	Reduce CO_2 per GDP by 20–25 % by 2020 from 2005
Japan	Reduce GHG emission by 25 % by 2020 from 1990 on condition that major emitting countries participate in international mitigation agreement
Republic of Korea	Reduce CO_2 emission by 30 % by 2020 compared with Business as Usual
Indonesia	Reduce CO_2 emission by 26 % by 2020 compared with Business as Usual. With international assistance, the target will be changed to 41 %
Singapore	Reduce CO_2 by 16 % by 2020 compared with Business as Usual
Papua New Guinea	Reduce GHG emission by 50 % by 2030 (base year not defined), and carbon neutral by 2050
New Zealand	Reduce GHG emission by 10–20 % by 2020 compared with 1990 levels by if the world agrees to a global goal of 2 °C.

the Convention applicable to all Parties." The new negotiation process is likely to review the gap between the long-term target and the short-term emission reduction target, and to negotiate ways to fill the gap.

(a) *Means to secure transparency*

The multilateral negotiations deal with other important elements. The schemes for monitoring, reporting and verifying (MRVs) have become another core element of negotiation since after the Copenhagen Accord.

Some industrialized countries hesitate to accept legally-binding emission reduction targets and they prefer to commit to voluntary, non-binding targets. Such relatively loose targets need to be monitored to ensure that countries make serious efforts to achieve their targets. It was therefore decided in the Cancun Agreement to conduct a series of processes to increase transparency of mitigation actions taken by industrialized countries. First, developed countries should submit annual GHG inventory reports and biennial reports on their progress in achieving emission reductions. The developed countries should also report on the provision of financial, technological and capacity-building support to developing country parties. A process for international assessment and review was established under the UNFCCC's Subsidiary Body for Implementation (SBI), with a view to promoting comparability and building confidence.

In many developing countries, methodologies for accurate data collection are needed to accumulate statistical data related to climate change. The MRV process is a way to secure emission limitation targets in each country. In the Cancun Agreement, developing countries are invited to submit to the UNFCCC secretariat information on nationally appropriate mitigation actions (NAMAs) for which they are seeking support, along with estimated costs and emission reductions, and the anticipated timeframe for implementation.

Developing countries are also requested to submit their national communications to the UNFCCC COP every 4 years. They are further requested to submit biennial update reports containing updates of national GHG inventories, including a national inventory report and information on mitigation actions, needs and support received. Internationally-supported mitigation actions will be reviewed domestically and will be subject to international MRV in accordance with guidelines to be developed under the UNFCCC. Domestically supported mitigation actions will be reviewed domestically in accordance with general guidelines to be developed under the UNFCCC. International consultation and analysis of biennial reports will be conducted by the SBI, in a manner that is non-intrusive, non-punitive and respectful of national sovereignty, aiming at increasing transparency of mitigation actions and efforts in developing countries.

(b) *Financial mechanisms*

Financial mechanisms are another key item under the current negotiations, because the level of emission mitigation efforts to be taken by developing nations depends on the amount of financial support by developed countries. Both short-term and mid-term financial support was agreed under the Copenhagen Agreement in 2009 (COP15). For the short term, US$30 billion for the period 2010–2012 was agreed, the allocation of which was to be balanced between adaptation and mitigation. For the mid-term, US$100 billion per year by 2020 was agreed to address the needs of developing countries.

The funds themselves may come from a wide variety of sources, public and private, bilateral and multilateral, including alternative sources. The Green Climate Fund (GCF) under the Cancun Agreement is to be designated as an operating entity of the financial mechanism of the UNFCCC under Article 11. A Transitional Committee is appointed to manage the Fund and, at the time of writing, a decision on how the existing funds under the UNFCCC and the Kyoto Protocol will be merged with the newly established GCF has yet to be made.

Although multilateral negotiations under the UNFCCC have not made substantial progress in the last several years, many countries acknowledge the importance of taking climate change mitigation action even without an agreement being reached at the international level. National actions toward LCD are prerequisite for domestic planning and economic development.

(c) *Co-benefits of climate mitigation policies*

Mitigation policies in many cases have co-benefits. In fact, co-benefits are becoming a major driving force for countries in Asia and the Pacific to set voluntary emission limitation and reduction targets.

First, many countries in Asia will benefit from energy-efficiency improvements. As explained earlier, most countries in the Asia-Pacific region import energy resources such as coal and oil. As demand for energy increases in these countries, improving energy efficiency is beneficial through minimizing costs of imports and improvements in energy security. Even for energy-exporting countries, saving energy at the household level will result in increases in exports, which is beneficial.

Second, mitigation can help improve people's health. For countries that use a lot of coal, installing clean coal technology will minimize local air pollution and health hazards such as respiratory diseases will be reduced due to improved local air quality. Indoor air pollution is also a serious issue in many developing countries. Traditional biomass fuel use has led to the destruction of forests. Shifting from traditional biomass fuel to renewable energy or electricity in households will reduce indoor air pollution in these countries.

Third, many countries in the region are interested in the recent debate on REDD (Reducing Emissions from Deforestation and Degradation in Developing Countries) and REDD-plus (Conservation and Sustainable Forest Management). Countries such as Indonesia and Papua New Guinea have incurred rapid deforestation and the REDD scheme has paved a way for those countries to be involved in emission mitigation policies. Chapter 6.3 provides more detail on the impact of REDD on ecosystem services.

Adaptation is also an important dimension of climate change policies, especially for countries in the Asia-Pacific region. Many small island states will be affected by sea-level rise. Some low-lying countries such as Bangladesh will be affected by floods. Nepal and India are concerned with the melting of glaciers in the Himalayas. Being fully prepared for extreme weather patterns will protect communities even if the extreme weather is not caused by climate change. Mainstreaming adaptation policies within development policies is imperative.

6.2.4.2 Regional Cooperation for LCD Pathways

While major international institutions to deal with climate change have been developed at multilateral levels under the auspices of the United Nations, cooperation at regional levels also has merit. First, each region has its regional circumstance that is hard to be shared with other regions. In the case of Asia, the region is different from other regions because of its rapid economic growth, as well as its rapid urbanization. As a consequence, Asia alone is responsible for more than 30 % of global GHG emissions. This means that any regional policy to reduce emissions in Asia could make a difference to one third of global emissions.

(a) *Asia and the Pacific Economic Cooperation (APEC)*
Asia and the Pacific Economic Cooperation (APEC) is an organization that mainly deals with economic cooperation in the region. Energy and climate change has become one of APEC's main issues in recent years. In the APEC Energy Ministerial Meeting held in May 2010, a joint ministerial declaration called "Fukui Declaration – *Low Carbon Paths to Energy Security: Cooperative Energy Solutions for a Sustainable APEC*" was agreed. The declaration included an aspirational energy-intensive reduction goal to reduce the ratio of energy use to economic output by at least 25 % from 2005 levels by 2030. The Energy Working Group (EWG), set up under APEC, has been instructed to increase analysis of the potential for further energy intensity improvement with a view to recommending an enhanced goal.

(b) *Association of South East Asian Nations (ASEAN)*
The Association of South East Asian Nations (ASEAN) plus three (China, Japan and Republic of Korea) is another group of countries in the Asian region that seeks economic cooperation. The 8th ASEAN+3 Ministers on Energy Meeting (AMEM+3) was held in Brunei Darussalam in September 2011. Ministers at this meeting noted good progress of ASEAN's aspirational goals of reducing regional energy intensity by 8 % and achieving its 15 % target for regional renewable energy in total power-installed capacity by 2015. The Ministers lauded the accomplishments of the CDM programme.

(c) *Asian Development Bank (ADB)*
The Asian Development Bank (ADB) is a financial organization established for the region. Expanding the use of clean energy, encouraging sustainable transport in urban areas, managing land-use and forests for carbon sequestration, promoting climate resilient development, and strengthening policies, governance and capacities are the areas of priority in ADB. Approved in 2008, ADB's Strategy 2020 reaffirms both ADB's vision of the region (freedom from poverty) and its mission to help its developing member countries improve their living conditions and quality of life (Asian Development Bank 2008). The ADB also started its "Pacific Climate Change Program" focusing on 3 main areas of small island countries:

- Climate proofing of on-going and planned ADB infrastructure projects with contributing development partners;
- Promoting renewable energy through new technology and research and development, and
- Working with partners to manage land, water, forests and costal and marine resources.

US$250 million was secured for the period 2010-2012. In addition, the ADB supports individual member countries by loans. In November 2011, the ADB decided to support Indonesia's drive to reduce GHG emissions and strengthen its resilience against climate change by providing a loan for US$100 M. Indonesia has pledged to cut its GHG emissions by 26 % over business-as-usual by 2020, and will aim to increase that to over 40 % with international assistance. Achieving the 26 % reduction will require an investment of billions of dollars between now and 2020. ADB's loan will be used to develop a national action plan to reduce GHG emissions, establish forest management units, establish a legal timber verification system, and promote geothermal energy.

(d) *Low Carbon Asia Research Network (LoCARNet)*
Having recently established a Low Carbon Initiative programme in 2012, the APN is networking with a new regional network – Low Carbon Asia Research Network (LoCARNet) based at the Institute of Global Environmental Studies (IGES), Japan under the guidance of the Ministry of Environment, Japan (MOEJ). LoCARNet (http://lcs-rnet.org/index.html) is an open network of researchers, research organizations that facilitates the formulation and implementation of science-based policies for low-carbon development in Asia. With the UNFCCC's advanced deliberations on a new framework for reducing GHG

Table 6.2 Wide range of disciplines required for research on low-carbon societies

Policy process	Discipline	Examples of application
Low-carbon goal setting	International relations, economics, planning, etc.	National development plan, long-term scenario goal setting, green development plan
Creation of low-carbon development policy	Sciences, engineering, energetics, agriculture and forestry, social infrastructure and urban engineering, economics, sociology, planning, public policy, law, integrated assessment	Preparation of inventories, development of scenarios, selection of technologies, formulation of roadmap, cost accounting, policy options, policy creation, integrated assessment, creation of low-carbon cities, lifestyle analysis
Low-carbon development policy assessment	Economics, public finance, assessment models	Mid- and long-term economic impact assessments, assessment of changes in industrial structure
Low-carbon development policy implementation	Public policy, law, sociology, behavioural science	Policy formulation, formulation of regional plan, consensus building, promotion of public participation
Feedback on assessment of policy outcomes	Public policy	Analysis of policy effects

Source: LoCARNet 2013

emissions in which all nations are expected to participate from 2020, LoCARNet is a timely addition to other ongoing networks, such as the International Research Network for Low Carbon Societies (LCS-RNet) established in 2009 following the G8 Environment Ministers Meeting in Kobe, Japan in 2008.

There are a wide range of disciplines required for research on low carbon societies and the aim of LoCARNet is to promote regional cooperation to facilitate the formulation and implementation of science-based policies for low-carbon growth in the Asian region, together with relevant stakeholders (Table 6.2).

LoCARNet aims to effectively promote research on low-carbon growth policy by enabling a sufficient amount of dialogue between scientists and policymakers and increase research capacity in the region through knowledge-sharing and information exchange in the context of not only north–south cooperation, but also south-south regional cooperation, as well.

6.2.5 Conclusion

As LCD involves every individual of the region, it is imperative that discussions for LCD need to involve multiple stakeholders. Especially when multilateral negotiation is not making much progress, multi-level activities should be implemented.

This calls for decision-making and the introduction of climate policies at various stages of governance, including international, regional, national and local. Integrated assessments will be needed across these levels of governance.

Recently, the increasing role of non-governmental actors has been observed. Cooperation from the private sector is necessary to achieve LCD. The business sector has voluntarily participated in LCD activities by providing society with energy efficient products. Studies on low-carbon societies are being undertaken by various research groups in many parts of the world and, in Asia, networks of research groups, such as LCS-RNet and, more recently, LoCARNet have been established to provide policy makers with up-to-date scientific findings on visions and pathways toward LCD. These networks serve as platforms for sharing research findings and facilitating collaboration among research institutions and various stakeholders who are interested in scientific research on low carbon development and societies. While these networks involve research institutions from all over the world, most studies consider Asia as one of the most crucial regions in the world that will affect global sustainable development.

6.3 Climate and Ecosystems Management

6.3.1 Natural Ecosystems and Sustainable Development

Natural ecosystems have a critical role to play in sustainable development. This realization is being expressed in the recent move towards a green economy where economic growth is balanced with the conservation of natural capital (UNEP 2011). The term "ecosystem" refers to *a dynamic complex of plant, animal and micro-organism communities and their non-living environment interacting as a functional unit* (CBD 1991). Ecosystems are utilized to meet various wellbeing needs that may be of monetary-economic significance or otherwise. Using the Millennium Ecosystems Assessment (MEA) framework, the role of ecosystems to human wellbeing can be viewed in terms of its provisioning, regulating, cultural, and supporting services roles (Millennium Ecosystem Assessment 2005). The provisioning role refers to the products people obtain from ecosystems, and includes food, fuel, fibre, freshwater, and genetic resources. Regulating services are the benefits people obtain from the regulation of ecosystem processes, including air quality maintenance, climate regulation, erosion control, regulation of human diseases, and water purification. Cultural services are the non-material benefits people obtain from ecosystems through spiritual enrichment, cognitive development, reflection, recreation, and aesthetic experiences. Supporting services are those that are necessary for the production of all other ecosystem services, such as primary production, production of oxygen, and soil formation.

While every ecosystem provides these different services, the exact nature and degree of the services vary with the type of ecosystem. Thus, for example, forest ecosystems that cover about 31 % of the land area of the earth are home to rich biodiversity.

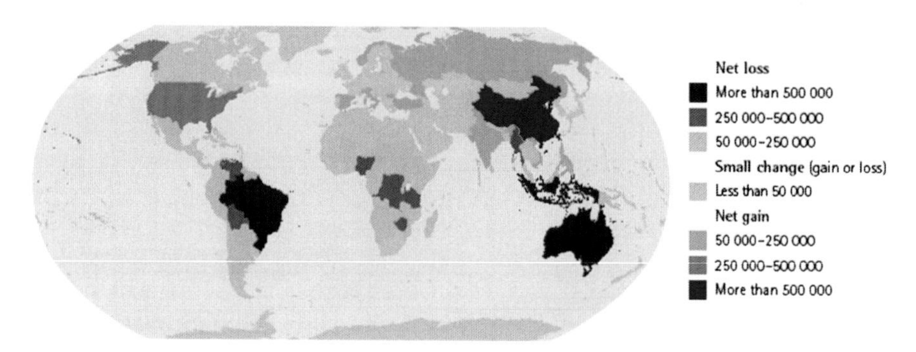

Fig. 6.5 Rate of forest loss and gain in Asia relative to other parts of the world (Source: FAO 2010a)

They contribute to soil formation and water regulation and are estimated to provide direct employment to at least ten million people, apart from being a source of livelihoods to millions more (FAO 2010a). It is estimated that about 410 million people are highly dependent on forests for subsistence and income, and 1.6 billion people depend on forest goods and services for some part of their livelihoods (Munang et al. 2011). Wood and manufactured forest products add more than US$450 billion to the world market economy annually, and the annual value of internationally-traded forest products is between US$150–200 billion.

At the regional and local levels, forests provide services in the form of water purification, flood and drought mitigation, waste decomposition and detoxification, soil generation and renewal, pollination, pest and disease control, seed dispersion, and moderation of weather extremes (Daily 1997). However, conversion of forests for other land use continues to rise, driven by better economic returns or the pressures of increasing space for "developmental" (infrastructure-related) activities (Millennium Ecosystem Assessment 2005; FAO 2010a; Braimoh et al. 2010). The rate of global deforestation is estimated to be around 5.6 million hectares per year (Fig. 6.5), primarily led by forest conversions in Africa (around 3 million hectares per year) and South America (around 3.5 million hectares per year), with Asia showing some gains in forests (around 1.7 million hectares per year) (FAO 2010a).

Similarly, fresh water ecosystems, though constituting only around 1 % of the world's surface area, contribute to food, essential water supply to human and other life for survival and production purposes, in addition to being a refuge for water-based biodiversity, performing various regulating functions such as nutrient recycling, power generation or being the basis of coastal livelihoods (Revenga et al. 2000; Millennium Ecosystem Assessment 2005). The value of these services is estimated to be trillions of dollars (Revenga et al. 2000).

Coastal and marine ecosystems are crucial to life on Earth because they support the livelihoods of billions of people and the economy of many nations (Harvey 2006). These ecosystems are highly productive and act as a repository of biological diversity which is vital to both human wellbeing and survival (Michel and Pandya 2010; UNEP 2006). The vast natural marine ecosystem is comprised of habitats from the productive near-shore regions up to the barren ocean floor. Thus, it includes oceans, estuaries and salt marshes, coral reefs and other tropical communities (mangrove forests) and coastal areas like lagoons, kelp and sea-grass beds and inter-tidal systems (for example, rocky, sandy and muddy shores). Coastal ecosystems exist at the interface between terrestrial and marine environments and include some of the most diverse and dynamic environments on earth (USAID 2010). It has great importance due to its ecological and socio-economic functions. It provides a number of livelihoods such as fisheries, ports, tourism, recreation, transportation and other industries (Michel and Pandya 2010; USAID 2010). Besides its economic benefits, it is essential in regulating atmospheric composition, cycling of nutrients and waste removal (Crooks et al. 2011).

More than one third of the world's population resides in coastal areas and they heavily rely on the goods and services provided by coastal and marine ecosystems. In the Asia-Pacific region about 60 % of the population live on or near coasts (Mimura 2006). Despite their great importance to human survival and wellbeing, these ecosystems are threatened by land-use change, over-fishing, pollution, invasion of alien (non-native) species and climate change (UNEP 2006; Millennium Ecosystem Assessment 2005). Productivity and biodiversity are greatly affected by these problems including the growing effects of climate change. This will, eventually, affect the pursuit of the Millennium Development Goals and marine-related goals, which both target sustainable development in the long term (UN 2010).

6.3.2 *Ecosystem Change and Impacts*

The rapid and extensive ecosystem change (under human influence) in the last century to advance economic development has caused rapid deterioration of natural ecosystems around the world (Millennium Ecosystem Assessment 2005). While this has led to a substantial rise in living standards, it has also caused irreversible loss in the diversity of life on the planet, which is expected to grow even worse in the first half of the present century. Most rivers have been totally restructured; oceans have been severely altered and depleted; coral reefs are near the tipping point of disappearing as functional ecosystems; over half of the land surface is devoted to livestock and crop agriculture, with little consideration for the ecosystem services that are being lost as a consequence (Mooney et al. 2009).

One region that continues to witness rapid change in natural ecosystems is Asia and the Pacific. Parts of Southeast Asia (the Indo-Malaysia and Melanesian land-mass) that are host to valuable tropical forests have been classified as biocultural diversity hotspots, based on the threats to their biological and social systems (Maffi 2007). While the FAO reports that forest area in this region is increasing, it is also evident that the threats of degradation are still high (FAO 2010a). This can partly be attributed to demographic changes (Table 6.3) but climate change is expected to further exacerbate these stresses (Braimoh et al. 2010; Fischlin et al. 2007). More specifically, the decline in its ability to perform regulating functions is of special concern because it could lead to its inability to provide other ecosystem services (Carpenter et al. 2009).

Table 6.3 Land area and population of Southeast Asia

Country	Land Area (1,000 ha)	Forest cover (%)	Percent annual rate of forest change (2000–2005)	Population in 2006 (1,000)	Percent urban population (2000)	Percent urban population (2025)	Human development index rank (2006)
Indonesia	181,157	49	−2.0	228,864	42	51	111
Philippines	29,817	24	−2.1	86,263	48	55	105
Vietnam	31,007	40	2.0	86,205	24	41	116
Thailand	51,089	28	−0.4	63,443	31	42	87
Malaysia	32,855	64	−0.7	26,113	62	81	66
Myanmar	65,755	49	−1.4	48,379	28	44	138
Singapore	69	3	0.0	4,381	100	100	23
Cambodia	17,652	59	−2.0	14,196	17	26	137
Lao PDR	23,080	70	−0.5	5,759	22	49	133
Timor-Leste	1,487	54	−1.3	1,113	24	36	162
Brunei Darussalam	527	53	−0.7	381	71	81	30
Southeast Asia	*434,495*	*47*	*−1.3*	*565,097*	*38*	*50*	**nn**

Source: Braimoh et al. 2010

6.3.3 Enhancing Ecosystems

There is increasing recognition of the importance of natural ecosystems as natural capital that provides essential services to humanity. For example, forest cover is increasing in many countries around the world. New forests are regenerating on former agricultural land, and forest plantations are being established for commercial and restoration purposes (Chazdon 2008; FAO 2010a). These artificially-established forests can improve ecosystem services and enhance biodiversity conservation, but they will not be the same as the composition and structure of the original forest cover (Sodhi et al. 2004). This is well illustrated by the proliferation of plantation forests that have arisen in parts of Southeast Asia such as in Malaysia and Indonesia. In 2008, 13.9 % and 60.2 % of the total agricultural land in Indonesia and Malaysia, respectively, were oil palm plantations (FAO 2010b). Such plantations, while providing a canopy cover, have adversely affected primary forest biodiversity (Fitzherbert et al. 2008; Danielsen et al. 2009). Still, there are numerous opportunities for combining forest restoration and regeneration goals to enhance sustainable rural livelihoods, community participation and development goals.

On a wider perspective, there are those who advocate shifting to "ecosystems stewardship." The central goal of this stewardship is to sustain the capacity to provide ecosystem services that support human wellbeing under conditions of uncertainty and change (Chapin et al. 2009).

Three broadly overlapping sustainability approaches are integrated:

- Reducing vulnerability to expected changes;
- Fostering resilience to sustain desirable conditions in the face of perturbations and uncertainty; and

– Transforming from undesirable trajectories when opportunities emerge. Its main strategies include maintaining diversity of options, enhancing social learning to facilitate adaptation, and adapting governance to implement potential solutions.

A key assumption is that the science of ecosystem stewardship is sufficiently mature to make important contributions to all social–ecological systems. The emphasis here is on the principle of adaptive management, which involves adapting to changes and learning to adapt through feedbacks and responses from the environment to better manage resources (Berkes et al. 2000). The concept inherently recognizes the value of integrating mainstream and traditional concepts of resource management to achieve desired objectives.

There will always be trade-offs in natural ecosystems management. In an increasingly resource-constrained world, increases in one ecosystem service or human activity typically result in the reduction in other services or activities (Carpenter et al. 2009). For example, in agriculture, humans have deliberately reduced genetic, stand and landscape diversity to attain greater productivity. Indeed, the general increase in provisioning services over the past century has been achieved at the expense of decreases in regulating and cultural services and biodiversity. However, win–win solutions in the conservation-and-development debate do exist. Table 6.4 illustrates what actions may be necessary to overcome barriers in sustainable management of ecosystems.

A promising approach to ecosystems management is through payments for environmental services (PES) or rewards for environmental services (RES). Efforts are under way to estimate the monetary value of the services that natural ecosystems provide. For example, Perrings (2010) used existing studies to estimate the mean value of both the macro-climatic regulation offered by terrestrial carbon sequestration, and the change in provisioning and cultural services offered by forest systems. The study showed that the mean values of forest ecosystem services, in US$/ha/year, are dominated by regulatory functions: specifically regulation of climate (US$1,965/ha/year), water flows (US$1,360/ha/year), and soil erosion (US$694/ha/year). Several businesses are also adopting environmentally friendly practices and are increasingly acquiring certification for good practices in resource use. The growing membership of certifying agencies that promote sustainable resource-use such as the Forest Stewardship Council, Marine Stewardship Council, Fair Trade Stewardship Council, among others, is testament to the growing realization among consumers and businesses on the need to abide by good and ethical practices of resource sourcing and use.

However, most efforts in environmental service management are not grounded on scientific evidence (Carpenter et al. 2009). While scientific understanding of ecosystem production functions is improving rapidly, it remains a limiting factor in incorporating natural capital into decisions, via systems of national accounting and other mechanisms (Daily and Matson 2008). There is a need for advances in ecosystem service production functions, trade-offs among multiple ecosystem services, and the design of appropriate monitoring programs for the implementation of conservation and development projects that will successfully advance both environmental and social goals (Tallis et al. 2008).

Table 6.4 Sustaining ecosystem services: Barriers, actions, and examples

Barriers	Actions	Examples
People fail to make the connection between healthy ecosystems and the attainment of social and economic goals	Develop and use information about ecosystem services	Perform regular monitoring and assessment Identify and manage tradeoffs Frame messages that resonate with the public Tailor information for citizens, producers, and purchasers
Local people often lack clear rights to use and make decisions about the ecosystem services they depend on for their livelihoods and well-being	Strengthen the rights of local people to use and manage ecosystem services	Ensure that individuals and communities have secure rights to ecosystem services Decentralize decisions about ecosystem services Bring local voices to the table to influence development projects and policies
The management of ecosystem services is fragmented among many different agencies and bodies that often work at cross-purposes and fail to coordinate across levels	Manage ecosystem services across multiple levels and timeframes	Establish the conditions for cooperation with communities Form bridging organizations Use co-management practices Raise priority of working across levels in national institutions
Government and business accountability mechanisms for decisions about ecosystem services are frequently absent or weak	Improve accountability for decisions that affect ecosystem services	Hold elected officials accountable Use public process to track ecosystem investments in meeting development goals Increase corporate transparency
Responsible management of ecosystem services does not always pay	Align economic and financial incentives with ecosystem stewardship	Eliminate perverse subsidies and reform taxation policies Include ecosystem risk in financial evaluations Support markets and payments for ecosystem services Incorporate ecosystem stewardship goals in managers' performance objectives

Source: Irwin and Ranganathan 2007

6.3.4 *Natural Ecosystems and Climate Change Adaptation*

Natural ecosystems services support substantial components of economies and social systems across Asia and the Pacific and they are keys to enhancing the resilience of local communities to climate change.

Asia and the Pacific harbors many of the world's most diverse and productive natural ecosystems, the world's deepest ocean floor, the world's largest mangroves, vast tropical rainforests, and the highest mountain peaks in the world. The regulating services provided by natural ecosystems are critical for climate change adaptation. These ecosystem services include climate and water regulation, protection from natural hazards such as floods and avalanches, water and air purification, carbon sequestration, and disease and pest regulation (UNEP 2009a). Appropriate protection and effective management of ecosystems are essential for cost-effective mitigation and adaptation for climate stabilization through use of natural carbon sequestration processes and secured delivery of essential ecosystem services; for example, clean air, food and water security (UNEP, n.d.). Therefore, it is important to adopt an ecosystem management approach in planning climate change adaptation and mitigation strategies.

The term "ecosystems-based adaptation" (EBA) is increasingly being used in the international arena and "*relates to the management of ecosystems within interlinked social-ecological systems to enhance ecological processes and services that are essential for resilience to multiple pressures, including climate change*" (Devisscher 2010). EBA includes a range of local- and landscape-scale strategies for managing ecosystems to increase resilience and maintain essential ecosystem services and reduce the vulnerability of people, their livelihoods and nature in the face of climate change (IUCN 2009; Colls et al. 2009). EBA addresses the role of ecosystem services in reducing the vulnerability of natural resource-dependent societies to climate change. EBA is a set of adaptation policies or measures that jointly addresses the vulnerability of ecosystems and the role of ecosystem services in reducing the vulnerability of society to climate change, using a multi-sectoral and multi-scale approach.

Natural ecosystems as natural capital provide provisioning services by enhancing rural livelihoods, especially in developing countries of Asia and the Pacific. Many of the rural communities in these countries rely on forest ecosystems for their livelihoods. In the past 25 years, many countries have overhauled their forest land use rights through some form of community forestry schemes, among them Nepal (Adhikari et al. 2007), the Philippines (Pulhin et al. 2007), and Indonesia (Hindra 2007). This is based on research findings that indicate greater access to forest resources is correlated with enhanced livelihoods and wellbeing in general. For example, in the western Himalayas, higher incidence of land poverty has been observed to be associated with lower forest access rates, while higher incidence of land-rich households is associated with higher forest access rates (Naidu 2011).

In China, poorer households derive greater benefits from non-timber forest products (NTFPs) than wealthier households (Fu et al. 2009). In Indonesia, larger forest areas are significantly correlated to the wellbeing of rural villages

(Dewi et al. 2005). In the Philippines, tenure reform and its associated financial, technical, and livelihood support have seen some promising socio-economic and environmental impacts through the transfer of certain rights to local communities that promote access, use and control of forest resources (Pulhin et al. 2008). In South Asia (India, Bhutan and Nepal), it was reported that higher forest biodiversity is positively correlated with livelihoods (Persha et al. 2010).

However, it must also be recognized that there are many barriers that constrain the full utilization of forest resources to provide livelihoods to the rural poor. Sunderlin et al. (2005) have shown the complex relationship between forests, poverty, livelihoods enhancement and conservation. Ineffective tenure reform, excessive regulatory barriers, poor market access, and weak community capacity, among others, limit the potential of forests to effectively contribute to poverty reduction in many developing countries (Larson et al. 2010). Even in a developed country like Republic of Korea, the benefits local people derive from forests can only accrue if certain conditions exist such as the presence of joint forest management agreements (Yeo-Chang 2009).

In any case, there seems ample evidence to at least suggest that forest ecosystems may provide safety nets in times of food and income scarcity such as may be expected as climate patterns change. More broadly, they provide indirect evidence that the health of natural ecosystems will be a critical ingredient in enhancing the resilience of local communities to climate variability and change. However, there is limited empirical data that provide direct support to this claim.

This is a gap that needs to be addressed by future research.

The importance of forest ecosystems to climate change adaptation coupled with numerous threats to their existence provides strong argument for redoubling efforts to conserve them. This will entail, among others, combining traditional knowledge and scientific knowledge; increasing participatory reform; maintaining and enhancing biodiversity; enhancing robust management strategies; improving inter-sectoral coordination; main-streaming forest adaptation into policy; and incorporating new actors and new modes of governance (Vickers et al. 2010).

At present, some key adaptation options and practices in the forestry sector have been documented in Asia and different parts of the world. These include reforestation and afforestation activities, establishment of early warning systems, use of appropriate silvicultural practices, various forest protection strategies, monitoring of degraded forests, establishment of forest corridors, adoption of soil and water conservation measures, agroforestry, and diversification of local economies and livelihoods through non-wood forest products. If effectively implemented, such adaptation options and practices can enhance community resilience to climate change (Pulhin et al. 2010).

Healthy mangrove ecosystems, which are part of the larger coastal and marine environment, can provide support through protection services from natural hazards like storms and flood, which are expected to increase in strength and frequency due to climate change (AIT/UNEP 2010). Mangroves act as natural revetments or dikes and can mitigate 70–90 % of the energy from wind-generated waves

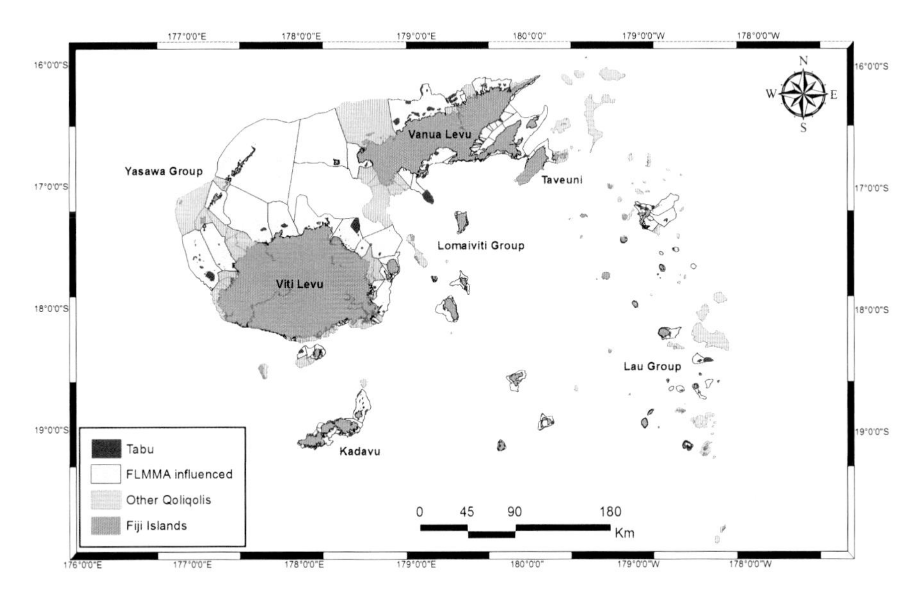

Fig. 6.6 Map showing boundaries of traditional fishing grounds in Fiji (Source: Fiji Locally Managed Marine Protected Area Network)

(UNEP-WCMC 2008). In addition, coastal wetlands can perform carbon sequestration and can transform carbon into sediments that can be stored for millennia (Crooks et al. 2011). Coral reefs provide offshore breakwaters which, reduce the impacts of sea surges and tropical storm waves before they reach the shoreline (UNEP-WCMC 2008).

The Locally Managed Marine Areas (LMMA) of the Pacific countries is a good example of adaptive management of coastal and marine ecosystems. Primarily governed by customary tenure systems, LMMAs are managed using tools, predominantly related to banning of resource sourcing (referred to as "tabu" – see Fig. 6.6) during certain periods and/or from certain areas (Govan 2009). A recent study highlights the successful management of coral, fish and other marine resources by the Pacific country communities, calling for a sensitive approach by scientists and policy bodies when designing interventions related to the ecosystem.

6.3.5 Natural Ecosystems and Climate Change Mitigation

Natural ecosystems in Asia and the Pacific have a critical role in climate change mitigation.

In Asia and the Pacific, natural ecosystems can both help exacerbate and mitigate greenhouse gas emissions. Forest ecosystems influence climate through a combination of physical, chemical, and biological processes that affect planetary energetics, the hydrologic cycle, and atmospheric composition (Bonan 2008). Tropical, temperate and boreal reforestation and afforestation mitigate climate change primarily through carbon sequestration. Tropical forests mitigate warming through evaporative cooling, but the low albedo of boreal forests is a positive climate forcing. The net climate forcing from these and other processes are not yet known.

Deforestation, degradation and poor forest management reduce carbon storage in forests, but sustainable forest management, planting and rehabilitation, can increase carbon sequestration (FAO 2006). From 1850 to 1995, 75 % of all carbon emissions from South and Southeast Asia were due to the conversion of forests to perennial crops (Vickers et al. 2010). This trend is expected to continue despite the reduction in deforestation rates in the region. Of key concern are the peatlands in Southeast Asia, which contribute 70 % of total emissions from peatlands while occupying a mere 15 % of the total area (Wetlands International 2009). This is equivalent to 1.3–3.1 % of global CO_2 emissions from the combustion of fossil fuel.

At the same time, forest ecosystems in the region provide significant opportunities to mitigate carbon emissions. Using figures from the 1990s, FAO estimates that between 170 and 660 Mt of carbon could be prevented each year if deforestation rates were reduced by 50 % (Vickers et al. 2010). The potential of forest lands to mitigate climate change has been estimated in a number of Asian countries such as India (Ravindranath et al. 2008), Indonesia (Boer 2001) China (Houghton and Hackler 2003) and the Philippines (Lasco and Pulhin 2001).

The emerging carbon market could offer significant financing for forest conservation in the region. Several government and non-government organizations are advocating financial mechanisms such as payments for avoiding deforestation in developing countries under REDD-plus scheme, perhaps in the post-2012 Kyoto Protocol. This is, in part, because it has long been recognized that deforestation, mainly in the tropics, accounts for nearly 20 % of all carbon-based greenhouse gas emissions (Denman et al. 2007). However, it has also been recognized that "the design and implementation of REDD policies will be neither simple nor straightforward, given the complexity of the social, economic, environmental and political dimensions of deforestation. Many of the underlying causes of deforestation are generated outside the forestry sector and alternative land uses tend to be more profitable than conserving forests" (Kanninen et al. 2007). REDD-plus could provide incentives to local communities to be more involved in forest conservation and rehabilitation. For example, in the Philippines, community-based forest management participants are exploring ways to take advantage of carbon finance (Lasco et al. 2010). However, experience with the Clean Development Mechanism (CDM) shows that there are many barriers in implementing them, foremost of which is the high transaction costs (Thomas et al. 2010).

Deforestation, forest degradation and peatlands conversion accounts for 60 % of Indonesia's GHG emissions (Brockhaus et al. 2011). As noted earlier in Sect. 6.2.4, in 2009, the Indonesian government pledged to unilaterally cut its emissions by 26 % by 2020 and by 41 % if given international support. Norway pledged to provide US$ 1 billion in funding to reduce emissions from forests and land-use change.

However, the impacts of REDD-plus projects on livelihoods of the rural poor, although likely to be positive, are still uncertain. A recent review of five pre-REDD projects show that the design, data collection, and analysis methods for understanding the impacts frequently lack sufficient rigor to inform future REDD-plus projects (Caplow et al. 2011). In Indonesia, land allocation policy and processes could negate the good intentions of REDD-plus projects (Brockhaus et al. 2011).

In general, ecosystems management to enhance biological carbon sequestration could lead to other co-benefits. For example, healthy ecosystems protect societies from disasters and improve their ability to cope with the impacts (UNEP 2009b). Agroforestry systems traditionally practiced in different parts of India, in accordance with respective agro-climatic conditions, have been shown to provide multiple benefits. Beyond a diverse income portfolio, such systems, which involve a combination of agricultural and forestry crops, are estimated to sequester about 12–228 Mt per hectare of carbon, which varies due to differences in biomass. These systems also contribute to soil fertility, sustenance of biodiversity, improvement in water-use efficiency, and provide various productive resources to the population (Pandey 2007).

Coastal and marine ecosystems also play a significant role in regulating CO_2 accumulation in the atmosphere. In a recent rapid response assessment of UNEP, it was found out that from all the biological carbon (or green carbon) captured in the world, 55 % was captured by living marine organisms. Blue carbon sinks and estuaries can capture and store from 870 to 1,650 Mt of carbon every year (UNEP n.d.). Oceans provide solutions to help mitigate climate change and opportunities for sustainable development. They generate oxygen and absorb carbon dioxide from the atmosphere, while at the same time provide essential goods and services for sustaining life on Earth. Coastal and marine ecosystems, which include mangroves, salt marshes and seagrass, store up to 70 % of the carbon in the marine environment (UN 2010).

6.3.6 Governance Issues

Successful management of natural ecosystems requires that appropriate governance structures are in place.

Managing ecosystems and building resilience ultimately refers to ensuring appropriate and well-functioning governance systems at the ecosystem level. These systems would be a combination of macro processes (including national, regional and international rights and obligations with attendant institutions and mechanisms) and sub-national processes that include ecosystem-level institutions of both formal and informal character. The non-formal institutions could include traditional forms of leadership institutions that still hold sway in the local contexts. As pointed out by Lebel and Daniels (2009), there are well defined power relations within an ecosystem context between different actors, and while participatory planning enhances better ecosystem outcomes, a system of regulations combined with adequate information and incentives enable better outcomes.

Further, it is important to note that ecosystems do not recognize political boundaries calling for enhanced trans-boundary co-operation in their governance. As highlighted by Badenoch (2002), through an example of flooding in Cambodia due to dam water spill in Viet Nam along the lower Mekong River system, the effects of ecosystem degradation in one part of an ecosystem can have highly damaging consequences in other parts. The same trans-boundary issue holds at the in-country level. In Viet Nam and Indonesia, flooding in the lowland areas has been attributed to deforestation and ecosystem degradation in the upper parts of the watersheds (see for example, Phong and Shaw 2010 for Viet Nam; Lasco and Boer 2006 for Indonesia). An integrated river basin management strategy has, therefore, been recommended to provide a framework for coordination among different stakeholders to tackle complex issues caused by conflicts resulting from multiple users and uses of natural resources (Phong and Shaw 2010).

To effectively respond to the complexity and dynamic changes confronting many natural ecosystems, which in essence constitute both social and ecological components, the notion of "adaptive governance" has recently emerged. Folke et al. (2005) describes adaptive governance as a form of governance that "connects individuals, organizations, agencies, and institutions at multiple organizational levels" where "key persons provide leadership, trust, vision, meaning, and they help transform management organizations toward a learning environment." It focuses on learning and managing resilience or building adaptive capacity where learning can take place at different levels and through various ways including interactions among different stakeholders (Lebel et al. 2010). It departs from the rigid structure traditionally imposed by central governments, but it is often loosely and self-organized as social networks composed of groups that draw on various knowledge systems and experiences. Crucial, however, to successful adaptive governance are enabling legislation, flexible institutions, and the presence of "bridging organisations" like non-government organization that will effectively link local actors and communities to other scales of organizations (Folke et al. 2005).

6.3.7 Outlook

From the foregoing discussion it is evident that some areas need immediate attention to enhance actions that improve resilience of ecosystems and populations deriving ecosystem services. Areas that the research community can contribute to include:

Integrating ecosystem management in climate change action planning.

Science had already proven the importance of ecosystem management in climate change adaptation and mitigation. However, corresponding policies and actions are not yet in place to support proper ecosystem management that also addresses climate change adaptation and disaster risk reduction (UNEP 2009b). Integration of ecosystem management into climate change adaptation and disaster risk reduction policy frameworks is important to enhance the adaptive capacity of stakeholders, particularly in developing countries vulnerable to climate change impacts.

Communication, education and capacity building.

A study conducted by Futerra Sustainability Communications revealed that different stakeholders respond to different messages. Hence a policy maker tends to weigh the opportunity costs and benefits of any activity, while a conservationist is moved by messages of responsibility to nature (Futerra Sustainability Communications 2010). What also came out was that people respond better when the messages are not playing on the guilt of their actions, but on the need for positive action. This implies the need for developing better social learning tools and educational materials that translate the knowledge on sustainable use of resources and opportunities for win-win scenarios between different stakeholders into user friendly formats. It also raises the need to address capacity gaps at various levels of governance from policy makers to people on the ground.

Increased networking and cross-learning among scientists and practitioners from different disciplines.

Clearly, developing implementable mechanisms to operationalize sustainable development involves the need for trans-disciplinary approaches. This requires fostering scientists and practitioners from different sectors and disciplines and countries to come together to develop solutions appropriate to ecosystems and to dependent populations.

Development and dissemination of tools and methods that capture co-benefits derivable from actions to mitigate and adapt to climate change and to manage sustainable use of ecosystems and resources are required. These need to complement efforts that enable better monitoring and assessment of the status of ecosystems and resource use. It would also be useful to examine how to integrate these approaches with various certification systems developed to ensure sustainable practice in business.

Strengthening inter-linkages among science, policy and practice.

Considering the importance and urgency to address ecosystem management problems as a way to achieve the goal of sustainable development, it is paramount to strengthen the linkages among the scientific and policy communities as well as the local communities to ensure informed decision-making processes at various levels. This requires adherence to the new research paradigm that engages different stakeholders at various levels in all the phases of the research process. Such stakeholder engagement has multiple benefits. It promises to educate the policy makers in current ecosystem issues, problems and solutions and hence to increase the chance of coming up with more scientifically-based policy prescriptions. Similarly, it can empower local communities whose lives are threatened by the adverse impacts of ecosystem degradation to take appropriate actions and avoid or reduce the risks associated with degradation.

6.4 Conclusion and Synthesis

The challenge posed by climate change will be keenly felt in the coming decades in Asia and the Pacific. In parallel, nations in the region will continue to aspire for sustainable development. Policy makers and development workers must find ways to ensure that both these concerns are addressed synergistically while avoiding negative outcomes.

One way to mitigate climate change while pursuing sustainable development is to pursue a low carbon development (LCD) pathway. It is clear that LCD requires negotiation across many stakeholders, including government, non-government agencies, industry and the broad community.

In Asia and the Pacific, natural ecosystems will continue to play a critical role in addressing climate change adaptation and mitigation. Nations in the region will have to find innovative ways to manage and rehabilitate natural ecosystems for a multiplicity of functions and services. This will involve greater collaboration and communication between scientists and policy makers as well as between natural and social scientists. In many developing countries, there is still very limited empirical information and research needs to be ramped up. North–south and South-South partnerships could help fill the gap.

References

Adger, W. N. (2003). Building resilience to promote sustainability: An agenda for coping with globalization and promoting justice. *IHDP Update, 2*, 1–3.

Adhikari, B., Williams, F., & Lovett, J. C. (2007). Local benefits from community forests in the middle hills of Nepal. *Forest Policy and Economics, 9*, 464–478.

AIT/UNEP. (2010). *Adaptation knowledge platform coastal ecosystem's role in climate change adaptation synthesis report*. Bangkok: AIT/UNEP Regional Resource Center for Asia & the Pacific.

Alcamo, J., Shaw, R., & Hordijk, L. (Eds.). (1990). *The RAINS model of acidification: Science and strategies in Europe*. Dordrecht: Kluwer.

Arico, S., & Valderrama, G. C. (2010). Traditional knowledge: From environmental management to territorial development. In S. M. Subramanian & B. Pisupati (Eds.), *Traditional knowledge in policy and practice: Approaches to development and human wellbeing* (pp. 208–225). Tokyo: UNU Press.

Asian Development Bank. (2008). *Strategy 2020: Working for an Asia and Pacific free of poverty*. Manila: ADB.

Badenoch, N. (2002). *Transboundary environmental governance principles and practice in mainland Southeast Asia*. Washington, DC: World Resources Institute.

Baumgartner, R. J. (2011). Critical perspectives of sustainable development research and practice. *Journal of Cleaner Production, 19*, 783–786.

Bélair, C., Ichikawa, K., Wong, B. Y. L., & Mulongoy, K. J. (Eds.). (2010). *Sustainable use of biological diversity in socio-ecological production landscapes. Background to the 'Satoyama Initiative for the benefit of biodiversity and human well-being'*. Montreal/Canada/Tokyo: Secretariat of the Convention on Biological Diversity, United Nations University-Institutes of Advanced Studies, and Ministry of Environment Japan.

Berkes, F. (2008). *Sacred ecology* (2nd ed.). New York/London: Routledge.

Berkes, F., Colding, J., & Folke, C. (2000). Rediscovery of traditional ecological knowledge as adaptive management. *Ecological Applications, 10*(5), 1251–1262.

Boer, R. (2001). Economic assessment of mitigation options for enhancing and maintaining carbon sink capacity in Indonesia. *Mitigation and Adaptation Strategies for Global Change, 6*, 257–290.

Bonan, G. B. (2008). Forests and climate change: Forcings, feedbacks, and the climate benefits of forests. *Science, 320*, 1444–1449.

Braimoh, A. K., Subramanian, S., Elliott, W., & Gasparatos, A. (2010). *Climate and human related drivers of biodiversity decline in Southeast Asia*. Yokohama: UNU-IAS Policy Report.

Brockhaus, M., Obidzinski, K., Dermawan, A., Laumonier, Y., & Luttrell, C. (2011). An overview of forest and land allocation policies in Indonesia: Is the current framework sufficient to meet the needs of REDD+? *Forest Policy and Economics, 18*, 30–37. doi:10.1016/j.forpol.2011.09.004.

Caplow, S., Jagger, P., Lawlor, K., & Sills, E. (2011). Evaluating land use and livelihood impacts of early forest carbon projects: Lessons for learning about REDD+. *Environmental Science & Policy, 14*(2), 152–167.

Carpenter, S. R., Mooney, H. A., Agard, J., Capistrano, D., DeFriese, R. S., Díaz, S., … & Whyte, A. (2009). Science for managing ecosystem services: Beyond the millennium ecosystem assessment. *PNAS, 106*(5), 1305–1312.

CBD. (1991). Article 2 of the convention on biological diversity. Available at http://www.cbd.int/convention/articles/?a=cbd-02

Chapin, F. S., III, Carpenter, S. R., Kofinas, G. P., Folke, C., Abel, N., Clark, W. C., … & Swanson, F. J. (2009). Ecosystem stewardship: Sustainability strategies for a rapidly changing planet. *Trends in Ecology & Evolution, 25*(4), 231–249.

Chazdon, R. L. (2008). Beyond deforestation: Restoring forests and ecosystem services on degraded lands. *Science, 320*(5882), 1458–1460. doi:10.1126/science.1155365.

Colls, A., Ash, N., & Ikkala, N. (2009). *Ecosystem-based adaptation: A natural response to climate change*. Gland: International Union for Conservation of Nature and Natural Resources (IUCN).

Crooks, S., Herr, D., Tamelander, J., Laffoley, D., & Vandever, J. (2011). *Mitigating climate change through restoration and management of coastal wetlands and near-shore marine ecosystems: Challenges and opportunities* (Environment department paper, Vol. 121). Washington, DC: World Bank.

Cruz, R. V., Harasawa, H., Lal, M., Wu, S., Anokhin, Y., Punsalmaa, B., & Ninh, N. H. (2007). Asia. In M. L. Parry, O. F. Canziani, J. P. Palutikof, P. J. van der Linden, & C. E. Hanson (Eds.), *Climate change 2007: Impacts, adaptation and vulnerability. Contribution of working group II to the fourth assessment report of the intergovernmental panel on climate change* (pp. 469–506). Cambridge, UK: Cambridge University Press.

Daily, G. C. (Ed.). (1997). *Nature's services: Societal dependence on natural ecosystems*. Washington, DC: Island Press.

Daily, G. C., & Matson, P. A. (2008). Ecosystem services: From theory to implementation. *PNAS, 105*(28), 9455–9456.

Danielsen, F., Beukema, H., Burgess, N. D., Parish, F., Bruhl, C. A., Donald, P. F., … & Fitzherbert, E. B. (2009). Biofuel plantations on forested lands: Double jeopardy for biodiversity and climate. *Conservation Biology, 23*, 348–358.

de Vries, B. J. M., & Petersen, A. C. (2009). Conceptualizing sustainable development: An assessment methodology connecting values, knowledge, worldviews and scenarios. *Ecological Economics, 68*, 1006–1019.

Denman, K. L., Brasseur, G., Chidthaisong, A., Ciais, P., Cox, P. M., Dickinson, R. E., … & Zhang, X. (2007). Couplings between changes in the climate system and biogeochemistry. In S. Solomon, D. Qin, M. Manning, Z. Chen, M. Marquis, K. B. Averyt, M. Tignor, & H. L. Miller (Eds.), *Climate change 2007: The physical science basis. Contribution of working group I to the fourth assessment report of the intergovernmental panel on climate change*. Cambridge, UK/New York: Cambridge University Press.

Devisscher, T. (2010). *Ecosystem-based adaptation in Africa. Rationale, pathways and cost estimates*. Stockholm Environment Institute, April 2010, p. 92.

Dewi, S., Belcher, B., & Puntodewo, A. (2005). Village economic opportunity, forest dependence, and rural livelihoods in East Kalimantan, Indonesia. *World Development, 33*, 1419–1434.

Edmonds, J., & Reilly, J. (1985). *Global energy: Assessing the future*. New York: Oxford University Press.

FAO. (2006). *Global forest resources assessment 2005* (Forestry paper, Vol. 147). Rome: United Nations Food and Agriculture Organization.

FAO. (2010a). *Global forest resources assessment 2010* (FAO forestry paper, No. 163). Rome. Retrieved from http://www.fao.org/docrep/013/i1757e/i1757e.pdf

FAO. (2010b). *Forestry trade flows – FAOSTAT*. Available from http://faostat.fao.org

Fischlin, A., Midgley, G. F., Price, J. T., Leemans, R., Gopal, B., Turley, C., & Velichko, A. A. (2007). Ecosystems, their properties, goods, and services. In M. L. Parry, O. F. Canziani, J. P. Palutikof, P. J. van der Linden, & C. E. Hanson (Eds.), *Climate change 2007: Impacts, adaptation and vulnerability. Contribution of working group II to the fourth assessment report of the intergovernmental panel on climate change* (pp. 211–272). Cambridge, UK: Cambridge University Press.

Fitzherbert, E. B., Struebig, M. J., Morel, A., Danielsen, F., Bruhl, C. A., Donald, P. F., & Phalan, B. (2008). How will oil palm expansion affect biodiversity? *Trends in Ecology & Evolution, 23*, 538–545.

Folke, C., Carpenter, S., Elmqvist, T., Gunderson, L., Holling, C. S., & Walker, B. (2002). Resilience and sustainable development: Building adaptive capacity in a world of transformations. *Ambio, 31*, 437–440.

Folke, C., Hahn, T., Olsson, P., & Norberg, J. (2005). Adaptive governance of social-ecological systems. *Annual Review of Environment and Resources, 30*, 441–473.

Friedlingstein, P., Houghton, R. A., Marland, G., Hackler, J., Boden, T. A., Conway, T. J., & Le Quéré, C. (2010). Update on CO_2 emissions. *Nature Geoscience, 3*, 811–812. doi:10.1038/ngeo_1022.

Fu, Y., Chen, J., Guo, H., Chen, A., Cui, J., & Hu, H. (2009). The role of non-timber forest products during agroecosystem shift in Xishuangbanna, southwestern China. *Forest Policy and Economics, 11*, 18–25.

Futerra Sustainability Communications. (2010). *Branding biodiversity: The new nature message*. London: Futerra.

Goma, H. C., Rahim, K., Nangendo, G., Riley, J., & Stein, A. (2001). Participatory studies for agroecosystem evaluation. *Agriculture Ecosystems and Environment, 87*, 179–190.

Govan, H. (2009). Status and potential of locally-managed marine areas in the South Pacific: Meeting nature conservation and sustainable livelihood targets through wide-spread implementation of LMMAs. SPREP/WWF/WorldFish-Reefbase/CRISP. 95pp + 5 annexes.

Gunderson, L. H., & Holling, C. S. (Eds.). (2002). *Panarchy: Understanding transformations in human and natural systems*. Washington, DC: Island Press.

Halsnaes, K., Markandya, A., & Shukla, P. (2011). Introduction: Sustainable development, energy, and climate change. *World Development, 39*, 983–986.

Harvey, N. (2006). *Global change and integrated coastal management: The Asia-Pacific region*. Dordrecht: Springer.

Hindra, B. (2007 August 6–8). Community forestry in Indonesia. Bangkok, Thailand: Paper presented on Asia Pacific Tropical Forest Investment Forum.

Houghton, R. A., & Hackler, J. L. (2003). Sources and sinks of carbon from land-use change in China. *Global Biogeochemical Cycles, 17*, 1034–1052. doi:10.1029/2002GB001970.

Indian Institute of Management Ahmedabad (IIMA), National Institute for Environmental Studies (NIES), Kyoto University, and Mizuho Information & Research Institute. (2009). *Low carbon society vision 2050 India*.

Irwin, F., & Ranganathan, J. (2007). *Restoring nature's capital: An action agenda to sustain ecosystem services*. Washington, DC: World Resources Institute. Retrieved from http://pdf.wri.org/restoring_natures_capital.pdf

IUCN. (2009). Ecosystem-based Adaptation (EBA), Gland, Switzerland: IUCN. Retrieved from http://cmsdata.iucn.org/downloads/iucn_position_paper_eba_september_09.pdf

Jabareen, Y. T. (2008). Toward participatory equality: Protecting minority rights under international law. *Israel Law Review, 41*, 635–676.

Kainuma, M., Matsuoka, Y., & Morita, T. (2003). *Climate policy assessment – Asia-Pacific integrated modeling*. Tokyo: Springer.

Kanninen, M., Murdiyarso, D., Seymour, F., Angelsen, A., Wunder, S., & German, L. (2007). *Do trees grow on money? The implications of deforestation research for policies to promote REDD*. Bogor: Center for International Forestry Research (CIFOR).

Kates, R. W., Clark, W. C., Corell, R., Hall, J. M., Jaeger, C. C., Lowe, I., & Svedin, U. (2000). *Sustainability science. Research and assessment systems for sustainability program discussion paper 2000–33*. Cambridge, MA: Harvard University.

Kates, R. W., Clark, W. C., Corell, R., Hall, J. M., Jaeger, C. C., Lowe, I., & Svedin, U. (2001). Sustainability science. *Science, 292*(5517), 641–642.

Kates, R., Parris, T. M., & Leiserowitz, A. A. (2005). What is sustainable development? *Environment, 47*(3), 9–21. Retrieved from http://www.environmentmagazine.org/Editorials/Kates-apr05-full.html

Kenter, J. O., Hyde, T., Christie, M., & Fazey, I. (2011). The importance of deliberation in valuing ecosystem services in developing countries—Evidence from the Solomon Islands. *Global Environmental Change, 21*(2), 505–521.

Larson, A., Barry, D., Dahal, R. G., & Colfer, C. (Eds.). (2010). *Forests for people: Community rights and forest tenure reform*. London: Earthscan.

Lasco, R. D., & Boer, R. (2006). *An integrated assessment of climate change impacts, adaptations and vulnerability in watershed areas and communities in Southeast Asia. A final report submitted to Assessments of Impacts and Adaptations to Climate Change (AIACC), project No. AS 21*. Washington, DC: The International START Secretariat.

Lasco, R. D., & Pulhin, F. B. (2001). Forestry mitigation options in the Philippines: Application of the COMAP model. *Mitigation and Adaptation Strategies for Global Change, 6*, 313–334.

Lasco, R. D., Pulhin, F. B., Sanchez, P. A. J., Delfino, R. J. P., Gerpacio, R., & Garcia, K. (2009). Mainstreaming adaptation in developing countries: The case of the Philippines. *Climate and Development, 1*, 130–146.

Lasco, R. D., Evangelista, R. S., & Pulhin, F. B. (2010). Potential of community-based forest management (CBFM) to mitigate climate change in the Philippines. *Small Scale Forestry Journal, 9*, 429–443.

LCS-RNet. (2009). *Time to act! Introduction to low carbon societies* [Brochure]. Retrieved from http://lcs-rnet.org/publications/pdf/2010LCS-RnetBro_EN.pdf

Lebel, L., & Daniels, R. (2009). The governance of ecosystem services from tropical upland watersheds. *Current Opinion in Environmental Sustainability, 1*(1), 61–68.

Lebel, L., Sinh, B. T., & Nikitina, E. (2010). Adaptive governance of risks: Climate, water and disasters. In R. Shaw, J. Pulhin, & J. Pereira (Eds.), *Climate change adaptation and disaster risk reduction: Issues and challenges* (Series: Community, environment and disaster risk management, Vol. 4, pp. 115–142). Bingley: Emerald Group Publishing Limited.

LoCARNet. (2013). Draft strategic action plan. Kanagawa: Institute for Global Environmental Strategies (IGES). Available from http://lcs-rnet.org/index.html

Maffi, L. (2007). Biocultural diversity and sustainability. In J. Pretty, A. Ball, T. Benton, J. Guivant, D. Lee, D. Orr, M. Pfeiffer, & H. Ward (Eds.), *The SAGE handbook of environment and society* (pp. 267–277). Los Angeles: Sage.

Meadows, D. H., Meadows, D. L., Randers, J., & Behrens, W. W. (1972). *The limits to growth*. New York: Universe Books.

Menon-Choudhury, D., Shukla, P. R., Biswas, D., & Nag, T. (2006). Electricity reforms, firm level responses and environmental implications. In P. K. Kalra & J. Ruet (Eds.), *Electricity act and technical choices for the power sector in India* (pp. 183–216). New Delhi: Manohar.

Mesarovic, M. D., & Pestel, E. (1974). *Mankind at the turning point: The second report to the club of Rome*. New York: Dutton.

Mexico City Workshop. (2002 May 20–23). Science and technology for sustainable development. Conclusions of a workshop to explore and synthesize findings from a two-year consultation process conducted by the International Council for Science, the InterAcademy Panel, the third world Academy of Sciences, and the Initiative on Science and Technology for Sustainability. Mexico City.

Michel, D., & Pandya, A. (2010). *Coastal zones and climate change*. Washington, DC: Stimson Center. Retrieved from http://www.stimson.org/books-reports/coastal-zones-and-climate-change/

Millennium Ecosystem Assessment. (2005). *Ecosystems and human well-being: Wetlands and water synthesis*. Washington, DC: World Resources Institute.

Mimura, N. (2006). Chapter 2: State of the environment in the Asia and Pacific coastal zones and effects of global change. In N. Harvey (Ed.), *Global change and integrated coastal management: The Asia-Pacific region* (pp. 17–38). Dordrecht: Springer.

Mimura, N., Nurse, L., McLean, R. F., Agard, J., Briguglio, L., Lefale, P., … & Sem, G. (2007). Small islands. In M. L. Parry, O. F. Canziani, J. P. Palutikof, P. J. van der Linden, & C. E. Hanson (Eds.), *Climate change 2007: Impacts, adaptation and vulnerability. Contribution of working group II to the fourth assessment report of the intergovernmental panel on climate change* (pp. 687–716). Cambridge, UK: Cambridge University Press.

Mooney, H., Larigauderie, A., Cesario, M., Elmquist, T., Hoegh-Guldberg, O., Lavorel, S., & Yahara, T. (2009). Biodiversity, climate change, and ecosystem services. *Current Opinion in Environmental Sustainability, 1*(1), 46–54.

Morita, T., Matsuoka, Y., Kainuma, M., Harasawa, H., & Kai, K. (1994). *AIM–Asian-Pacific integrated model for evaluationg GHG emissions and global warming impacts* (pp. 254–273). Bangkok, Thailand: In Global Warming Issues in Asia. Asian Institute of Technology.

Munang, R., Thiaw, I., Thompson, J., Ganz, D., Girvetz, E., & Rivington, M. (2011). *Sustaining forests: Investing in our common future* (UNEP policy series, Vol. 5). Nairobi: UNEP.

Munasinghe, M. (2010). Addressing the sustainable development and climate change challenges together: Applying the sustainomics framework. *Procedia Social and Behavioral Sciences, 41*, 6634–6640.

Naidu, S. C. (2011). Access to benefits from forest commons in the Western Himalayas. *Ecological Economics, 71*, 201–210. doi:10.1016/j.ecolecon.2011.09.007.

National Institute for Environmental Studies (NIES), Kyoto University, Ritsumeikan University, and Mizuho Information and Research Institute. (2008). *Japan scenarios and actions towards Low-Carbon Societies (LCSs)*, Global Environmental Research Fund (GERF/S-3-1), Japan-UK Joint Research Project "a Sustainable Low-Carbon Society (LCS)".

Nordhaus, W. D. (1979). *The efficient use of energy resources*. New Haven: Yale University Press.

Pandey, D. P. (2007). Multifunctional agroforestry systems in India. *Current Science, 92*(4), 455–463.

Perrings, C. (2010). *Biodiversity, ecosystem services, and climate change: The economic problem* (Environmental economics series, Vol. 120). Washington, DC: The World Bank.

Persha, L., Fischer, H., Chhatre, A., Agrawal, A., & Benson, C. (2010). Biodiversity conservation and livelihoods in human-dominated landscapes: Forest commons in South Asia. *Biological Conservation, 143*, 2918–2925.

Phong, T., & Shaw, R. (2010). River basin management for effective disaster risk reduction in the face of changing climate. In R. Shaw, J. Puhin, & J. Pereira (Eds.), *Climate change adaptation and disaster risk reduction: Issues and challenges* (Series: Community, environment and disaster risk management, Vol. 4, pp. 265–289). Bingley: Emerald Group Publishing Limited.

Pulhin, J. M., Inoue, M., & Enters, T. (2007). Three decades of community-based forest management in the Philippines: Emerging lessons for sustainable and equitable forest management. *International Forestry Review, 19*(4), 865–883.

Pulhin, J. M., Dizon, J. T., Cruz, R. V. O., Gevaña, D. T., & Dahal, G. R. (2008). *Tenure reform on Philippine forest lands: Socio-economic and environmental impacts*. UP Los Baños: CFNR-UPLB, CIFOR, and RRI.

Pulhin, J. M., Lasco, R. D., Pulhin, F. B., Ramos, L., & Peras, R. J. J. (2010). Climate change adaptation and community forest management. In R. Shaw, J. Puhin, & J. Pereira (Eds.), *Climate change adaptation and disaster risk reduction: Issues and challenges* (Series: Community, environment and disaster risk management, Vol. 4, pp. 237–258). Bingley: Emerald Group Publishing Limited.

Ravindranath, N. H., Chaturvedi, R. K., & Murthy, I. K. (2008). Forest conservation, afforestation and reforestation in India: Implications for forest carbon stocks. *Current Science, 95*(2), 216–222.

Revenga, C., Brunner, J., Henninger, N., Kassem, K., & Payne, R. (2000). *Pilot analysis of global ecosystems freshwater systems*. Washington, DC: World Resources Institute.

Robinson, J., Bradley, M., Busby, P., Connor, D., Murray, A., Sampson, B., & Soper, W. (2006). Climate change and sustainable development: Realizing the opportunity. *Ambio, 35*, 2–8.

Schipper, E. L. F. (2007). Climate change adaptation and development: Exploring the linkages. Tyndall centre working paper No. 107. Tyndall Centre for Climate Change Research. Available from http://www.preventionweb.net/files/7782_twp107.pdf

Shukla, P. R., Chandler, W., Ghosh, D., & Logan, J. (1999). *Developing countries & global climate change, electric power options in India*. Washington, DC: Pew Center on Global Climate Change.

Shukla, P. R., Nag, T., & Biswas, D. (2005). Electricity reforms and firm level responses: Changing ownership, fuel choices and technology decisions. *International Journal of Global Energy Issues, 23*(2–3), 260–279.

Sneddon, C., Howarth, R. B., & Norgaard, R. B. (2006). Sustainable development in a post-Brundtland world. *Ecological Economics, 57*, 253–268.

Sodhi, N. S., Koh, L. P., Brook, B. W., & Ng, P. K. L. (2004). Southeast Asian biodiversity: An impending disaster. *Trends in Ecology & Evolution, 19*, 654–660.

SRES. (2000). Special report on emissions scenarios prepared for policy makers by the Intergovernmental Panel on Climate Change (IPCC) working group III. ISBN 92-9169-113-5.

Statement of the Friibergh Workshop on Sustainability Science. (2000 October 11–14). Friibergh, Sweden. Retrieved from http://sustsci.aaas.org/content.html?contentid=774

Sunderlin, W. D., Angelsen, A., Belcher, B., Burgers, P., Nasi, R., Santoso, L., & Wunder, S. (2005). Livelihoods, forests, and conservation in developing countries: An overview. *World Development, 33*, 1383–1402.

Tallis, H., Kareiva, P., Marvier, M., & Chang, A. (2008). An ecosystem services framework to support both practical conservation and economic development. *PNAS, 105*(28), 9457–9464.

Thomas, S., Dargusch, P., Harrison, S., & Herbohn, J. (2010). Why are there so few afforestation and reforestation clean development mechanism projects? *Land Use Policy, 27*, 880–887.

UN (2010). Oceans and climate change. Office of Legal Affairs. Division for Ocean Affairs and the Law of the Sea United Nations.

UNEP. (2006). *Marine and coastal ecosystems and human wellbeing: A synthesis report based on the findings of the millennium ecosystem assessment*. Nairobi: UNEP.

UNEP. (2009a). Ecosystem management: Part of the climate change solution. UNEP Research Brief. Retrieved from www.macaulay.ac.uk/copenhagen/documents/UNEP-CC-EM-5-page-brief.pdf

UNEP. (2009b, June). The role of ecosystem management in climate change adaptation and disaster risk reduction. Copenhagen discussion series paper 2. Retrieved from http://www.unep.org/climatechange/Portals/5/documents/UNEP-DiscussionSeries_2.pdf

UNEP. (2011). Towards a green economy: Pathways to sustainable development and poverty eradication: A synthesis for policy makers. Retrieved from http://www.unep.org/greeneconomy

UNEP (n.d). Ecosystem management: Part of the climate change solution. UNEP Research Brief. www.macaulay.ac.uk/copenhagen/documents/UNEP-CC-EM-5-page-brief.pdf

UNEP-WCMC. (2008). *National and regional networks of marine protected areas: A review of progress*. Cambridge, UK: UNEP World Conservation Monitoring Centre (UNEP-WCMC).

USAID. (2010). Adapting to coastal climate change: A guidebook for development planners. Retrieved from http://www.crc.uri.edu/download/CoastalAdaptationGuide.pdf

van Vuuren, D. P., Edmonds, J., Kainuma, M., Riahi, K., Thomson, A., Hibbard, K., … & Rose, S. K. (2011). The representative concentration pathways: An overview. *Climatic Change, 109*, 5–31. doi:10.1007/s10584-011-0148-z.

Vickers, B., Kant, P., Lasco, R., Bleany, A., Milne, S., Suzuki, R., … & Pohnan, E. (2010). *Forests and climate change in the Asia Pacific region* (Forests and climate change working paper, Vol. 7). Rome: FAO.

Wetlands International. (2009). *The global peatland CO_2 picture. Peatland status and drainage related emissions in all countries of the world*. The Netherlands: Wetlands International, Ede.

World Commission on Environment and Development (WCED). (1987). Our common future. Report of the Buntland Commission/WCED. Published as annex to general assembly document A/42/427, Development and International Co-operation: Environment.

Xiulian, H., & Kejun, J. (2008). Country Scenarios toward Low-Carbon Society (LCS). The Third workshop of Japan-UK joint research project. National Institute of Environmental Studies (NIES), Japan.

Yeo-Chang, Y. (2009). Use of forest resources, traditional forest-related knowledge and livelihood of forest dependent communities: Cases in South Korea. *Forest Ecology and Management, 257*(10), 2027–2034.

Yohe, G. W., Lasco, R. D., Ahmad, Q. K., Arnell, N. W., Cohen, S. J., Hope, C., … & Perez, R. T. (2007). Perspectives on climate change and sustainability. In M. L. Parry, O. F. Canziani, J. P. Palutikof, P. J. van der Linden, & C. E. Hanson (Eds.), *Climate change 2007: Impacts, adaptation and vulnerability. Contribution of working group II to the fourth assessment report of the intergovernmental panel on climate change* (pp. 811–841). Cambridge, UK/New York: Cambridge University Press.

Chapter 7
Future Directions for Climate Research in Asia and the Pacific

Michael J. Manton and Linda Anne Stevenson

Abstract There are clear trends of increasing temperature in the Asia-Pacific region. There are observed trends in extreme climate events and evidence of changes in large-scale climate systems including the monsoon and the associated Hadley circulation. Modeling the climate of the region provides opportunities for improved understanding and prediction, but there remain challenges especially for mountainous terrain and small islands. Current projections for future climate indicate that existing stresses are likely to be exacerbated.

Urbanization is expected to continue and better understanding of the interactions between climate and urban areas is essential. Further work is needed to improve our understanding of adaptation and mitigation both in urban areas and in small communities. Significant challenges exacerbated by climate variability and change need to be overcome so that future needs for rice and wheat can be met. Management strategies need to be implemented globally so that fisheries will be able to provide necessary food for the region. Local management strategies are also needed to ensure water security.

Regional and international cooperation is providing initial support for integrated assessments that can investigate pathways towards low carbon development (LCD) across the region. Natural ecosystem services support substantial components of economies across the region and new strategies are being developed to enhance the resilience of natural ecosystems impacted by climate change. Natural ecosystems in Asia and the Pacific can contribute significantly to the mitigation of climate change.

M.J. Manton (✉)
School of Mathematical Sciences, Monash University, VIC 3800, Australia
e-mail: michael.manton@monash.edu

L.A. Stevenson
Asia-Pacific Network for Global Change Research, APN Secretariat,
4F, East Building, 1-5-2 Wakinohama Kaigan Dori, Chuo-ku, Kobe 651-0073, Japan
e-mail: lastevenson@apn-gcr.org; clothears_2008@yahoo.co.jp

M.J. Manton and L.A. Stevenson (eds.), *Climate in Asia and the Pacific: Security, Society and Sustainability*, Advances in Global Change Research 56,
DOI 10.1007/978-94-007-7338-7_7,

Communities, particularly poor and remote communities, are vulnerable to climate change and there is a need for capacity building in research, policy development and implementation to reduce these vulnerabilities. International cooperation exists in the development of mechanisms to promote systematic observations of geophysical variables. Further cooperation is needed to ensure that consistent high-quality socio-economic data are collected, archived and accessible. Continuous monitoring of the geophysical environment and associated socio-economic variables, and developing and analyzing indicators of climate interactions with natural ecosystems and human societies is needed to fully interpret and respond to the complex socio-economic interactions with the Earth's climate.

Keywords Asia-Pacific • Climate research • Security • Society • Sustainability • Urbanization

7.1 Findings and Future Directions

In this book we have considered the current status of understanding of climate and its interactions across the region under the topics:

- Climate variability and change
- Climate and urbanization
- Climate and security
- Climate and society
- Climate and sustainability.

Key questions were raised in Chap. 1 to focus the analysis of each topic. The following text summarizes the findings and the need for future work related to each topic; the headings in italics provide headline findings.

There are clear trends of increasing temperature across Asia and the Pacific.

In line with most of the world, there are clear signs of rising temperatures across Asia and the Pacific. These trends tend to be larger at higher latitudes, with for example winter temperatures in Mongolia rising by more than 3.5 °C over the past 60 years (Bohannon 2008). Because of the inherent spatial and temporal variability of precipitation, it is much harder to identify sustained trends in precipitation than in temperature, and both increases and decreases can be seen in the mean daily precipitation over the last 20 years over Asia and the Pacific (Levizzani and Gruber 2007).

A consequence of economic and industrial progress in much of the world has been increases in the atmospheric concentrations of not only greenhouse gases (GHGs) but also aerosols. There is evidence of these aerosols reducing the direct solar radiation at the surface, leading to "global dimming" over some decades of the last half of the Twentieth Century. Although this effect is declining over much of the world, there has been a continuing decrease in surface solar radiation in China and India (Wild et al. 2009). There is both observational and modeling evidence that aerosols in Asia, arising from biomass burning and consumption of fossil fuels, are affecting both temperature and precipitation (Nakajima et al. 2007).

The glaciers of the Himalaya Tibetan Plateau (HTP) are important features of the geography and ecology of the whole region. There have been a number of field studies of the mass balances of these glaciers, but increasingly satellite-based observations are used to monitor changes in glacier properties. While difficulties in monitoring the huge number of glaciers across the HTP lead to uncertainties, there is strong evidence of retreating glaciers in Nepal (Fujita and Nuimura 2011) and India (Kumar et al. 2009).

Continuing evidence is emerging of changes in the climate of Asia and the Pacific. However, much of the region is difficult to access owing to remoteness, harsh climate or steep terrain. These difficulties mean that systematic monitoring is not straightforward and may not be seen as a priority for communities with limited resources. While satellite-based measurements can provide the required spatial and temporal coverage for main environmental variables, direct in situ observations are also needed to provide benchmarks for remotely sensed data and to provide local detail that cannot be captured from satellites. Direct observations can also be used as a resource by local communities for understanding and managing their environment.

There are observed trends in extreme climate events across Asia and the Pacific, with decreases in the frequency of cold nights and increases in the frequency of hot days.

For more than a decade, countries of Asia and the Pacific have been cooperating in the analysis of daily temperature and rainfall records to identify trends in extreme climate events on a regional as well as local basis. For example, the analysis of Kwon (2007), which is also complemented by the work of Sheikh (2008) and Islam et al. (2009), finds that there have been consistent decreases in the frequency of cool nights in Asia, accompanied by increases in hot days. Although there have been suggestions of trends in temperature extremes being larger at higher elevations, Revadekar et al. (2012) find no consistent behavior of temperature trends with elevation across South Asia.

The inherent variability of rainfall means that, as with trends in mean values, it is very difficult to observe statistically significant trends in extreme precipitation on a regional basis. On the other hand, analysis of global trends in temperature and rainfall suggests that there has been an increase in the intensity of heavy rainfall both locally (Taiwan) and globally (Liu et al. 2009).

Tropical cyclones are a major aspect of the climate of much of Asia and the Pacific. While there is some suggestion of increases in intensity and frequency of tropical cyclones (typhoons) in the Pacific, detailed analysis of global trends indicates that there is real uncertainty in separating trends from natural variability (Knutson et al. 2010). On the other hand, future projections of tropical cyclone activity suggest that greenhouse warming will lead to more intense storms. Moreover, analysis of the human risks from tropical cyclones (Peduzzi et al. 2012) implies that the greatest risks, both currently and into the future, are with the communities of Asia and the Pacific.

There is evidence of changes in the large-scale climate systems of Asia and the Pacific including the monsoon and the associated Hadley circulation.

The regular seasonal variations of the Asian-Australian monsoon system are a major influence on the human communities and natural ecosystems of the region, and so any consistent change in these variations will have significant impacts. There are indications of decreasing trends in annual precipitation in northern China and south eastern Australia, with increasing trends in the Tibetan Plateau, Southeast Asia, Republic of Korea and Northwest Australia. These trends may suggest an intensification of the summer monsoon in these regions (Kwon 2007). On the other hand, model simulations of the South Asia monsoon (Ramanathan et al. 2005) suggest a weakening of the monsoon owing to anthropogenic influences of aerosols and GHGs.

The global-scale Hadley circulation is found to be widening in both the northern (Johanson and Fu 2009) and southern hemispheres (Lucas et al. 2012), and this trend is consistent with the expected effects of human activities. However, modeling studies of the South Asia summer monsoon (Fan et al. 2010) and the Hadley circulation over Australia (Kent et al. 2011) find substantially smaller trends than those observed.

It is apparent that the interactions between the monsoon and the global circulation are being influenced by anthropogenic activities. However, our current observations and models are not sufficiently detailed to capture and explain all these interactions.

While modeling the climate of Asia and the Pacific provides opportunities for improved understanding and prediction, there remain challenges especially related to the small-scale topographic features of the region, such as mountainous terrain and small islands.

Modeling provides the means to assimilate a range of observations from different platforms to develop a consistent picture of global and regional climate. It also supports detailed studies of the impacts of specific drivers of climate, such as GHGs or anthropogenic aerosols, on climate on many scales, and so it is an essential element in studies aimed at identifying the causes of observed changes and variations in climate.

There has been effective collaboration amongst modeling groups in Asia and the Pacific to develop and evaluate global and regional climate models; for example, the Regional Model Inter-comparison Project (RMIP) has promoted modeling studies of extreme weather patterns and the seasonal cycle across the region (Feng and Fu 2006). Under the auspices of the World Climate Research Programme (WCRP), the Coordinated Regional Climate Downscaling Experiment (CORDEX) is promoting regional downscaling activities aimed at supporting communities involved in climate vulnerability and adaptation studies (Whitehall et al. 2012), which is a key development in using science to cater to the practical needs of climatically-vulnerable communities. Downscaling is the process of extending the usefulness of the output of global climate models to local scales using statistical and regional modeling methods, and groups in Asia and the Pacific are directly involved in these studies (for example, Lee et al. 2007). Other important work that utilizes tailored downscaled models is also being undertaken directly in decision-making of local governments; for example, the recent vulnerability and adaptation work by Pulhin (2009) using simplified climate models.

While we can expect growing and continuing progress on climate modeling, the Asia and Pacific region provides a range of challenges associated with its geography. The need for very fine scale results arises in the mountainous areas of the HTP, where the steep terrain drives local climate variations, and in the vast Pacific and Indian Oceans where the specific characteristics of small isolated islands should be resolved by the models. Similarly, the features of urban areas that interact strongly with local climate should be resolved in models for Asia where urbanization is a dominant feature of human settlement. Much work is needed to ensure that the specific challenges of Asia and the Pacific are priorities of the international modeling community.

Current projections for the future climate across Asia and the Pacific under climate change indicate that existing regional stresses are likely to be exacerbated.

Although there remain considerable uncertainties in the projections of future climate change in Asia and the Pacific, current projections provide a useful indication of the climate regimes that will need to be managed in the decades ahead. Some of the uncertainties arise from the difficulty in modeling the details of large-scale features such as the El Niño–Southern Oscillation and of the South Pacific Convergence Zone, which strongly influence the climate of the Pacific and other parts of Asia.

Projections of future climate change suggest that warming will be greatest at higher latitudes and higher elevations. There are also indications that the monsoon will be affected in South Asia (Kumar et al. 2011) and in East Asia (Ninomiya 2011). A substantial research program (Power et al. 2011) has focused on the development of a better understanding of past and future climate change across developing island countries of the Pacific. Under the program, an interactive web-based system has been developed (Pacific Climate Futures) to provide information on potential climate change for 15 countries in the South Pacific (http://www.pacificclimatefutures.net/).

Urbanisation is expected to continue across Asia and the Pacific with more than 40 % of the urban population by 2025 living in cities of less than one million people, and so better understanding of the interactions between climate and urban areas is essential.

The process of urbanization is very likely to continue with almost three billion people living in urban areas in Asia and the Pacific by 2050. While the current trend of mega-cities is expected to also continue, the majority of people will be living in cities with populations less than one million; by 2050 that fraction will be almost 60 %.

Continuing urbanization means that there needs to be greater focus on understanding and managing interactions between the urban environment and climate. These interactions have impacts at regional as well as local scales. The urban heat island (UHI) effect has been recognized for many years (Oke 1973), but more work is needed to mitigate UHI phenomena through careful urban design. An example of effective action to reduce the UHI effect (with concomitant benefits for recreation and tourism) is the reclaiming of the Cheonggyecheon River in the city of Seoul (Cho 2010).

There is considerable evidence of urban areas affecting precipitation in the local region, with both increases (Inamura et al. 2011) and decreases (Kaufmann et al. 2007) being found. We noted earlier the suggestion that global warming itself is driving a trend towards heavier rainfall across the world (Liu et al. 2009).

A consequence of urbanization and industrialization has been increased air pollution, which, in turn, impacts human health as well as a range of other features of society and the environment (Molina and Molina 2004). Motor vehicles are a substantial source of air pollution in cities across Asia and the Pacific (Walsh 2003). In the short term in developing countries, demand for private transportation leads to a range of undesirable consequences, such as congestion and air pollution (Vasconcellos 2001). However, it is likely that increasing costs of car ownership, advanced technology, improved public transport and improved urban design will lead to significant changes in urban transport systems over the longer term (Garnaut 2008), in turn leading to improved urban air quality.

Urban areas are major contributors to GHG emissions, but the per capita contributions vary with factors such as population density and wealth.

The growth of urbanization in Asia and the Pacific is generally associated with economic growth thus presenting a major challenge by way of providing water, food and energy to urban areas while enhancing the social and economic well-being of communities. Consequently urban areas are directly and indirectly responsible for most of the GHG emissions in the region.

Energy consumption is a major source of GHG emissions across Asia and the Pacific. For example, energy production accounts for 30–40 % of GHG emissions in four large cities in China (Dhakal 2009). Emissions associated with transportation vary from city to city across the region, but it is likely at least in the short term that these emissions will rise with increasing urban infrastructure (Marcotullio and Marshall 2007). Industrial processes also account for about 10 % of GHG emissions in Asia, but a growing issue for the region is emissions from the management of waste, including waste water (Jha et al. 2008).

While the largest cities have the highest GHG emissions, per capita emissions vary with a number of factors. The overall efficiency of a city varies with population density, and low density urban areas in Asia tend to have much higher GHG emissions per capita than more dense cities; medium density cities have the lowest emissions (Marcotullio et al. 2012). There is also strong evidence that per capita emissions tend to be higher in communities with greater wealth and in colder climates. These results suggest that emission mitigation policies should recognize the inherent efficiencies of urban areas.

The variety in types of urban areas in Asia and the Pacific means that policies for adaptation and mitigation of climate change should not be constrained to a common solution.

The generally high population densities of urban areas of Asia and the Pacific mean that these communities are vulnerable to climate hazards, including tropical cyclones and severe storms, flooding and storm surges, and heat waves. Moreover, these risks are likely to increase under future climate change. One study suggests that excess mortality due to heat stress could increase fourfold in the future (Takahashi et al. 2007). Much urban development in Asia and the Pacific is along coastlines and these regions are vulnerable to the impacts of sea level rise (Hanson et al. 2011). Flooding, which is already a major hazard for Asia cities, is also expected to become increasingly costly for urban areas. In general, the impacts of weather-related disasters on both infrastructure and people are likely to increase over time across the region.

The vulnerability of both individuals and communities to climate change varies with a number of factors, including financial, social and natural resources. Across all urban areas, there are particular groups, such as the poor, the elderly and the very young that have limited capacity to adapt to the impact of climate change (Satterthwaite et al. 2007). The largest number of those living in poverty are in the Asia-Pacific region (Wan and Sebastian 2011).

Further work is needed to improve our understanding of the costs of adaptation and mitigation, especially for small communities and the poor.

A growing number of municipalities, NGOs and civil society organizations are involved in actions to reduce GHG emissions and to adapt to climate change. However, the costs and impacts of options for sector-wide mitigation strategies at the urban level are not fully understood at this time. There needs to be further evaluation of not only sector-based strategies but also of the potential for trade-offs by measures that act across sectors for both mitigation and adaptation.

Adaptation depends upon the quality of local knowledge, local capacity and willingness to act (Satterthwaite et al. 2007), and so local communities need to participate in the development and implementation of adaptation strategies. Further research is needed to enhance our understanding of governance related to mitigation and adaptation policies, including the power relationships influencing outcomes at local levels.

Significant challenges, exacerbated by climate variability and change, need to be overcome so that the future needs for rice and wheat across Asia can be met.

Most of the world's smallholder farmers are in Asia, and their livelihoods are threatened by climate-related factors including floods, droughts, tropical cyclones, heat waves and wild fires (Sivakumar and Motha 2007). These natural hazards are exacerbated by changing economic factors, such as rising market prices for seed, fertilizer and animal feed. At the same time, these communities have limited capacity to adapt to changing circumstances, owing to the limits of their socio-economic resources. Government and business policies are therefore needed to lead any substantial focus on climate change adaptation in agriculture in Asia and the Pacific.

Rice and wheat are the main crops for Asia and the Pacific. About 90 % of the world production of rice is in Asia, and demand is increasing with population. The main hazards for increasing rice production come from heat, drought, floods and salinity intrusion (Wassmann et al. 2009). Asia produces about 40 % of the global production of wheat.

Changes in agronomic practices, such as modifying planting dates, cultivar substitution and improved water management, are basic means for adapting to climate change and variability. Basic mitigation policies can include enhancement of carbon sequestration by establishing permanent land cover, using conservation tillage, incorporating rotations of forage, and improving nutrient management with fertilizers.

The development and implementation of improved practices for agricultural production in Asia and the Pacific need to be underpinned by enhancing the knowledge and resources of both agribusiness and individual farmers. This in turn depends upon better communication links between farmers, policy makers and agriculture researchers to ensure the effective transfer of scientific and technical advances into practice. The development of strategies to adapt to climate variability and change needs to incorporate local socio-economic factors as well as climate and agricultural science.

Strict management strategies need to be implemented globally so that fisheries will be able to provide necessary food for Asia and the Pacific in future under climate change.

The world's oceans play an important role in the response of the global climate system to enhanced levels of GHGs by absorbing significant fractions of the associated heat and CO_2. The interactions between the atmosphere, ocean and ocean ecosystems are therefore very complex. As most marine animals are cold-blooded, they are very sensitive to environmental temperature. It is likely that fish behavior, such as spawning times and locations, adapts to the prevailing physical and biological conditions (Heath and Gallego 1998). Similarly variations and changes in the environment can lead to changes in fish production, as measured by fish catch (Sakurai et al. 2000).

While fish is a major component of the diets of communities across Asia and the Pacific, there are regional differences in the amount and type of fish consumed. East Asia consumes about one third of the total global consumption of fish, and annual consumption is increasing with the increasing wealth of those communities. Regional food security is therefore dependent on the sustainability of ocean fisheries while accounting for climate variability and change.

Modeling studies have been carried out to determine the sensitivity of fisheries of the Pacific and Indian Oceans to climate change. For example, Kishi et al. (2010) analyzed the potential impact of climate change on chum salmon. Such studies provide the foundation for the development of management strategies for sustainable fish yields. However, bearing in mind the life span of many fish species, these strategies need to be implemented now so that the gradual effects of climate change can be managed effectively. Moreover, because fish migrate over large distances, international cooperation is needed for the development and implementation of these policies.

Local management strategies and multi-national agreements will be needed to ensure water security across Asia and the Pacific in the future.

Ready access to potable water is essential for the maintenance of a sustainable economy, and so access to water is a critical issue for the communities of Asia and the Pacific under the impacts of growing populations and economies as well as climate change. The Himalayas – Tibetan Plateau (HTP) is a critical region for the water security of Asia, as it provides water for about 20 % of the world's population living in a dozen countries. Based on current changes and projections of future climate, it is clear that the hydrology of the HTP will be substantially affected by climate change (Immerzeel et al. 2010). These effects are being felt not just in the HTP but also in all the downstream countries dependent upon the annual glacial melt from Asia's "water tower." Uncertainties in both future climate and the hydrological variations across the region mean that it is difficult to specify these downstream impacts. This is shown by Thayyen and Gergan (2010) who note that interannual variations in stream flow can be more sensitive to annual precipitation than glacier melt.

An international challenge to regional water security is arising from the large number of hydro-electric power stations planned for rivers that flow from the HTP. While hydro-electricity is seen as a "green" alternative to the combustion of fossil fuels, the local and downstream impacts of damming rivers can be profound and a fine balance is needed to avoid adverse impacts, particularly in situations that are trans-boundary in nature requiring cross-border dialogue and cooperation.

It is clear that regional water security is dependent upon international understanding and cooperation. However, at present there are no effective processes to promote dialogue and mechanisms for the development of adaptive responses to the trans-boundary issues related to the rivers of the HTP (Morton 2011). On the other hand, there have been some successful bilateral and even multilateral water treaties to resolve specific issues.

Because the nations of the Pacific Ocean are composed of thousands of islands, water security tends to be a national issue. Moreover, the sources of fresh water for island communities are rainfall and its associated run-off, together with groundwater. As populations increase, these sources can be limiting and water quality can be compromised (White and Falkland 2010). The current challenges of water security are expected to be exacerbated by climate change in the future.

The local supply and demand characteristics for freshwater can vary from island to island across the Pacific, and so strategies for water security need to vary. More research is needed to properly account for the impacts of climate change on island hydrology, and that research should inform the development of effective policies for water management. Advances in water technologies and infrastructure are needed to provide feasible solutions that are appropriate to the technical capabilities of local communities.

A range of technological, policy and educational strategies need to be developed to meet future energy demands across Asia and the Pacific while constraining GHG emissions.

For several decades there has been a linear relationship between national living standards (measured by GDP) and energy consumption. Moreover energy production has tended to be associated with the combustion of fossil fuels, and so there has also been a linear relationship between energy and GHG emissions. Thus, the steady improvement of wellbeing has been linked to the emission of GHGs. In order to underpin the continuing improvement in the wellbeing of communities across Asia and the Pacific while mitigating GHG emissions, it will be necessary to break the link between GDP and GHG. This link can be broken by reducing the connection between wellbeing and energy consumption or that between GHG and energy production.

The link between energy and well-being is reduced by increasing the efficiency of energy use in the community, through strategies such as more energy efficient machinery and buildings. The energy-GHG link is reduced by either more efficient consumption of fossil fuels or by alternative energy sources. Each country across the region is developing policies to promote a reduced dependence on fossil fuels.

China has implemented plans to increase energy intensity by 16 % by 2015 and GHG emission per unit GDP by 17 % by 2015. The plans include strategies to increase forest cover, increase the energy efficiency of industry, invest in public transport systems, and promote energy generation non-fossil fuel sources. Targets have been set for cities, and monitoring systems are being established to track progress towards the goals.

In India policies are being implemented to reduce the risks to energy security and the environment by opening the coal industry to the private sector, promoting oil and gas production, supporting overseas investment in energy sources, and promoting the use of renewable and nuclear energy sources. India's National Action Plan on Climate Change includes policies on energy efficiency, expanded forest cover, and sustainable agricultural practices.

Japan is dependent upon the importation of energy from overseas. However, demand for energy has been declining for a decade owing to a reduction in its manufacturing industries and in its population. In response to the Kyoto Protocol, Japan has had a focus on improved energy efficiency and on nuclear power plants. However, the impacts of the tsunami of March 2011 have led to a revision of national policy on nuclear energy. One consequence of the national energy crisis arising from the tsunami was a reduction of more than 10 % in demand for energy over the summer of 2011. This feature, along with a reluctance to depend upon nuclear power, has led to an increased emphasis on energy demand management and on the promotion of renewable energy sources.

It is clear that, while there is a common interest across Asia and the Pacific in reducing the carbon emissions, each country is developing its own strategies that take into account national interests and characteristics. Comparative studies of the individual policy options will be important to assess progress at national and regional levels.

While there are a number of international agreements relating to climate change governance, the complexity of the issues at all levels means that strategies for Asia and the Pacific are often fragmented.

The causes and impacts of climate change are global in nature, and so any attempts to manage these features require international actions, with ramifications at the regional, national and local levels. Even natural climate variability, resulting in floods, wild fires, droughts and heat waves, has impacts that require government actions at the local and national levels to both manage and mitigate the impacts of such natural disasters. As the impacts of climate change become more apparent, the overlap in governance arrangements at the national and local levels for managing natural disasters becomes clear.

Climate governance should be incorporated into a broader framework of governance for sustainable development. Such arrangements recognize the effects of climate change, extending across social and economic aspects of communities as well as the environment. The UN Framework Convention on Climate Change (UNFCCC) is a concrete measure of international recognition of the significance of climate change at the global level. However, its implementation has been limited, and other multilateral environmental agreements are being enhanced and developed to support decision makers involved with climate governance.

The leaders of the Association of South East Asian Nations (ASEAN) have prepared a Road Map for an ASEAN Community 2009–2015 that links climate change with sustainable development. The Road Map lays out goals, strategies and actions to underpin political and security cooperation, economic cooperation and socio-cultural cooperation. The ASEAN actions cover a range of issues, including trans-boundary environmental pollution, environmental education, environmental technology, harmonization of databases, and freshwater resources.

Under the auspices of the Secretariat of the Pacific Regional Environment Programme (SPREP), the Pacific Islands Framework for Action on Climate Change 2006–2015 has been prepared (SPREP 2011). The Pacific Island leaders recognize the special concerns and interests of small low-lying island countries, particularly arising from sea-level rise. The Framework contains principles relating to adaptation measures, improved understanding and awareness of climate change, mitigation of GHG emissions, and governance.

Supporting these international and regional agreements are a range of national and local governance arrangements to mitigate and adapt to the impacts of climate change. Given the diversity of socio-economic and cultural backgrounds, these governance arrangements tend to be fragmented. However, there is scope for a more programmatic approach to the introduction of climate adaptation and poverty alleviation into national and local development plans. The international framework for climate change governance also needs to be strengthened to provide a clearer context for regional, national and local actions across Asia and the Pacific.

The multiple hazards confronting remote mountain and small island communities across Asia and the Pacific, together with limited capacity for adaptation, mean that those communities are very vulnerable to the impacts of climate change.

Remote mountain communities and small island communities of Asia and the Pacific are vulnerable to a range of natural and social hazards. The highland areas of the HTP have a population of about 30 million people with a wide range of cultures and languages. Although their resources are limited, these communities have adapted to their environment over hundreds of years. The relatively rapid changes associated with global warming bring additional stresses. Warming itself is leading to shifting tree lines and biodiversity loss, as well as increased hazards from glacial lake outburst flooding (GLOF) and changes in seasonal stream flow.

In the eastern Tibetan Himalayas of Yunnan province of China, a climate field school has been established to improve understanding of the relationship between climate change and traditional medicine, including the growing of traditional crops and herbs. Through this process, it is expected that indigenous knowledge can be adapted to the changes occurring through climate change.

The Karen community of Huay Hin Lad, Thailand cooperated with other local groups to protect their lands as commercial logging begun operation in the 1980s. Actions included the analysis of the carbon footprint of their community in order to demonstrate both their basic efficiency and their vulnerability to external changes. Similar case studies demonstrate that indigenous knowledge has generally underpinned sustainable living conditions for remote mountain communities. However, their capacity for adaptation is limited and they are vulnerable to the additional stresses of climate change.

The small island developing states of the Pacific and Indian Oceans are vulnerable to a range of natural hazards, such as storm surges and tropical cyclones, as well as the additional stresses of climate change, especially through sea level rise. Social relationships are typically based on kinship in these island states and this factor is important for economic activity, local politics and land tenure. The associated cooperative processes also contribute to the resilience of the communities. The economies of the communities tend to be small and open, with adverse terms of trade.

Small island communities are vulnerable to the impacts of climate change, including sea level rise, more intense storms, drought and heat stress. However, their traditional life styles had evolved to be resilient to extreme environmental conditions, through strategies for food security, community cooperation, and well-designed settlements (Campbell 2006). However, these practices have declined since colonization and their exposure to the global economy. On the other hand, the small island nations have been very active in negotiations under the UNFCCC, and they have worked together to support activities aimed to enhancing their capacity to adapt to climate change at both regional and community levels.

Most of the settlements of the small island states are coastal, often only a few meters above sea level. It follows that many of these communities may not be sustainable as sea level rises. Migration, both within and from these states, is likely to become a substantial strategy for adaptation to climate change. International cooperation will be a necessary aspect of the successful implementation of such approaches in the longer term.

The vast number of islands in the Pacific is a significant constraint on the implementation of adaptation strategies. Moreover activities tend to be financed as projects rather than long-term programmes, which further limit the overall effectiveness of investments. For both remote mountain communities and small island communities, it is important for the international community to recognize the special vulnerabilities of these groups and to ensure that informed and sustained programs of adaptation are prepared and implemented. Such programs need to be informed by research into the specific hazards and vulnerabilities of each community.

Climate change is exacerbating both communicable and non-communicable diseases across Asia and the Pacific, including malnutrition that limits child growth and development

Climate extremes, such as cold spells and heat waves, have direct impacts on human health. Air quality and the presence of vector-borne diseases are also closely linked to climate. The basic necessities for human health, nutritious food and clean water, are also affected by climate. Many communities are slowly learning to mitigate the impacts of such climate fluctuations, but some socio-economic groups, such as the poor, are more vulnerable than most.

Climate change is exacerbating the hazards of climate for human health. However, it is not clear that the health sector across Asia and the Pacific recognizes the increasing risks, and more capacity building is needed to ensure that public health programs are adequate. The impacts of climate change will be greatest on the currently vulnerable groups of society: the poor, the elderly and the very young. More work is needed to identify these groups, to assess their specific vulnerabilities, and to implement strategies to reduce their risks.

Action is needed in both rich and poor countries; in particular, currently fragmented health systems need to be replaced by more coherent systems that can respond to public heath threats related to climate change in addition to delivering basic clinical services. The interactions between human health and urban planning need to be recognized, as most of the populations of the region are moving to urban areas.

Regional and international cooperation is providing initial support for integrated assessments that can investigate pathways towards low carbon development across Asia and the Pacific

Low carbon development (or green growth) is now seen as a policy aim to mitigate and adapt to the impacts of climate change. Integrated assessment models (for example, Kainuma et al. 2003) provide a mechanism for evaluating different strategies for economic development, by simulating a variety of factors associated with the interactions between climate and the economy. These models are used to test the effectiveness of alternative strategies in progress towards low carbon development, which has the long-term goal of reducing GHG emissions while ensuring the well-being of the community.

Country studies can be carried out using integrated assessment models to investigate the impacts of alternative socio-economic scenarios on GHG emissions. Such scenarios can vary from the application of high-technology to reduce emissions to the adoption of simple life-styles and reduced economic growth. Studies have been conducted for Japan (Kainuma et al. 2003) and India (IIMA et al. 2009). Case studies of this type provide a basis for policy-makers and communities to decide on future strategies for socio-economic progress.

Reduction of GHG emissions is negotiated internationally under the UNFCCC, and many countries of Asia and the Pacific have agreed to voluntary emission reductions. International negotiations continue towards a legal instrument that can be applied to all parties under the UNFCCC. Independent of formal commitments to emission reductions, processes are being developed for monitoring, reporting and verifying progress towards voluntary targets by individual countries. There has also been international work on the development of financial mechanisms to assist developing countries implement both mitigation and adaptation strategies.

There are a number of co-benefits for countries of Asia and the Pacific in joining these international arrangements, ranging from increased energy efficiency to improved community health. There are also benefits in countries cooperating on a regional as well as international level. The Association of South East Asian Nations (ASEAN) and the Asia and Pacific Economic Cooperation (APEC) provide contexts for such agreements in the region.

Actions toward low carbon development have foundations in international and regional arrangements. However, detailed implementation needs to come through at national and local levels. Integrated assessments can be applied at all levels to support decision-making. In order to promote progress in the development of integrated assessment models and to assure their quality, an international network has been established involving the key research groups around the world. Progress towards low carbon development requires cooperation across all relevant stakeholders, including government, non-government organizations, industry and the broad community.

Natural ecosystem services support substantial components of economies across the Asia and the Pacific, and new strategies are being developed to enhance the resilience of natural ecosystems impacted by climate change

There is increasing recognition that sustainable development is critically linked to natural ecosystems. For example, about 400 million people are dependent upon forests for subsistence and income, and the livelihood of about 1.6 billion people is linked to forest goods and services (Munang et al. 2011). Similarly, the value of freshwater ecosystems is estimated to be trillions of dollars (Revenga et al. 2000). Ecosystem services like these examples have been jeopardized by economic development that did not recognize their value. Moreover, climate change is imposing additional stresses on natural ecosystems.

A more sustainable approach to economic development can be achieved by recognizing natural ecosystems as natural capital providing essential services to human communities. Ecosystems stewardship aims to sustain ecosystem services that support human well-being under conditions of uncertainty and change (Chapin et al. 2009). This approach implies that there will be trade-offs between services and related activities. However, more work is needed to improve our understanding of the underpinning science, such as ecosystem production functions and the design of monitoring programs (Tallis et al. 2008).

Many diverse and productive ecosystems are in Asia and the Pacific, and these ecosystems act to regulate the climate. Strategies for climate change adaptation and mitigation therefore need to include ecosystem management. In particular, such strategies should include the combination of indigenous and scientific knowledge, the maintenance of biodiversity, and the enhancement of coordination among stakeholders.

Natural ecosystems in Asia and the Pacific can contribute significantly to the mitigation of climate change

Natural ecosystems of Asia and the Pacific can mitigate the effects of GHG emissions through for example carbon sequestration by forests. This means that sustainable forest management can increase carbon sequestration, while the conversion of natural forests to perennial crops contributes to GHG emissions. The potential of forests in Asia to mitigate climate change has been estimated for a number of countries including the Philippines (Lasco and Pulhin 2001).

Ecosystem management strategies aimed at enhancing carbon sequestration can generally have co-benefits. Such benefits include protection from natural disasters, improved soil fertility and better water efficiency.

It is clear from the science that management of natural ecosystems is a key strategy for both mitigating and adapting to climate change. However, the policy frameworks to support these practices are not yet in place. It is necessary to integrate ecosystem management into planning for climate change actions at all levels of governance. Associated with this integration is the implementation of education and capacity building programs to ensure that the basis of ecosystem policy is understood by all stakeholders.

7.2 Overarching Issues

In Sect. 7.1, we summarized the key issues and future directions arising from the individual chapter of this book. It is now useful to capture some of the important overarching issues that cut across several of the topics discussed earlier.

Communities across the region have generally adapted over many years to natural variations in climate. However, the combination of increasing population and economies together with climate change is imposing new stresses on both natural ecosystems and human societies, and new strategies are needed to manage these stresses. In particular, there is a real possibility that relevant traditional knowledge will be lost as communities migrate from rural to urban areas.

The ongoing process of natural climate variability has driven communities to understand its impacts and the need for strategies to mitigate or adapt to those impacts. On the other hand, the relative slowness of discernible impacts of climate change together with the need to change long-standing strategies mean that special education and communication efforts (including the acceptance of ethical requirements associated with global problems) are needed in all communities. Most sectors have at least some established practices that are no longer optimal under climate change conditions, and changing such practices requires concerted and informed education policies.

The stresses on communities in Asia and the Pacific vary with locality and often with time as the climate changes. Thus there is a continuing need for vulnerability studies and risk assessments of climate impacts across many sectors of society. Indeed the complexity of especially urban areas means that assessments should be integrated across sectors and should lead towards low-carbon development.

The analysis in the book has found that many communities, especially poor and remote communities, are vulnerable to the risks of climate change. To reduce these vulnerabilities, there is a need for capacity building on all aspects of climate interactions with society. Capacity building is needed not only in research but also in policy development and implementation. Especially in policy development and implementation, there is evidence that strong and ethical leadership, often from the local level, is an essential element in obtaining successful outcomes. The identification and support of such leaders will be important in ensuring that societies across Asia and the Pacific can react appropriately to the future challenges of climate change.

The underpinning of understanding and progress is through observation. Internationally there has been much cooperation in the development of mechanisms to promote systematic observations of geophysical variables. Further cooperation at international and regional levels is needed to ensure that consistent high-quality socio-economic data are also collected, archived and accessible. Free and open exchange of all data relevant to understanding and managing climate and its impacts is essential to the future progress of all communities across Asia and the Pacific. Particular emphasis should be on the collection of data that can be analyzed to provide indicators of the interactions between climate, natural ecosystems and human societies.

Continuous monitoring of the geophysical environment and associated socio-economic variables; developing and analyzing indicators of climate interactions with natural ecosystems and human societies is needed to fully interpret and respond to the complex socio-economic interactions with the Earth's climate.

References

Bohannon, J. (2008). The big thaw reaches Mongolia's pristine north. *Science, 319*, 567–568.

Campbell, J. R. (2006). *Rational disaster reduction in Pacific Island communities* (GNS science report, Vol. 2006/038). Wellington: Institute of Geological and Nuclear Sciences Limited.

Chapin, F. S., III, Carpenter, S. R., Kofinas, G. P., Folke, C., Abel, N., Clark, W. C., Swanson, F. J., et al. (2009). Ecosystem stewardship: Sustainability strategies for a rapidly changing planet. *Trends in Ecology & Evolution, 25*(4), 231–249.

Cho, M. R. (2010). The politics of urban nature restoration: The case of Cheonggyecheon restoration in Seoul, Korea. *International Development Planning Review, 32*, 145–165.

Dhakal, S. (2009). Urban energy use and carbon emissions from cities in China and policy implications. *Energy Policy, 37*, 4208–4219.

Fan, F., Mann, M. E., Lee, S., & Evans, J. L. (2010). Observed and modeled changes in the South Asian summer monsoon over the historical period. *Journal of Climate, 23*, 5193–5205.

Feng, J., & Fu, C. (2006). Inter-comparison of 10-year precipitation simulated by several RCMs for Asia. *Advance in Atmospheric Science, 23*, 531–542.

Fujita, K., & Nuimura, T. (2011). Spatially heterogeneous wastage of Himalayan glaciers. *PNAS, 108*(34), 14011–14014.

Garnaut, R. (2008). *The Garnaut climate change review*. Cambridge, UK: Cambridge University Press.

Hanson, J., Nicholls, R., Ranger, N., Hallegatte, S., Corfee-Morlot, J., Herweijer, C., & Chateau, J. (2011). A global ranking of port cities with high exposure to climate extremes. *Climatic Change, 104*(1), 89–111.

Heath, M. R., & Gallego, A. (1998). Bio-physical modelling of the early life stages of haddock, melanogrammus aeglefinus, in the North Sea. *Fisheries Oceanography, 7*, 110–125.

Immerzeel, W. W., van Bleck, L. H., & Bierkens, M. F. P. (2010). Climate change will affect the Asian water towers. *Science, 328*, 1382–1385.

Inamura, T., Izumi, T., & Matsuyama, H. (2011). Diagnostic study of the effects of a large city on heavy rainfall as revealed by an ensemble simulation: a case study of central Tokyo, Japan. *Joural of Applied Meteorology Climatology, 50*, 713–728.

Indian Institute of Management Ahmedabad (IIMA), National Institute for Environmental Studies (NIES), Kyoto University, and Mizuho Information & Research Institute. (2009). *Low carbon society vision 2050 India.*

Islam, S., Rehman, N., & Sheikh, M. M. (2009). Future change in the frequency of warm and cold spells over Pakistan simulated by the PRECIS regional climate model. *Climatic Change, 94*, 35–45. doi:10.1007/s10584-009-9557-7.

Jha, A. K., Sharma, C., Singh, N., Ramesh, R., Purvaja, R., & Gupta, P. K. (2008). Greenhouse gas emissions from municipal solid waste management in Indian mega-cities: A case study of Chennai landfill sites. *Chemosphere, 71*, 750–758.

Johanson, C. M., & Fu, Q. (2009). Hadley cell widening: Model simulations versus observations. *Journal of Climate, 22*, 2713–2725.

Kainuma, M., Matsuoka, Y., & Morita, T. (2003). *Climate policy assessment – Asia-Pacific integrated modeling*. Tokyo: Springer.

Kaufmann, R. K., Seto, K. C., Schneider, N., Liu, Z., Zhou, L., & Wang, W. (2007). Climate response to rapid urban growth: Evidence of a human-induced precipitation deficit. *Journal of Climate, 20*, 2299–2306.

Kent, D. M., Kirono, D. G. C., Timbal, B., & Chiew, F. H. S. (2011). Representation of the Australian sub-tropical ridge in the CMIP3 models. *International Journal of Climatology, 33*(1), 48–57. doi:10.1002/joc.3406.

Kishi, M. J., Kaeriyama, M., Ueno, H., & Kamezawa, Y. (2010). The effect of climate change on the growth of Japanese chum salmon (*Oncorhynchus keta*) using a bioenergetics model coupled with a three-dimensional lower tropic ecosystem model (NEMURO). *Deep-Sea Research II, 57*, 1257–1265.

Knutson, T. R., McBride, J. L., Chan, J., & Emanuel, K. (2010). Tropical cyclones and climate change. *Nature Geoscience, 3*, 157–163.

Kumar, R., Areendran, G., & Rao, P. (2009). *Witnessing change: Glaciers in the Indian Himalayas*. Pilani: WWF-India and Birla Institute of Technology.

Kumar, K. K., Patwardhan, S. K., Kulkarni, A., Kamala, K., Koteswara Rao, K., & Jones, R. (2011). Simulated projections for summer monsoon climate over India by high-resolution regional climate model (PRECIS). *Current Science, 101*, 312–326.

Kwon, W. T. (2007). *Development of indices and indicators for monitoring trends in climate extremes and its application to climate change projection* (Final research report for APN project: ARCP2007-20NSG). Retrieved from APN E-Lib http://www.apn-gcr.org/resources/items/show/1537

Lasco, R. D., & Pulhin, F. B. (2001). Forestry mitigation options in the Philippines: Application of the COMAP model. *Mitigation and Adaptation Strategies for Global Change, 6*, 313–334.

Lee, D. K., Gutowski, W. J., Kang, H. S., & Kim, C. J. (2007). Intercomparison of precipitation simulated by regional climate models over East Asia in 1997 and 1998. *Advances in Atmospheric Science, 24*, 539–554.

Levizzani, V., & Gruber, A. (2007). The international precipitation working group: A bridge towards operational applications. In V. Levizzani, P. Bauer, & F. J. Turk (Eds.), *Measuring precipitation from space: EURAINSAT and the future* (pp. 705–712). Dordrecht: Springer.

Liu, S. C., Fu, C., Shiu, C.-J., Chen, J.-P., & Wu, F. (2009). Temperature dependence of global precipitation extremes. *Geophysical Research Letters, 36*, L17702. doi:10.1029/2009GL040218.

Lucas, C., Nguyen, H., & Timbal, B. (2012). An observational analysis of Southern Hemisphere tropical expansion. *Journal of Geophysical Research, 117*, D17112. doi:10.1029/2011JD017033.

Marcotullio, P. J., & Marshall, J. D. (2007). Potential futures for road transportation CO_2 emissions in the Asia Pacific. *Asia Pacific Viewpoint, 48*, 355–377.

Marcotullio, P. J., Sarzynsky, A., Albrecht, J., & Schulz, N. B. (2012). The geography of urban greenhouse gas emissions in Asia: A regional approach. *Global Environmental Change, 22*, 944–958.

Molina, M. J., & Molina, L. T. (2004). Megacities and atmospheric pollution. *Journal of Air Waste Management Association, 54*, 644–680.

Morton, K. (2011). Climate change and security at the third pole. *Survival, 53*, 121–132.

Munang, R., Thiaw, I., Thompson, J., Ganz, D., Girvetz, E., & Rivington, M. (2011). *Sustaining forests: Investing in our common future* (UNEP policy series, Vol. 5). Nairobo: UNEP.

Nakajima, T., Yoon, S.-C., Ramanathan, V., Shi, G.-Y., Takemura, T., Higurashi, A., Schutgens, N., et al. (2007). Overview of the atmospheric brown cloud East Asian regional experiment 2005 and a study of the aerosol direct radiative forcing in East Asia. *Journal of Geophysical Research, 112*(D24), D24S91.

Ninomiya, K. (2011). Characteristics of the Meiyu and Baiu frontal precipitation zone in the CMIP3 20th century simulation and 21st century projection. *Journal of Meteorological Society of Japan, 89*, 151–159.

Oke, T. R. (1973). City size and the urban heat island. *Atmospheric Environment, 7*, 769–799.

Peduzzi, P., Chatenoux, B., Dao, H., De Bono, A., Herold, C., Kossin, J., Mouton, F., & Nordbeck, O. (2012). Global trends in tropical cyclone risk. *Nature Climate Change, 2*, 289–294.

Power, S. B., Schiller, A., Cambers, G., Jones, D., & Hennessy, K. (2011). The Pacific climate change science program. *Bulletin of the American Meteorological Society, 92*, 1409–1411. doi:10.1175/BAMS-D-10-05001.1.

Pulhin, J. M. (2009). *Capacity development on integration of science and local knowledge for climate change impacts and vulnerability assessments* (APN project CIA2009-02-Pulhin). Available from APN E-Lib http://www.apn-gcr.org/resources/items/show/1699

Ramanathan, V., Chung, C., Kim, D., Bettge, T., Buja, L., Kiehl, J. T., & Wild, M. (2005). Atmospheric brown clouds: Impacts on South Asian climate and hydrological cycle. *PNAS, 102*(15), 5326–5333.

Revadekar, J. V., Hameed, S., Collins, D., Manton, M., Sheikh, M., Borgaonkar, H. P., & Shrestha, M. L. (2012). Impact of altitude and latitude on changes in temperature extremes over South Asia during 1971–2000. *International Journal of Climatology, 33*(1), 199–209. doi:10.1002/joc.3418.

Revenga, C., Brunner, J., Henninger, N., Payne, R., & Kassem, K. (2000). *Pilot analysis of global ecosystems: freshwater systems*. Washington, DC: World Resources Institute.

Sakurai, Y., Kiyofuji, H., Saitoh, S., Goto, T., & Hiyama, Y. (2000). Changes in inferred spawning areas of *Todarodes pacificus* (Cephalopoda: Ommastrephidae) due to changing environmental conditions. *ICES Journal of Marine Science, 57*, 24–30.

Satterthwaite, D., Huq, S., Pelling, M., Reid, H., & Lankao, P. R. (2007). *Adapting to climate change in urban areas, the possibilities and constraints in low- and middle-income nations* (Human settlements discussion paper series, Theme: Climate change and cities, Vol. 1). London: International Institute for Environment and Development.

Sheikh, M. M. (2008). Development and application of climate extreme indices and indicators for monitoring trends in climate extremes and their socio-economic impacts in South Asian

Countries. APN project ARCP2008-10CMY-Sheikh. Available from APN E-Lib http://www.apn-gcr.org/resources/items/show/1550

Sivakumar, M. V. K., & Motha, R. P. (Eds.). (2007). *Managing weather and climate risks in agriculture*. Dordrecht: Springer.

SPREP. (2011). Pacific islands framework for action on climate change 2006–2015 2nd edition, Reprinted by Secretariat of the Pacific Regional Environment Programme (SPREP), Samoa. Available from http://www.sprep.org

Takahashi, K., Honda, Y., & Emori, S. (2007). Estimation of changes in mortality due to heat stress under changed climate. *Risk Research, 10*, 339–354.

Tallis, H., Kareiva, P., Marvier, M., & Chang, A. (2008). An ecosystem services framework to support both practical conservation and economic development. *PNAS, 105*(28), 9457–9464.

Thayyen, R. J., & Gergan, J. T. (2010). Role of glaciers in watershed hydrology: A preliminary study of a 'Himalayan catchment'. *The Cryosphere, 4*, 115–128.

Vasconcellos, E. A. (2001). *Urban transportation, environment and equity: The case for developing countries*. London: Earthscan.

Walsh, M. P. (2003). Vehicle emission and health in developing countries. In G. McGranahan & F. Murry (Eds.), *Air pollution and health in rapidly developing countries* (pp. 146–175). London: Earthscan.

Wan, G., & Sebastian, I. (2011) *Poverty in Asia and the Pacific: An update* (Asian development bank economics working paper No. 267). http://dx.doi.org/10.2139/ssrn.1919973

Wassmann, R., Jagadish, S. V. K., Sumfleth, K., Pathak, H., Howell, G., Ismail, A., & Heuer, S. (2009). Regional vulnerability of climate change impacts on Asian rice production and scope for adaptation. *Advances in Agronomy, 102*, 91–133.

White, I., & Falkland, A. (2010). Management of freshwater lenses on small Pacific islands. *Hydrogeology Journal, 18*, 227–246.

Whitehall, K., Mattmann, C., Waliser, D., Kim, J., Goodale, C., Hart, A., & Hewitson, B. (2012). Building model evaluation and decision support capacity for CORDEX. *WMO Bulletin, 61*(2), 29–34.

Wild, M., Trüssel, B., Ohmura, A., Long, C. N., König-Langlo, G., Dutton, E. G., & Tsvetkov, A. (2009). Global dimming and brightening: An update beyond 2000. *Journal of Geophysical Research, 114*(D10), doi:10.1029/2008JD011382.

Index

A
Adaptation policy framework (APF), 211–212
ADB. *See* Asian Development Bank (ADB)
Aeroallergens, 241
AFOLU. *See* Agriculture, forestry and other land use (AFOLU)
Agriculture
adaptation and mitigation, 140–141
climate variability and change, 137–139
dominance of rice, 132
environmental challenges and opportunities, 136
farming community, 149–150
humid and sub-humid tropics, 133
research-based activities, 149
rice, 133–135
scale challenges, 136
socio-economic systems, 133
supply challenges, water, 134, 136
wheat, Asia, 139, 140
Agriculture, forestry and other land use (AFOLU), 78, 81
AIM. *See* Asia-Pacific Integrated Model (AIM)
Air pollution
"atmospheric brown clouds", 77
chemicals, 76
dispersion and dilution, 75–76
gases, aerosols and particles, 75
GHG emissions, 78
heat stress, 87
mega-cities, 76
motor vehicles, 76
and respiratory disorders, 240–241
temperatures, 86
APEC. *See* Asia and the Pacific Economic Cooperation (APEC)
APF. *See* Adaptation policy framework (APF)
APMN. *See* Asia-Pacific Mountain Network (APMN)
Architecture and agents, climate governance
ASEAN, 205
Australia, 206–207
decision-making, 206
global level, 205
Malaysia, 207
Nepal, 207–208
PIFACC, 206
rural women, 208–209
youth empowerment, 209, 210
ASEAN. *See* Association of South East Asian Nations (ASEAN)
Asia and the Pacific Economic Cooperation (APEC), 266, 302
Asian climate change projections
GCMs, 43–44
RCMs, 45–47
Asian Development Bank (ADB), 267
Asia-Pacific Integrated Model (AIM), 258
Asia-Pacific Mountain Network (APMN), 209, 210
Asia-Pacific region
agronomic practices, 296
Asian-Australian monsoon, 292
balancing energy demands, reduced GHG emissions, 12
carbon development, 301
China, 261–262
climate
adaptation and poverty alleviation, 299
hazards and vulnerabilities, 8
and interactions, 290
mitigation and adaptation strategies, 8–9
modeling, 5–6
cold spells and heat waves, 301

M.J. Manton and L.A. Stevenson (eds.), *Climate in Asia and the Pacific: Security, Society and Sustainability*, Advances in Global Change Research 56,
DOI 10.1007/978-94-007-7338-7, © Springer Science+Business Media Dordrecht 2014

Asia-Pacific region (*cont.*)
economic development, 302
ecosystem management strategies, 303
ecosystem services, 15
effects, human health, 14
ENSO, 293
extreme climate events, 5
fisheries (*see also* Fisheries)
food, 9–10
global consumption, 296
geographical variations, 3
GHG emissions and global warming, 294
global-scale Hadley circulation, 292
HTP (*see* Himalayan Tibetan Plateau (HTP))
India, 262–263
individual policy options, 298
institutional arrangements, 12
interactive web-based system, 293
international modeling community, 293
IOD, 2
Karen community, 300
Kyoto Protocol, 298
large-scale climate systems, 3–4
low carbon development pathways, 14
mountain and small island communities, 299–300
national living standards, 297–298
natural and human systems, 3
NGOs and civil society organizations, 295
policy frameworks, 3
projections, climate, 6
region of interest, 2
remote communities, 13–14
rice and wheat, 9, 295
satellite-based measurements, 291
smallholder farmers, 295
society, security and sustainability, 6–7
socio-economic factors, 296
spatial and temporal variability, 290
temperature and rainfall records, 291
time scale, 3
trends, mean climate, 4
urban areas and climate, 7–8
urbanization and industrialization, 7, 294
water security, 10–12
Association of South East Asian Nations (ASEAN)
and APEC, 302
architecture and agents, 205
Community 2009–2015, 299
LCD pathways, 267
Road Map, 299

B
Balancing energy demands, 12
Bangkok Metropolitan Region (BMR), 88
Biofuels, 9, 101, 179
Bleaching, SST, 229
Bus Rapid Transit, 102

C
Carbon capture and storage (CCS), 258, 259
Chiang Rai (Karen Community), 223–224
China
eleventh 5-year plan, 175–177
energy policies and twelfth 5-year plan, 177–179
Clean Development Mechanism (CDM), 278
Climate and ecosystems management
change and impacts, 271–272
and climate change adaptation (*see* Climate change)
communication, education and capacity building, 281
enhancement, 272–274
governance issues, 279–280
integration, 280–281
natural and sustainable development, 269–271
networking and cross-learning, 281
strengthening inter-linkages, 281
Climate and energy security
China (*see* China)
CO_2 emissions, 173, 174
economic growth and GHG, 173
India, 179–182
Japan (*see* Japan)
mitigation policies, 174
nuclear policy, 174
Climate and food security
agriculture (*see* Agriculture)
Asia-Pacific region, 132
cereal stocks-to-utilization ratio, 131
FAO input price index, 132
fisheries (*see* Fisheries)
smallholder farms, 131
Climate and sustainability
adaptation and mitigation, 282
change mitigation, 256
economic development, 256
and economy (*see* Low carbon development (LCD))
and ecosystems management (*see* Climate and ecosystems management)
Millennium Development Goals, 255

national and international development discourse., 254
participatory planning, 255
social and economic development, 255
Climate and water security
flood disaster management (*see* Flood disaster management)
HTP (*see* Himalayan Tibetan Plateau (HTP))
in Pacific Islands (*see* Pacific Island Countries (PICs))
Climate change
agricultural production, 133
boundaries of traditional fishing grounds, Fiji, 277
carbon sequestration, 278
CDM, 278
and climate variability (*see* Climate variability and change)
coastal and marine ecosystems, 279
EBA, 275
ecosystem services, 275
energy renewable and efficiency, 180–182
and fish ecology change (*see* Fish ecology change)
food and income scarcity, 276
forest conservation, 278
forest ecosystems, 276
forestry sector, 276
healthy mangrove ecosystems, 276
Japan's recent energy policy, 184
NTFPs, 275
oceanic environment, 141–144
utilization, forest resources, 276
water security, 158–162
Climate change trends. *See* Climate variability trends
Climate governance
architecture and agents, 205–209
CO_2 emissions, 201–202
complexity, 215
establishments, 201
human activities, 200–201
institutional frameworks
agencies, 203
carbon emissions, 204–205
Copenhagen Accord, 204
development and CDM contributions, 203
politics, mitigation and adaptation, 203
regional development, 203
knowledge-base (*see* Knowledge-based climate governance)
resilient community, 216
socio-economic challenges, 202
Climate hazards
flooding
Bangkok and Bangladesh, 89
impacts, 89–91
sea-level rise and storm surges, 88
increased temperatures, 85–86
landslides, 89, 91
monsoon rains, 88
potential risks, 84–85
sea-level rise, 87–88
tropical cyclones, 92
and vulnerabilities, 8
warming, physical changes, 84
water shortage, 91–93
Climate modeling, 5–6
Climate Model Intercomparison Project Phase 3 (CMIP3), 6
Climate research, Asia-Pacific. *See* Asia-Pacific region
Climate resilience, 230–231
Climate society
governance, 200–216
human health, 236–246
remote communities, 216–236
Climate variability and change
CO_2 and photosynthesis, 139
HYV, 137
modeling techniques, temperature, 138
non-productive water losses, 137
rainfall and water supply, 137
rice-based systems, 138
water scarcity, 139
Climate variability trends
annual total precipitation day amount, 36
Asia-Pacific surface climates, 18
attribution, change, 38–39
cool nights, Southeast Asia, 33, 34
ENSO, 22–25
glaciers (*see* Glaciers, climate variability trends)
global warming, 38
heavy rainfall events, 38
HTP, 18
IOD, 25
large scale circulation and monsoon system, 19–22
magnitude, cool days, 33, 35
PDO/IPO, 25–27
precipitation intensity, 38
surface temperature and precipitation, 27–29
tropical cyclones, 28
very wet days, 36, 37

Climate vulnerabilities
capacity building, 109
energy transmission and efficiency, 107
food system, 106
insurance and financing, 110
urban poverty, 98–99
water resources, 105
CMIP3. *See* Climate Model Intercomparison Project Phase 3 (CMIP3)
Coordinated Regional Downscaling Experiment (CORDEX)
at IITM, 48
and KMA, 49
Coral Triangle, 229

D
Dynamical and statistical downscaling methods
Asian climate change projections, 43–47
climate modeling, 39
coordinated projects, regional modeling, 47–49
regional model downscaling, 40–43

E
Earth System Governance (ESG) project, 210–211
East Asian summer monsoon index (EASMI), 20
Eastern Tibetan Himalayas, 221–222
Ecosystem-based adaptation (EBA), 275
Ecosystem services, 15
El Niño–Southern Oscillation (ENSO)
climate anomalies, 24
correlations, SOI, 22, 23
effects, climate change, 25
inter-annual global climate variability, 22
phases, 24
projections, future climate change, 293
tropical cyclone activity, 24
ESG project. *See* Earth System Governance (ESG) project

F
Fish ecology change
air temperature, 146
chum salmon, 148
common squid, 147–148
Pacific tuna, 146–147
walleye pollack, 148
Fisheries
climate and ecosystem considerations, 144–146
decadal variability, 150
distribution and abundance, 150
economic situations, 150
fish ecology and climate change, 146–148
human activities, 141
management-based, ecosystem model, 148–149
physical characteristics, climate change, 141–144
Total Allowable Catch (TAC), 150
Flood disaster management
definition, 164
disaster management and societal vulnerabilities, 162–164
domestic institutions, 169–170
flood affected countries in Asia, 162, 163
framing disaster, 165
fund recovery and rehabilitation operations, 165
institutional capacities, 168–169
performance evaluation, 167–168
reduction, risk, 172
regional cooperation, 170–172
risks, 166–167
State agencies, 173
structural measures, 172
Food and Agriculture Organization of the United Nations (FAO)
cereal stocks-to-utilization, 131
fish, 132
input price index, 132
Food systems
adaptation response, 106
distribution, 96
nutritional transition, 96
peri-urban agriculture, 106
production, 95–96

G
GCMs. *See* Global climate models (GCMs)
GHG emissions
and AFOLU, 81
capacity building, 109
communities, 108
density, 82
description, 61, 78
"direct" and "indirect", 77
electricity production, 102
energy production, 78–79
growth rates, 82–83

industrial processes and product use, 80
population size, 81–82
power production and waste disposal, 77
residential, 81
transportation, 79–80, 101
waste and wastewater, 80–81
wealth and climate, 83–84
GHGs. *See* Greenhouse gases (GHGs)
Glacial-lake outburst floods (GLOFs), 11, 153, 219–220
Glaciers, climate variability trends
India Himalayas, 31–32
mass balance studies, 28, 30
Nepal Himalayas, 30–31
New Zealand, 32–33
Global climate models (GCMs)
Asian climate, 43–44
global warming, 39
Meiyu-Baiu-Changma front, 44
projections, future climate, 5
RCMs (*see* Regional climate models (RCMs))
GLOFs. *See* Glacial-lake outburst floods (GLOFs)
Greenhouse gases (GHGs)
and aerosols, 290
anthropogenic influences, 292
emissions (*see* GHG emissions)
energy-GHG link, 298
and GDP, 298
and UNFCCC, 302

H

High yielding varieties (HYV), 137
Himalayan Tibetan Plateau (HTP)
altitude, 18
Asia's water resources, 155
climate change, 297
earthquakes, 216
geography and ecology, 291
glacial ice, 151
glaciation, 220
Gobi desert, 216
highland areas, 218
hydro-electric power stations, 297
hydrology, 153–155
mass balance, glacier, 4
population growth and urbanization, 151
retreating glaciers, 291
temperature and precipitation changes, 152
water policy and governance, 155–156
water source, 220
water tower, 151, 217–218, 297
HTP. *See* Himalayan Tibetan Plateau (HTP)
Human health
aeroallergens, 241
air pollution and respiratory disorders, 240–241
capacities, skills and education needs, 244
contamination, drinking water, 242–243
cross-cutting challenges, 243
description, 245–246
impacts of weather, 238–239
infectious diseases, 241–242
IPCC Fourth Assessment Report, 236–237
malnutrition, 241
management, health risks, 245
public health and health care infrastructure, 243–244
susceptible, vulnerable and displaced populations, 243
trends, 238
weather and climate events, 239–240
Hydrology
agricultural production, 155
atmospheric temperatures, 152
glacial melt, 153
GLOFs, 153
Himalayan glaciers, 153
hydroelectric power generation, 154
inter-annual variability, 154
rainwater harvesting, 155
snow melt and rainfall contribution, 153
HYV. *See* High yielding varieties (HYV)

I

India
climate change, renewable energy and energy efficiency, 180–182
energy challenges and prospects, 182
energy supply trends, 179–180
Indian Institute of Tropical Meteorology (IITM), 48
Indian Ocean Dipole (IOD), 4, 25
Industry and services, 98
Integrated assessment model (IAM)
and AIM, 258
carbon tax scenario, India's, 260, 261
and CCS, 258, 259
communication tool, 258
emission scenarios, 258
high-tech, carbon-intense society, 258, 259
and IPCC, 257
and RCPs, 257
sustainability scenario, India's, 260

Interdecadal Pacific Oscillation (IPO). *See* Pacific Decadal Oscillation (PDO)
Intergovernmental Panel on Climate Change (IPCC), 257
Inter-tropical convergence zone (ITCZ), 19
IOD. *See* Indian Ocean Dipole (IOD)

J
Japan
earthquake and nuclear power plant, 184–186
energy and climate change policies, 184
energy supply trends, 182–183
Jawaharlal Nehru National Urban Renewal Mission (JNNURM), 104

K
KMA. *See* Korea Meteorological Administration (KMA)
Knowledge-based climate governance
agricultural systems, 214
APF, 211–212
CAPaBLE Capacity Development Programme, 213
carbon emission, 214
climate variability, 213
decision-making process, 213
description, 209–210
ESG project, 210–211
scientific tools, 212
training workshops, NIWA, 213–214
United Nation's Conference on Human Environment, 214–215
Korea Meteorological Administration (KMA), 49

L
Land and energy
Chinese cities, 72
development, 72
electricity consumption, 72–73
populations, 71
transportation fuel consumption, 73
vehicle usage, 73
Land-cover change (LCC), 42
LCD pathways
ADB, 267
adoption of, 14
APEC, 266
ASEAN, 267
CO_2 emissions, 256
LoCARNet, 267–268
low-carbon societies, 268
Low Carbon Asia Research Network (LoCARNet), 267–268
Low carbon development (LCD)
Asia and Pacific, 261–263
climate change policies, 256
climate policies, 269
IAM, 257–261
pathways (*see* LCD pathways)
and societies, 269
society, 256
UNFCCC, 263–266

M
Mahatma Gandhi National Rural Employment Guarantee Act (MGNREGA), 209
Marine ecosystems, 9, 10
Mega-cities
air pollution, 76
sea-level rise, 87–88
1950–1990 trends, 62–65
1990–2010 trends, 64–68
2010–2050 trends, 68–71
Millennium Ecosystems Assessment (MEA), 269
Monsoon Asia
Asian monsoon region:, 19
EASMI, 20
equatorial Walker circulation, 20
global warming, 20
zonal-mean updraft, 20–21
Monsoon circulation
Asia, 19–21
description, 19
ITCZ, 19
Pacific, 21–22
Mountain communities
Chiang Rai (Karen Community), 223–224
Eastern Tibetan Himalayas, 221–222
ecosystems and biodiversity, 219
glacial melt and GLOF, 219–220
HTP (*see* Himalayan Tibetan Plateau (HTP))
Xishuangbanna (Dai People), 222–223

N
National adaptation programs of action (NAPAs), 104
Natural and human systems, 3

Natural ecosystems
and climate change (*see* Climate change)
coastal and marine, 270
definition, 269
demographic changes, 271
ecological and socio-economic functions., 271
governance issues, 279–280
humanity, 272
land area and population, Southeast Asia, 271, 272
MEA, 269
monitoring programs, 273
natural ecosystems, 271
oil palm plantations, 272
PES, 273
rate of forest loss and gain, 270
science of ecosystem stewardship, 273
sustainability approaches, 272–273
sustaining ecosystem services, 273, 274
Natural variability, monsoon, 3
Non-timber forest products (NTFPs), 275
Numerical weather prediction (NWP) models, 40
Nutritional transition, 96

P

Pacific
and Asia (*see* Asia-Pacific region)
PNA oscillation, 21–22
SPCZ, 21
surface climates, 21
Pacific Centre for Environment and Sustainable Development (PACE-SD), 203, 213, 233, 234
Pacific Climate Change Science Program (PCCSP), 6
Pacific Decadal Oscillation (PDO)
climate variability, 25
negative phase, 27
SST, 25, 26
Pacific Island Countries (PICs)
climate change, water security, 158–162
current water security issues, 157–158
water management and governance, 162
Pacific Islands Framework for Action on Climate Change (PIFACC), 203, 206
Payments for environmental services (PES), 273
PCCSP. *See* Pacific Climate Change Science Program (PCCSP)
PICs. *See* Pacific Island Countries (PICs)
Public health and health care infrastructure, 243–244

R

RCPs. *See* Representative concentration pathways (RCPs)
Regional climate models (RCMs)
Asia-Pacific region, 41, 42
climate modeling, 5
climatological performance, indices, 41
and CORDEX, 48–49
impacts and adaptation communities, 41
internal forcing, 40
lateral boundary forcing, 40
projection, Asia-Pacific, 45–47
and RMIP, 47–48
stand-alone variable-resolution global models, 40–41
statistical downscaling, 43
strongly-stretched variable-resolution global models, 41
Regional Model Intercomparison Project (RMIP), 6, 47–48
Remote communities
description, 235–236
"geographic hotspots", 216, 217
mountain communities, 216–224
small island (*see* Small island developing states (SIDS))
Representative concentration pathways (RCPs), 257
Resilient cities
policy needs, 113
research gaps, 112
socio-economic and biophysical sub-systems, 111
uncertainties, 111–112
Rewards for environmental services (RES), 273
Rice and wheat, asia, 9
RMIP. *See* Regional Model Intercomparison Project (RMIP)

S

Sea surface temperature (SST)
and ENSO, 25
and PDO, 25, 26
topographical features, 2
Secretariat of the Pacific Regional Environment Programme (SPREP), 299

Security
  Asia, 297
  coal industry, 298
  food, 296
  hydro-electric power stations, 297
Small island developing states (SIDS)
  adaptation activities, 232–233
  atolls, 227
  barriers to adaptation, 235
  bleaching, coral reefs, 229
  challenges, 228
  Coral Triangle, 229
  countries, 224, 225
  definition, 224–225
  forced migration, 233–235
  human health issues, 230
  isolation, 226–227
  mitigation, 231–232
  'oceanic' islands, 225–226
  PACE-SD Integrated Project Cycle, 233, 234
  resilience and vulnerability, 230–231
  sea-level rise, 229
  social, political and economic characteristics, 228
  trade winds, 227
  tropical cyclones, 230
  volcanic islands, 227
Society
  climate interactions, 304
  NGOs and civil organizations, 295
  urbanization and industrialization, 294
South Pacific Convergence Zone (SPCZ), 21
SPREP. *See* Secretariat of the Pacific Regional Environment Programme (SPREP)
SST. *See* Sea surface temperature (SST)
Sustainability, 7, 296

T
Total Allowable Catch (TAC), 150
Tropical cyclones, 92

U
UHI. *See* Urban heat island (UHI)
UNDP. *See* United Nations Development Programme (UNDP)
UN Framework Convention on Climate Change (UNFCCC)
  co-benefits, climate mitigation policies, 265–266
  emission targets, 263–264
  financial mechanisms, 265
  GHG emissions, 302
  global level, 299
  regional and community levels, 300
  secure transparency, 264–265
United Nations Development Programme (UNDP), 66
Urban adaptation strategies
  description, 99–100
  energy transmission and efficiency, 107
  food, 106
  NAPAs and JNNURM, 104
  policy and governance (*see* Urban climate governance)
  transportation, 106–107
  types, 104
  water resources, 105
Urban climate
  air pollution, 75–77
  description, 73
  precipitation, 74–75
  UHI, 73–74
Urban climate governance
  capacity building, 109
  description, 107–108
  development and disaster risk management, 108
  insurance and financing, 110
  opportunities, 110–111
  participation, 109–110
Urban design, 100–101
Urban heat island (UHI), 73–75, 86, 87, 100–101
Urbanization
  adaptation strategies (*see* Urban adaptation strategies)
  climate hazards (*see* Climate hazards)
  description, 60–61
  energy production, transmission and distribution, 96–97
  environment and climate, 293
  "flying geese" model, 61
  food, 95–96
  GHG emissions (*see* GHG emissions)
  human settlement, 293
  and industrialization, 294
  industry and services, 98
  land and energy use, 71–73
  mitigation strategies (*see* Urban mitigation strategies)
  resilient cities, 111–113
  transportation, 93–94
  1950–1990 trends, 62–65

1990–2010 trends, 64–68
2010–2050 trends, 68–71
UHI, 293
urban climate, 73–77
vulnerability, 98–99
water supply and sanitation, 94–95
Urban mitigation strategies
description, 99–100
development, 103–104
energy production and demand, 102–103
policy and governance (*see* Urban climate governance)
transportation, 101–102
urban design, 100–101

V

Vector-borne diseases (VBD), 241
Vulnerability
APF, 212
health risks and population, 245
PACE-SD Integrated Project Cycle, 233
and resilience, 230–231
social, 216

W

Water security
agriculture and domestic uses, 13
climate, topography and culture variations, 12
GLOF, 11
growing populations and economies, 10
Kathmandu Valley, Himalayas, 11
population growth and climate change, 11–12
World Climate Research Programme (WCRP), 6

X

Xishuangbanna (Dai People), 222–223

Printed by Publishers' Graphics LLC
LMO131026.15.13.38